Practical Data Acquisition for Instrumentation and Control Systems

Titles in the series

Practical Cleanrooms: Technologies and Facilities (David Conway)

Practical Data Acquisition for Instrumentation and Control Systems (John Park, Steve Mackay)

Practical Data Communications for Instrumentation and Control (John Park, Steve Mackay, Edwin Wright)

Practical Digital Signal Processing for Engineers and Technicians (Edmund Lai)

Practical Electrical Network Automation and Communication Systems (Cobus Strauss)

Practical Embedded Controllers (John Park)

Practical Fiber Optics (David Bailey, Edwin Wright)

Practical Industrial Data Networks: Design, Installation and Troubleshooting (Steve Mackay, Edwin Wright, John Park, Deon Reynders)

Practical Industrial Safety, Risk Assessment and Shutdown Systems (Dave Macdonald)

Practical Modern SCADA Protocols: DNP3, 60870.5 and Related Systems (Gordon Clarke, Deon Reynders)

Practical Radio Engineering and Telemetry for Industry (David Bailey)

Practical SCADA for Industry (David Bailey, Edwin Wright)

Practical TCP/IP and Ethernet Networking (Deon Reynders, Edwin Wright)

Practical Variable Speed Drives and Power Electronics (Malcolm Barnes)

Practical Data Acquisition for Instrumentation and Control Systems

John Park ASD, IDC Technologies, Perth, Australia

Steve Mackay CPEng, BSc(ElecEng), BSc(Hons), MBA, IDC Technologies, Perth, Australia

AMSTERDAM • BOSTON • HEIDELBERG • LONDON • NEW YORK • OXFORD
PARIS • SAN DIEGO • SAN FRANCISCO • SINGAPORE • SYDNEY • TOKYO

Newnes is an imprint of Elsevier

Newnes

Newnes
An imprint of Elsevier
Linacre House, Jordan Hill, Oxford OX2 8DP
200 Wheeler Road, Burlington, MA 01803

First published 2003

British Library Cataloguing in Publication Data
A catalogue record for this book is available from the British Library

ISBN   07506 57960

For information on all Newnes publications, visit
our website at www.newnespress.com

Typeset and Edited by Vivek Mehra, Mumbai, India
(vivekmehra@tatanova.com)

Transferred to digital printing 2006

Contents

Appendix F Number systems   **389**

Appendix G GPIB (IEEE-488) mnemonics & their definitions **398**

Index   **403**

Preface

In less than a decade, the PC has become the most widely used platform for data acquisition and control. The main reasons for the popularity of PC-based technology are low costs, flexibility and ease of use, and, last but not the least, performance. This solid and dependable trait is all thanks to the use of 'off-the-shelf' components. Data acquisition with a PC enables one to display, log and control a wide variety of real world signals such as pressure, flow, and temperature. This ability coupled with that of easy interface with various stand-alone instruments makes the systems ever more desirable.

Until the advent of the PC, data acquisition and process monitoring were carried out by using dedicated data loggers, programmable logic controllers and or expensive proprietary computers. Today's superb software-based operator interfaces make the PC an increasingly attractive option in these typical applications:

- Laboratory data acquisition and control
- Automatic test equipment (ATE) for inspection of components
- Medical instrumentation and monitoring
- Process control of plants and factories
- Environmental monitoring and control
- Machine vision and inspection

The key to the effective application of PC-based data acquisition is the careful matching of real world requirements with appropriate hardware and software. Depending on your needs, monitoring data can be as simple as connecting a few cables to a plug-in board and running a menu-driven software package. At the other end of the spectrum, you could design customized sensing and conversion hardware, or perhaps develop application software to optimize a system.

This book gives both the novice and the experienced user a solid grasp of the principles and practical implementation of interfacing the PC and stand-alone instruments with real world signals. The main objective of this book is to give you a thorough understanding of PC-based data acquisition systems and to enable you to design, specify, install, configure, and program data acquisition systems quickly and effectively.

After reading this book, we believe you will be able to:

- Demonstrate a sound knowledge of the fundamentals of data acquisition (with a focus on PC-based work)
- Competently install and configure a simple data acquisition system
- Choose and configure the correct software
- Avoid the common pitfalls in designing a data acquisition system

This book is intended for engineers and technicians who are:

- Electronic engineers
- Instrumentation and control engineers
- Electrical engineers
- Electrical technicians
- Systems engineers
- Scientists working in the data acquisition area
- Process control engineers
- System integrators
- Design engineers

A basic knowledge of electrical principles is useful in understanding the outlined concepts, but this book also focuses on the fundamentals; hence, understanding key concepts should not be too onerous.

The structure of the book is as follows.

Chapter 1: Introduction. This chapter gives a brief overview of what is covered in the book with an outline of the essentials and main hardware and software components of data acquisition.

Chapter 2: Analog and digital signals. This chapter reviews analog and digital inputs to the data acquisition system, through such techniques as temperature measurement and the use of strain gauges.

Chapter 3: Signal conditioning. This chapter discusses how signals are conditioned before the data acquisition system can accurately acquire it.

Chapter 4: The PC for real time work. This chapter considers the various PC related issues to make it suitable for real time work such as software and hardware.

Chapter 5: Plug-in data acquisition boards. This chapter assesses the wide range of methods of using plug-in data acquisition boards such as analog inputs/ outputs, digital inputs/outputs and counter/timer configurations.

Chapter 6: Serial data communications. This chapter reviews the fundamental definitions and basic principles of digital serial data communications with a focus on RS-232 and RS-485.

Chapter 7: Distributed and stand-alone loggers/controllers. This chapter discusses the hardware and software configurations of stand-alone logger/controllers.

Chapter 8: IEEE 488 standard. This chapter reviews the IEE 488 standard with a reference to the IEEE 488.2 and SCPI approaches.

Chapter 9: Ethernet and fieldbus systems. This chapter briefly outlines the essentials of Ethernet and Fieldbus systems.

Chapter 10: The universal serial bus (USB). This chapter reviews the key features of the universal serial bus, which will have a major impact on PC-based data acquisition.

Chapter 11: Specific techniques. This chapter discusses how the PC can be used for process control applications.

Chapter 12: The PCMCIA card. This chapter discusses the essentials of the PCMCIA card as applied to data acquisition systems.

1

Introduction

In 1981, when IBM released its first personal computer or PC (as it became widely known) its open system design encouraged the development of a wide range of compatible add-on products by independent third party developers. In addition, the open system design has encouraged the proliferation of IBM compatible PCs in the market place, resulting in a rapid increase in the speed and power of the PC, as competitors vie for a market edge.

Accompanied by a significant drop in cost and a rapid expansion in software, which utilizes the increased power of the processor, the PC is now the most widely used platform for digital signal processing, image processing, data acquisition, and industrial control and communication applications. In many applications, indeed for data acquisition and process control, the PCs power and flexibility allow it to be configured in a number of ways, each with its own distinct advantages. The key to the effective use of the PC is the careful matching of the specific requirements of a particular data acquisition application to the appropriate hardware and software available.

This chapter reviews the fundamental concepts of data acquisition and control systems and the various system configurations, which make use of the PC.

1.1 Definition of data acquisition and control

Data acquisition is the process by which physical phenomena from the real world are transformed into electrical signals that are measured and converted into a digital format for processing, analysis, and storage by a computer.

In a large majority of applications, the data acquisition (DAQ) system is designed not only to acquire data, but to act on it as well. In defining DAQ systems, it is therefore useful to extend this definition to include the control aspects of the total system. Control is the process by which digital control signals from the system hardware are convened to a signal format for use by control devices such as actuators and relays. These devices then control a system or process. Where a system is referred to as a data acquisition system or DAQ system, it is possible that it includes control functions as well.

1.2 Fundamentals of data acquisition

A data acquisition and control system, built around the power and flexibility of the PC, may consist of a wide variety of diverse hardware building blocks from different equipment manufacturers. It is the task of the system integrator to bring together these individual components into a complete working system.

The basic elements of a data acquisition system, as shown in the functional diagram of Figure 1.1, are as follows:

- Sensors and transducers
- Field wiring
- Signal conditioning
- Data acquisition hardware
- PC (operating system)
- Data acquisition software

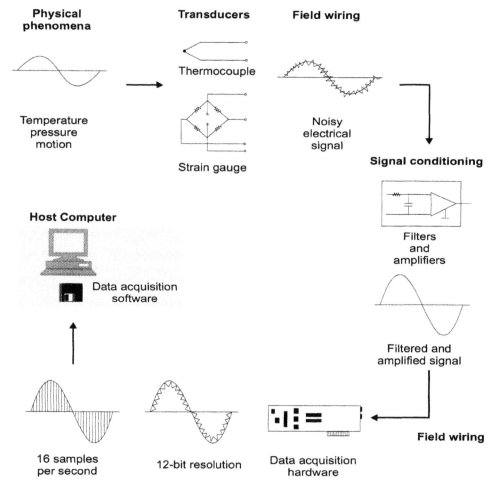

Figure 1.1
Functional diagram of a PC-based data acquisition system

Each element of the total system is important for the accurate measurement and collection of data from the process or physical phenomena being monitored, and is discussed in the following sections.

1.2.1 Transducers and sensors

Transducers and sensors provide the actual interface between the real world and the data acquisition system by converting physical phenomena into electrical signals that the signal conditioning and/or data acquisition hardware can accept.

Transducers available can perform almost any physical measurement and provide a corresponding electrical output. For example, thermocouples, resistive temperature detectors (RTDs), thermistors, and IC sensors convert temperature into an analog signal, while flow meters produce digital pulse trains whose frequency depends on the speed of flow.

Strain gauges and pressure transducers measure force and pressure respectively, while other types of transducers are available to measure linear and angular displacement, velocity and acceleration, light, chemical properties (e.g. CO concentration, pH), voltages, currents, resistances or pulses. In each case, the electrical signals produced are proportional to the physical quantity being measured according to some defined relationship.

1.2.2 Field wiring and communications cabling

Field wiring represents the physical connection from the transducers and sensors to the signal conditioning hardware and/or data acquisition hardware. When the signal conditioning and/or data acquisition hardware is remotely located from the PC, then the field wiring provides the physical link between these hardware elements and the host computer. If this physical link is an RS-232 or RS-485 communications interface, then this component of the field wiring is often referred to as communications cabling.

Since field wiring and communications cabling often physically represents the largest component of the total system, it is most susceptible to the effects of external noise, especially in harsh industrial environments. The correct earthing and shielding of field wires and communications cabling is of paramount importance in reducing the effects of noise. This passive component of the data acquisition and control system is often overlooked as an important integral component, resulting in an otherwise reliable system becoming inaccurate or unreliable due to incorrect wiring techniques.

1.2.3 Signal conditioning

Electrical signals generated by transducers often need to be converted to a form acceptable to the data acquisition hardware, particularly the A/D converter which converts the signal data to the required digital format. In addition, many transducers require some form of excitation or bridge completion for proper and accurate operation.

The principal tasks performed by signal conditioning are:

- Filtering
- Amplification
- Linearization
- Isolation
- Excitation

Filtering

In noisy environments, it is very difficult for very small signals received from sensors such as thermocouples and strain gauges (in the order of mV), to survive without the sensor data being compromised. Where the noise is of the same or greater order of magnitude than the required signal, the noise must first be filtered out. Signal conditioning equipment often contains low pass filters designed to eliminate high frequency noise that can lead to inaccurate data.

Amplification

Having filtered the required input signal, it must be amplified to increase the resolution. The maximum resolution is obtained by amplifying the input signal so that the maximum voltage swing of the input signal equals the input range of the analog-to-digital converter (ADC), contained within the data acquisition hardware.

Placing the amplifier as close to the sensor as physically possible reduces the effects of noise on the signal lines between the transducer and the data acquisition hardware.

Linearization

Many transducers, such as thermocouples, display a non-linear relationship to the physical quantity they are required to measure. The method of linearizing these input signals varies between signal conditioning products. For example, in the case of thermocouples, some products match the signal conditioning hardware to the type of thermocouple, providing hardware to amplify and linearize the signal at the same time.

A cheaper, easier, and more flexible method is provided by signal conditioning products that perform the linearization of the input signal using software.

Isolation

Signal conditioning equipment can also be used to provide isolation of transducer signals from the computer where there is a possibility that high voltage transients may occur within the system being monitored, either due to electrostatic discharge or electrical failure. Isolation protects expensive computer equipment from damage and computer operators from injury. In addition, where common-mode voltage levels are high or there is a need for extremely low common mode leakage current, as for medical applications, isolation allows measurements to be accurately and safely obtained.

Excitation

Signal conditioning products also provide excitation for some transducers. For example: strain gauges, thermistors and RTDs, require external voltage or current excitation signals.

1.2.4 Data acquisition hardware

Data acquisition and control (DAQ) hardware can be defined as that component of a complete data acquisition and control system, which performs any of the following functions:

- The input, processing and conversion to digital format, using ADCs, of analog signal data measured from a system or process – the data is then transferred to a computer for display, storage and analysis
- The input of digital signals, which contain information from a system or process

- The processing, conversion to analog format, using DACs, of digital signals from the computer – the analog control signals are used for controlling a system or process
- The output of digital control signals

Data acquisition hardware is available in many forms from many different manufacturers. Plug-in expansion bus boards, which are plugged directly into the computer's expansion bus, are a commonly utilized item of DAQ hardware. Other forms of DAQ hardware are intelligent stand-alone loggers and controllers, which can be monitored, controlled and configured from the computer via an RS-232 interface, and yet can be left to operate independently of the computer.

Another commonly used item of DAQ hardware, especially in R&D and test environments, is the remote stand-alone instrument that can be configured and controlled by the computer, via the IEEE-488 communication interface. Several of the most common DAQ system configurations are discussed in the section **Data acquisition and control system configuration p. 6**

1.2.5   Data acquisition software

Data acquisition hardware does not work without software, because it is the software running on the computer that transforms the system into a complete data acquisition, analysis, display, and control system.

Application software runs on the computer under an operating system that may be single-tasking (like DOS) or multitasking (like Windows, Unix, OS2), allowing more than one application to run simultaneously.

The application software can be a full screen interactive panel, a dedicated input/output control program, a data logger, a communications handler, or a combination of all of these.

There are three options available, with regard to the software required, to program any system hardware:

- Program the registers of the data acquisition hardware directly
- Utilize low-level driver software, usually provided with the hardware, to develop a software application for the specific tasks required
- Utilize off-the-shelf application software – this can be application software, provided with the hardware itself, which performs all the tasks required for a particular application; alternatively, third party packages such as LabVIEW and Labtech Notebook provide a graphical interface for programming the tasks required of a particular item of hardware, as well as providing tools to analyze and display the data acquired

1.2.6   Host computer

The PC used in a data acquisition system can greatly affect the speeds at which data can be continuously and accurately acquired, processed, and stored for a particular application. Where high speed data acquisition is performed with a plug-in expansion board, the throughput provided by bus architectures, such as the PCI expansion bus, is higher than that delivered by the standard ISA or EISA expansion bus of the PC.

Depending on the particular application, the microprocessor speed, hard disk access time, disk capacity and the types of data transfer available, can all have an impact on the speed at which the computer is able to continuously acquire data. All PCs, for example, are capable of programmed I/O and interrupt driven data transfers. The use of Direct Memory Access (DMA), in which dedicated hardware is used to transfer data directly into the computer's memory, greatly increases the system throughput and leaves the computer's microprocessor free for other tasks. Where DMA or interrupt driven data transfers are required, the plug-in data acquisition board must be capable of performing these types of data transfer.

In normal operation the data acquired, from a plug-in data acquisition board or other DAQ hardware (e.g. data logger), is stored directly to System Memory. Where the available system memory exceeds the amount of data to be acquired, data can be transferred to permanent storage, such as a hard disk, at any time. The speed at which the data is transferred to permanent storage does not affect the overall throughput of the data acquisition system.

Where large amounts of data need to be acquired and stored at high speed, disk-streaming can be used to continuously store data to hard disk. Disk-streaming utilizes a terminate-and-stay-resident (TSR) program to continuously transfer data acquired from a plug-in data acquisition board and temporarily held in system memory, to the hard disk. The limiting factors in the streaming process may be the hard disk access time and its storage capacity. Where the storage capacity is sufficient, the amount of contiguous (unfragmented) free hard disk space available to hold the data, may affect the system performance, since the maximum rate at which data can be streamed to the disk is reduced by the level of fragmentation.

If real-time processing of the acquired data is needed, the performance of the computer's processor is paramount. A minimum requirement for high frequency signals acquired at high sampling rates would be a 32-bit processor with its accompanying co-processor, or alternatively a dedicated plug-in processor. Low frequency signals, for which only a few samples are processed each second, would obviously not require the same level of processing power. A low-end PC would therefore be satisfactory. Clearly, the performance requirements of the host computer must be matched to the specific application. As with all aspects of a data acquisition system the choice of computer is a compromise between cost and the current and future requirements it must meet.

One final aspect of the personal computer that should be considered is the type of operating system installed. This may be single-tasking (e.g. MS-DOS) or multitasking (e.g. Windows 2000). While the multitasking nature of Windows provides many advantages for a wide range of applications, its use in data acquisition is not as clear-cut. For example, the methods employed by Windows to manage memory can provide difficulties in the use of DMA. In addition, interrupt latencies introduced by the multitasking nature of Windows can lead to problems when interrupt driven data transfers are used. Therefore, careful consideration must be given to the operating system and its performance in relation to the type of data acquisition hardware and the methods of data transfer, especially where high-speed data transfers are required.

1.3 Data acquisition and control system configuration

In many applications, and especially for data acquisition and process control, the power and flexibility of the PC, allows DAQ systems to be configured in a number of ways, each with its own distinct advantages. The key to the effective use of the PC is the careful matching of the specific requirements of a particular data acquisition application to the appropriate hardware and software available.

The choice of hardware, and the system configuration, is largely dictated by the environment in which the system will operate (e.g. an R&D laboratory, a manufacturing plant floor or a remote field location). The number of sensors and actuators required and their physical location in relation to the host computer, the type of signal conditioning required, and the harshness of the environment, are key factors.

Several of the most common system configurations are as follows:

- Computer plug-in I/O
- Distributed I/O
- Stand-alone or distributed loggers and controllers
- IEEE-488 instruments

1.3.1   Computer plug-in I/O

Plug-in I/O boards are plugged directly into the computers expansion bus, are generally compact, and also represent the fastest method of acquiring data to the computers memory and/or changing outputs. Along with these advantages, plug-in boards often represent the lowest cost alternative for a complete data acquisition and control system and are therefore a commonly utilized item of DAQ hardware.

As shown in Figure 1.2, examples of plug-in I/O boards are, multiple analog input A/D boards, multiple analog output D/A boards, digital I/O boards, counter/timer boards, specialized controller boards (such as stepper/servo motor controllers) or specialized instrumentation boards (such as digital oscilloscopes).

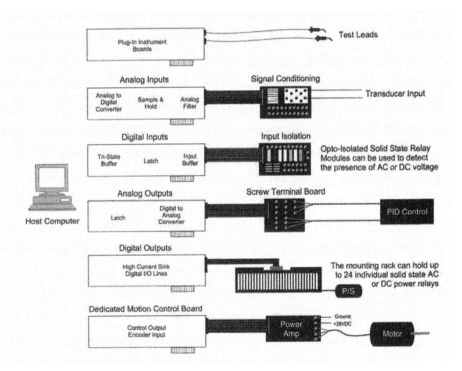

Figure 1.2
Example of computer plug-in I/O boards

Multi-function DAQ boards, containing A/D converters (ADCs), D/A converters (DACs), digital I/O ports and counter timer circuitry, perform all the functions of the equivalent individual specialized boards. Depending on the number of analog inputs/outputs and digital inputs/outputs required for a particular application, multi-function boards represent the most cost effective and flexible solution for DAQ systems.

Plug-in expansion boards are commonly used in applications where the computer is close to the sensors being measured or the actuators being controlled. Alternatively, they can be interfaced to remotely located transducers and actuators via signal conditioning modules known as two-wire transmitters. This system configuration is discussed in the following section on Distributed I/O.

1.3.2   Distributed I/O

Often sensors must be remotely located from the computer in which the processing and storage of the data takes place. This is especially true in industrial environments where sensors and actuators can be located in hostile environments over a wide area, possibly hundreds of meters away. In noisy environments, it is very difficult for very small signals received from sensors such as thermocouples and strain gauges (in the order of mV) to survive transmission over such long distances, especially in their raw form, without the quality of the sensor data being compromised.

An alternative to running long and possibly expensive sensor wires, is the use of distributed I/O, which is available in the form of signal conditioning modules remotely located near the sensors to which they are interfaced. One module is required for each sensor used, allowing for high levels of modularity (single point to hundreds of points per location). While this can add reasonable expense to systems with large point counts, the benefits in terms of signal quality and accuracy may be worth it.

One of the most commonly implemented forms of distributed I/O is the digital transmitter. These intelligent devices perform all required signal conditioning functions (amplification, filtering, isolation etc), contain a micro-controller and A/D converter, to perform the digital conversion of the signal within the module itself. Converted data is transmitted to the computer via an RS-232 or RS-485 communications interface. The use of RS-485 multi-drop networks, as shown in Figure 1.3, reduces the amount of cabling required, since each signal-conditioning module shares the same cable pair. Linking up to 32 modules, communicating over distances up to 10 km, is possible when using the RS-485 multi-drop network. However, since very few computers have built in support for the RS-485 standard, an RS-232 to RS-485 converter is required to allow communications between the computer and the remote modules.

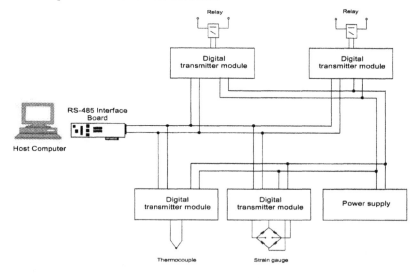

Figure 1.3
Distributed I/O – digital transmitter modules

1.3.3   Stand-alone or distributed loggers/controllers

As well as providing the benefits of intelligent signal conditioning modules, and the ability to make decisions remotely, the use of stand-alone loggers/controllers increases system reliability. This is because once programmed, the stand-alone logger can continue to operate, even when the host computer is not functional or connected. In fact, stand-alone loggers/controllers are specifically designed to operate independently of the host computer. This makes them especially useful for applications where the unit must be located in a remote or particularly hostile environment, (e.g. a remotely located weather station), or where the application does not allow continuous connection to a computer (e.g. controlling temperatures in a refrigerated truck).

Stand-alone loggers/controllers are intelligent powerful and flexible devices, easily interfaced to a wide range of transducers, as well as providing digital inputs and digital control outputs for process control.

The stand-alone logger/controller and logging data are programmed either by a serial communications interface or by using portable and reusable PCMCIA cards. The credit card size PCMCIA card is especially useful when the stand-alone logger/controller is remotely located, but requires an interface connected to the computer. This is shown in Figure 1.4.

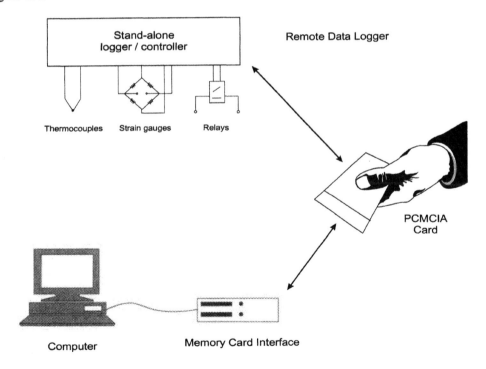

Figure 1.4
Using PCMCIA cards to program and log data from a stand-alone logger/controller

The most commonly used serial communications link for direct connection between the computer and the stand-alone logger/controller is the RS-232 serial interface. This allows programming and data logging up to distances of 50 meters, as shown in Figure 1.5. Where the stand-alone unit must be located remotely, a portable PC can be taken to the remote location or communications performed via a telephone or radio communications network using modems, as shown in Figure 1.6.

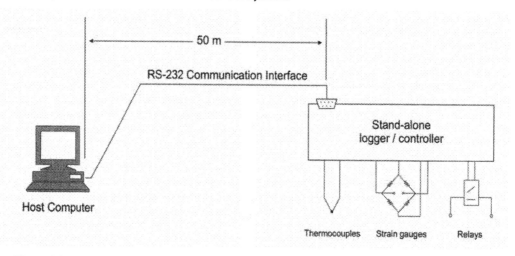

Figure 1.5
Direct connections to a stand-alone logger/controller via an RS-232 serial interface

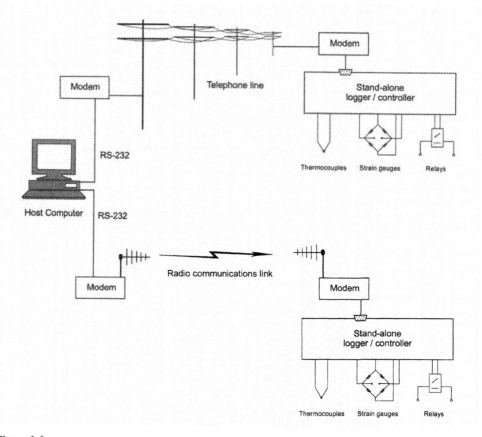

Figure 1.6
Remote connection to a stand-alone logger/controller via a telephone or radio communications network

Where an application requires more than one logger/controller, each unit is connected within an RS-485 multi-drop network. A signal unit, deemed to be the host unit, can be connected directly to the host computer via the RS-232 serial interface, as shown in Figure 1.7, thus avoiding any requirement for an RS-232 to RS-485 serial interface card.

The same methods of programming or logging data from each logger/controller are available either via the serial communications network or via using portable and reusable memory cards.

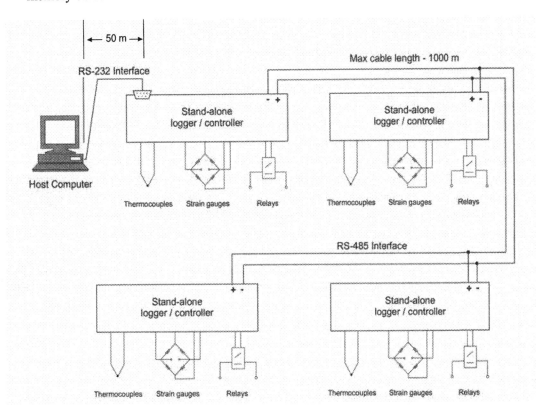

Figure 1.7
Distributed logger/controller network

1.3.4   IEEE-488 (GPIB) remote programmable instruments

The communications standard now known as GPIB (General Purpose Interface Bus), was originally developed by Hewlett-Packard in 1965 as a digital interface for interconnecting and controlling their programmable test instruments. Originally referred to as the Hewlett Packard Interface Bus (HPIB), its speed, flexibility and usefulness in connecting instruments in a laboratory environment led to its widespread acceptance, and finally to its adoption as a world standard (IEEE-488). Since then, it has undergone improvements (IEEE-488.2) and SCPI (Standard Commands for Programmable Instruments), to standardize how instruments and their controllers communicate and operate.

Evolving from the need to collect data from a number of different stand-alone instruments in a laboratory environment, the GPIB is a high-speed parallel communications interface that allows the simultaneous connection of up to 15 devices or instruments on a short common parallel data communications bus. The most common configuration requires a GPIB controller, usually a plug-in board on the computer, which addresses each device on the bus and initiates the devices that will communicate to each other. The maximum speed of communications, the maximum length of cable, and the maximum cable distance between each device on the GPIB is dependent on the speed and processing power of the GPIB controller and the type of cabling used. Typical transfer speeds are of the order of 1 Mbyte/s, while the maximum cable length at this data transfer rate is 20 m. This makes GPIB remote instruments most suited to the research laboratory or industrial test environment.

Thousands of GPIB-compatible laboratory and industrial instruments, such as data loggers and recorders, digital voltmeters and oscilloscopes are available on the market for a wide range of applications and from a wide range of manufacturers. A typical system configuration is shown in Figure 1.8.

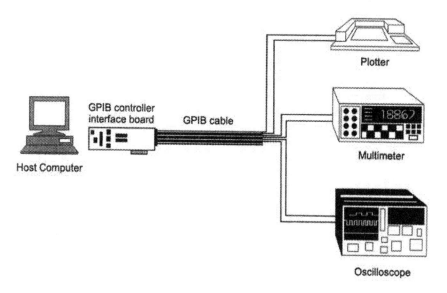

Figure 1.8
A typical GPIB system configuration

2

Analog and digital signals

2.1   Classification of signals

In the real world, physical phenomena, such as temperature and pressure, vary according to the laws of nature and exhibit properties that vary continuously in time; that is they are all analog time-varying signals.

Transducers convert physical phenomena into electrical signals such as voltage and current for signal conditioning and measurement within DAQ systems. While the voltage or current output signal from transducers has some direct relationship with the physical phenomena they are designed to measure, it is not always clear how that information is contained within the output signal. For example, in the case of a flow meter, the output is a digital pulse train whose frequency is directly proportional to the rate of flow. While the change in the flow rate of a fluid may be varying slowly with time, the output signal is a digital pulse train that may vary quickly in time, dependent on the flow rate, and not on the speed of change in the flow rate. This is shown in Figure 2.1.

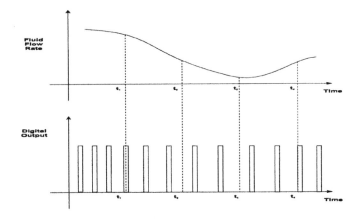

Figure 2.1
The rate of fluid flow and sign at output from a flow meter transducer

This leads us to the need for the classification of signals in DAQ systems, because it is the information contained within a signal that determines its classification, and therefore the method of signal measurement and or the type of hardware required to produce that signal. The classification of signals that may be encountered in data acquisition and control systems are defined in the sections below.

2.1.1   Digital signals binary signals

A digital, or binary, signal can have only two possible specified levels or states; an 'on' state, in which the signal is at its highest level, and an 'off' state, in which the signal is at its lowest level. This is shown in Figure 2.2.

For example, the output voltage signal of a transistor-to-transistor logic (TTL) switch can only have two states – the value in the 'on' state is 5 V, while the value in the 'off' state is 0 V. Control devices, such as relays, and indicators such as LEDs, require digital output signals like those provided on digital I/O boards.

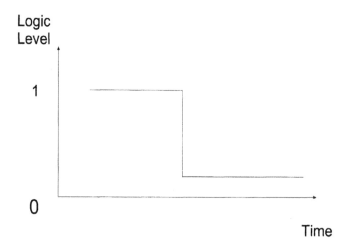

Figure 2.2
A binary digital signal

Digital pulse trains

A digital pulse train is a special type of digital signal, comprising a sequence of digital pulses as shown in Figure 2.3. Like all digital signals, a digital pulse can have only two defined levels or states. It is defined as a pulse because it remains in a non-quiescent state for a short period. A positive going pulse is one that makes a transition from its lowest logic state to its highest logic state, remains at the high logic state for a short duration, and then returns to the low logic state. A negative going pulse makes a transition from its highest logic state to the low logic state, remains there for a short duration, and then returns to the high logic state. The information conveyed in a digital pulse train is conveyed in the number of pulses that occur, the rate at which pulses occur and or the time between pulses.

The output signals from a flow meter or from an optical encoder mounted on a rotating shaft are examples of a digital pulse train. It is also possible for a DAQ system to be required to output a digital pulse train as part of the control process. A stepper motor, for example, requires a series of digital pulses to control its speed and position. While input and output digital pulse trains can be practically measured or produced using digital I/O boards, counter/timer I/O boards are more effective in performing these functions.

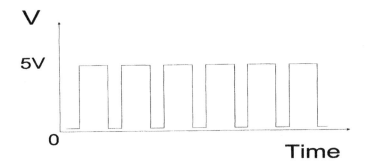

Figure 2.3
Digital pulse train signal

2.1.2   Analog signals

Analog signals contain information within the variation in the magnitude of the signal with respect to time. The relevant information contained in the signal is dependent on whether the magnitude of the analog signal is varying slowly or quickly with respect to time, or if the signal is considered in the time or frequency domains.

Analog DC signals

Analog DC signals are static or slowly varying DC signals. The information conveyed in this type of signal is contained in the level or amplitude of the signal at a given instant in time, not in how this level varies with respect to time. This is shown in Figure 2.4.

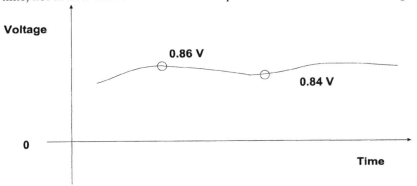

Figure 2.4
An analog DC signal

As the timing of the measurements made of slowly varying signals is not critical, the DAQ hardware would only be required to convert the signal level to a digital form for processing by the computer using an analog-to-digital converter (ADC). Low speed A/D boards would be capable of measuring this class of signal. Temperature and pressure monitoring are just two examples of slowly varying analog signals in which the DAQ system measures and returns a single value indicating the magnitude of the signal at a given instant in time. Such signals can be used as inputs to digital displays and gauges or processed to indicate a control-action (e.g. turn on a heater or open a valve) required for a particular process.

For example, control hardware like a valve actuator, requires only a slowly varying analog signal; the magnitude at a given point in time determining the control setting. DAQ hardware that could perform this task would only be required to convert the digital control setting to an analog form using a digital-to-analog converter (DAC) at the required instant in time. A low-speed general purpose D/A board could perform this function.

The most important parameters to consider for low speed A/D boards and D/A boards are the accuracy and resolution in which the slowly varying signal can be measured or output respectively.

Analog AC signals

The information conveyed in analog AC signals is contained not only in the level or amplitude of the signal at a given instant in time, but also how the amplitude varies with respect to time. The shape of the signal, its slope at a given point in time, the frequency, and location of signal peaks, can all provide information about the signal itself. An analog AC signal is shown in Figure 2.5.

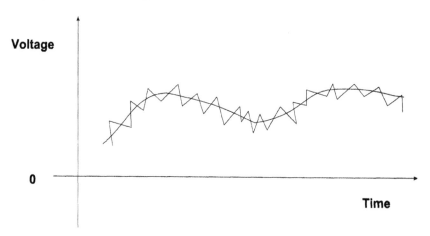

Figure 2.5
An analog AC signal

Since an analog AC signal may vary quite quickly with respect to time, the timing of measurements made of this type of signal may be critical. Hence, as well as converting the signal amplitude to a useful digital form for processing by the computer using an ADC, the DAQ hardware would be required to take the measurements close enough together to reproduce accurately the shape, and therefore the information, contained in the signal. Further to this, the information extracted from the signal may vary depending on when the measurement of the signal started and ended. DAQ hardware used to measure these signals would require an ADC, a sample clock, to time the occurrence of each A/D conversion, and a trigger to start and/or stop the measurements at the proper time, according to some external event or condition, so that the relevant portion of the signal can be obtained. A high-speed A/D board would be capable of performing these functions.

As all time varying signals can be represented by the summation of a series of sinusoidal waveforms of different magnitudes and frequencies, another useful way of extracting information is through the frequency spectrum of a signal. This indicates the magnitudes and frequencies of each of the sinusoidal components that comprise the signal rather than the time-based characteristics of the signal (i.e. shape, slope at a given point etc). This is shown in Figure 2.6.

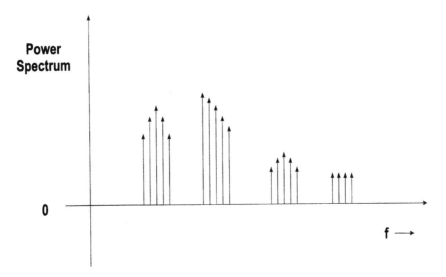

Figure 2.6
An analog AC signal in the frequency domain

Analysis in the frequency domain allows for easier detection and extraction of the wanted signal by filtering out unwanted noise components having frequencies much higher than the desired signal. The digital signal processing (DSP) required to convert the time-measured signal into frequency information and possibly perform analysis on the frequency spectrum, can be achieved with software or with special DSP hardware.

2.2   Sensors and transducers

A transducer is a device that converts one form of energy or physical quantity into another, in accordance with some defined relationship. Where a transducer is the sensing element that responds directly to the physical quantity to be measured and forms part of an instrumentation or control system, then the transducer is often referred to as a sensor.

In data acquisition systems, transducers sense physical phenomena and provide electrical signals that the system can accept. For example, thermocouples, resistive temperature detectors (RTDs), thermistors, and IC sensors convert temperature into an analog voltage signal, while flow transducers produce digital pulse trains whose frequency depends on the speed of flow.

Two defined categories of transducer exist:

- Active transducers convert non-electrical energy into an electrical output signal. They do not require external excitation to operate. Thermocouples are an example of an active transducer.
- Passive transducers change an electrical network value, such as resistance, inductance or capacitance, according to changes in the physical quantity being measured. Strain gauges (resistive change to stress) and LVDTs (inductance change to displacement) are two examples of this. To be able to detect such changes, passive devices require external excitation.

2.3   Transducer characteristics

Transducers are classified according to the physical quantity they measure (e.g. temperature, force etc).

Beyond the obvious selection of the type of transducer required to measure a particular physical quantity and any cost considerations, the characteristics that are most important in determining a transducer's applicability for a given application are as follows:

- Accuracy
- Sensitivity
- Repeatability
- Range

Accuracy

When a range of measurements is made of any process it is essential to know the accuracy of the readings and whether the same is maintained over the entire range or not. The accuracy of a transducer describes how close a measurement is to the actual value of the process variable being measured. It describes the maximum error that can be expected from a measurement taken at any point within the operating range of the transducer. Manufacturers usually provide the accuracy of a transducer as a percentage error over the operating range of the transducer, such as ± 1% between 20°C and 120°C, or as a rating (i.e. ± 1°C) over the operating range of the transducer.

Sensitivity

Sensitivity is defined as the amount of change in the output signal from a transducer to a specified change in the input variable being measured. Highly sensitive devices, such as thermistors, may change resistance by as much as 5% per °C, while devices with low sensitivity, such as thermocouples, may produce an output voltage that changes by only 5μV per °C.

Repeatability

If two or more measurements are made of a process variable at the identical state, a transducer's repeatability indicates how close the repeated measurements will be. The ability to generate almost identical output responses to the same physical input throughout its working life is an indication of the transducer's reliability and is usually related to the cost of the transducer.

Range

A transducer is usually constructed to operate within a specified range. The range is defined as the minimum and maximum measurable values of a process variable between which the defined limits of all other specified transducer characteristics (i.e. sensitivity, accuracy etc) are met. A thermocouple, for example, could well work outside its specified operating range of 0°C to 500°C, however its sensitivity outside this range may be too small to produce accurate or repeatable measurements.

Several variables affect the accuracy, sensitivity, and repeatability of the measurements being made.

In the process of measuring a physical quantity, the transducer disturbs the system being monitored. As an example, a temperature measuring transducer lowers the temperature of the system being monitored, while energy is used to heat its own mass.

Transducers are responsive to unwanted noise in the same way that a record player's magnetic cartridge is sensitive to the alternating magnetic field of the mains transformer (giving rise to 'mains hum').

Some transducers are subject to excitation signals that alter their response to the input physical quantity being measured. As an example, an RTD's excitation current can result in self-heating of the device, thereby changing its resistance.

2.4 Resistance temperature detectors (RTDs)

2.4.1 Characteristics of RTDs

Resistance temperature detectors (RTDs) are temperature sensors generally made from a pure (or lightly doped) metal whose resistance increases with increasing temperature (positive resistance temperature coefficient).

Most RTD devices are either wire wound or metal film. Wire wound devices are essentially a length of wire wound on a neutral core and housed in a protective sleeve. Metal film RTDs are devices in which the resistive element is laid down on a ceramic substrate as a zig-zag metallic track a few micrometers thick. Laser trimming of the metal track precisely controls the resistance. The large reduction in size with increased resistance that this construction allows, gives a much lower thermal inertia, resulting in faster response and good sensitivity. These devices generally cost less than wire wound RTDs.

The most popular RTD is the platinum film PT100 (DIN 43760 Standard), with a nominal resistance of 100 Ω ± 0.1 Ω at 0°C. Platinum is usually used for RTDs because of its stability over a wide temperature range (–270°C to 650°C) and its fairly linear resistance characteristics. Tungsten is sometimes used in very high temperature applications. High resistance (1000 Ω) nickel RTDs are also available. If the RTD element is not mechanically stressed (this also changes the resistance of a conductor), and is not contaminated by impurities, the devices are stable over a long period, reliable and accurate.

2.4.2 Linearity of RTDs

In comparison to other temperature measuring devices such as thermocouples and thermistors, the change in resistance of an RTD with respect to temperature is relatively linear over a wide temperature range, exhibiting only a very slight curve over the working temperature range. Although a more accurate relationship can be calculated using curve fitting – the Callendar-Van Dusen polynomial equations are often used – it is not usually required. Since the error introduced by approximating the relationship between resistance and temperature as linear is not significant, manufacturers commonly define the temperature coefficient of RTDs, known as alpha (α), by the expression:

$$Alpha \ (\alpha) = \frac{R_{100} - R_0}{100 \times R_0} \ \ \Omega / \Omega / °C$$

Where:

R_0 = Resistance at 0°C
R_{100} = Resistance at 100°C

This represents the change in the resistance of the RTD from 0°C to 100°C, divided by the resistance at 0°C, divided by 100°C.

From the expression of alpha (α) it is easily derived that the resistance R_T of an RTD, at temperature T can be found from the expression:

$R_T = R_0(1 + \alpha T)$

Where:

R_0 = Resistance at 0°C

For example, a PT100 (DIN 43760 Standard), with nominal resistance of 100 $\Omega \pm 0.1$ Ω at 0°C has an alpha (α) of 0.00385 Ω / Ω / °C. Its resistance at 100°C will therefore be 138.5 Ω.

2.4.3   Measurement circuits and considerations for RTDs

Two-wire RTD measurement

Since the RTD is a passive resistive device, it requires an excitation current to produce a measurable voltage across it. Figure 2.7 shows a two-wire RTD excited by a constant current source, I_{EX} and connected to a measuring device.

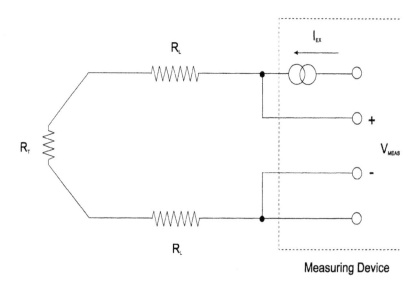

Figure 2.7
Two-wire RTD measurement

Any resistance, R_L, in the lead wires between the measuring device and the RTD will cause a voltage drop on the leads equal to ($R_L \times I_{EX}$) volts. The voltage drop on the wire leads will add to the voltage drop across the RTD, and depending on the value of the lead wire resistance compared to the resistance of the RTD, may result in a significant error in the calculated temperature.

Consider an example where the lead resistance of each wire is 0.5 Ω. For a 100 ω RTD with an alpha (α) of 0.385 Ω / °C, the lead resistance corresponds to a temperature error of 2.6°C (1 .0 Ω / 0.385 Ω / °C).

This indicates that if voltage measurements are made using the same two wires which carry the excitation current, the resistance of the RTD must be large enough, or the lead wire resistances small enough, that voltage drops due to the lead wire resistances are negligible. This is usually true where the leads are no longer than a few (<3) meters for a 100 Ω RTD.

Four-wire RTD measurement

A better method of excitation and measurement, especially when the wire lead lengths are greater than a few meters in length, is the four-wire RTD configuration shown in Figure 2.8.

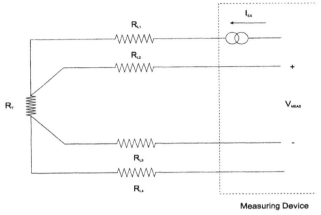

Figure 2.8
Four-wire RTD measurements

RTDs are commonly packaged with four (4) leads, two current leads to provide the excitation current for the device, and two voltage leads for measurement of the voltage developed. This configuration eliminates the voltage drops caused by excitation current through the lead resistances (R_{L1} and R_{L4}). Since negligible current flows in the voltage lead resistances, (R_{L2} and R_{L3}) only the voltage drop across the resistance R_T of the RTD is measured.

Three-wire RTD measurement

A reduction in cost is possible with the elimination of one of the wire leads. In the three-wire configuration shown in Figure 2.9, only one lead R_{L1} adds an error to the RTD voltage measured.

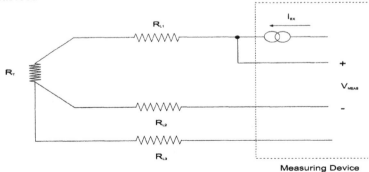

Figure 2.9
Three-wire RTD measurements

Self-heating

Another consequence of current excitation of the RTD is the possible effect that internal heating of the device may have on the accuracy of the actual temperature measurements being made. The degree of self-heating depends on the medium in which the RTD is being used, and is typically specified as the rise in temperature for each mW of power dissipated for a given medium (i.e. still air).

For a PT100 RTD device, the self-heating coefficient is 0.2°C/mW in still air, although this will vary depending on the construction of the RTD housing and its thermal properties. With an excitation current of 0.75 mA the power to be dissipated by the device is 56 μW $[(0.75 \times 10^{-3})^2 \times 100]$ corresponding to a rise in the temperature of the device due to self-heating of 0.011°C (56 μW × 0.2).

Inaccuracies in the temperature measurement due to self-heating problems, can be greatly reduced by:

- Minimizing the excitation power
- Exciting the RTDs only when a measurement is taken
- Calibrating out steady state errors

2.5 Thermistors

A cheap form of temperature sensing is provided by the thermistor, which is a thermally sensitive semiconductor resistor formed from the oxides of various metals. The type and composition of the semiconductor oxides used (i.e. manganese, nickel, cobalt etc) depend on the resistance value and temperature coefficient required.

More commonly used thermistor devices exhibit a negative temperature coefficient and have a high degree of sensitivity to small changes in temperature, typically 4% / °C.

Their accuracy is typically ten times better than thermocouples but not as accurate as RTDs. Thermistors are non-linear devices and directly useful over typical temperature ranges of –80°C up to 250°C. With regard to this, modern microprocessor based systems (either PCs or stand-alone data loggers) can be used to relieve some of the limitations caused by non-linearities, by modeling the non-linearities with quadratic equations.

Thermistors exhibit a high resistance, typically 3 kΩ, 5 kΩ, 6 kΩ and 10 kΩ at 25°C, although values as low as 100 Ω are available. High resistance means that the lead resistances of wires used to excite thermistors are usually negligible, requiring only two wire measurement schemes.

One of the attractions of thermistors is the wide range of shapes in the form of beads, discs, rods and probes that can be easily manufactured. Their small size means they have a fast thermal response, but can be quite fragile compared to RTDs that are more robust.

Just as excitation currents for RTDs can cause self-heating problems, this is even more the case for thermistors due to the higher device resistance values.

Self-heating problems can be greatly reduced by:

- Minimizing the excitation power
- Exciting the RTDs only when a measurement is taken
- Calibrating out steady state errors. Some authorities state that the temperature rise, in °C, due to self-heating can be calculated by dividing the proposed internal power dissipation by 8 mW.

2.6 Thermocouples

A thermocouple is two wires of dissimilar metals that are electrically connected at one end (measurement junction) and thermally connected at the other end (the reference junction). This is shown in Figure 2.10 below.

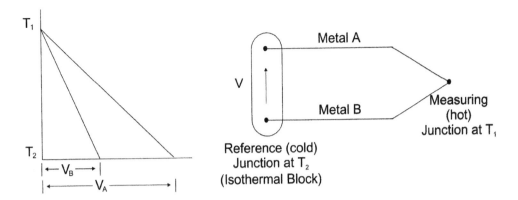

Figure 2.10
Thermocouple measurement

Its operation is based on the principle that temperature gradients in electrical conductors generate voltages in the region of the gradient.

Different conductors will generate different voltages for the same temperature gradient. Therefore, a small voltage, equal to the difference between the voltages generated by the thermal gradient in each of the wires ($V = V_A - V_B$), can then be measured at the reference junction.

Note that this voltage is produced by the temperature gradient along the wires and not by the junction itself. As long as the conductors are uniform along their lengths, then the output voltage is only affected by the temperature difference between the measurement (hot) junction and tile reference (cold) junction, and not the temperature distribution along the conductor between them.

2.6.1   Reference junction compensation

Calculations determining the temperature corresponding to a given measured voltage of a thermocouple assume that this voltage corresponds to a temperature gradient that is referenced to 0°C. Clearly, where the reference junction is allowed to follow ambient temperature, this is not the case.

Where ambient temperature variations of the reference junction would cause significant errors in the temperature calculation from the voltage output of the thermocouple, two methods of reference junction compensation exist:

Maintain the reference junction at a constant known temperature such as an ice bath (0°C). This is where the term 'cold junction' was originally derived.

Measure the temperature of the reference junction and add the reference junction voltage. The reference junction voltage is equal to the voltage, which would be generated by the same thermocouple if its measurement junction was at ambient temperature and its reference junction was at 0°C.

Obviously the second option is far easier to implement and has led to the design of many cold junction compensation circuits. The necessary voltage correction can be carried out with software, hardware, or a combination of both.

Hardware compensation

Hardware compensation requires dedicated circuitry to generate a compensation voltage according to the ambient temperature of the isothermal block, and add this voltage to the voltage measured at the measuring junction. As the voltage vs temperature relationship varies between thermocouples, each thermocouple type must have a separate compensation circuit that operates over the required working range of ambient temperatures. This makes hardware compensation circuitry for thermocouples complex and expensive, and by their nature, prone to inherent errors.

Software compensation

Software compensation requires only that an additional direct reading temperature sensor, such as a thermistor or silicon sensor, be used to measure the isothermal block temperature of the reference junction. Software is then used to calculate the equivalent reference junction voltage, either by polynomial equations, or look-up tables, for the thermocouple type being used. Once calculated, this value is added to the measured output voltage from the thermocouple. The resulting voltage is converted back to a temperature, representing the true thermocouple temperature.

Note: It is not always the case that changes in the ambient temperature lead to significant errors in determining the thermocouple temperature, as shown by the example below.

Example: Consider a type S thermocouple used to measure temperatures of 1500°C within a furnace. The ambient temperature of the reference junction is 25°C ± 15°C.

Since the sensitivity of the thermocouple is 12 μV / °C at 1500°C and a change from 10°C to 40°C at the reference junction produces a change of 180 μV in the net output voltage, the equivalent change in temperature at the measuring junction is 15°C.

This represents at most a 1% error of 1500°C over the operating temperature range of the reference junction. In this case, the error introduced by changes in the reference junction temperature might be ignored.

2.6.2   Isothermal block and compensation cables

Quite often thermocouples, especially those used in industrial applications, are at a considerable distance from the measuring points and require extension leads and connectors. Conventional copper wire and connectors cannot be used for the extensions as unwanted thermocouples are created. Wire and connectors of the same material as the thermocouple must be used. The use of extension cables made of similar but less pure metals than the actual thermocouple, is an economical way of extending the thermocouple circuit.

This wire, though considerably cheaper, has a limited temperature range of typically 0°C to 100°C and must not be used where temperatures exceed this range.

Where inline connectors are used these must also be of the same material as the thermocouples. Color-coded and polarized connectors (to prevent alloy reversal) are available.

References junctions are held at the same temperature by an 'isothermal block', a physical arrangement that ensures good thermal conductivity between the ends of the thermocouple cable. It is advisable to protect the isothermal block from rapid ambient temperature changes.

2.6.3   Thermocouple linearization

In addition to requiring cold-junction compensation, thermocouples are also highly non-linear, and thus require linearization. For example, a J type thermocouple has a thermal coefficient of 22 μV per °C at – 200°C, but 64 μV per °C at 750°C.

For most purposes, some form of software-based linearization is used. Two techniques of linearization are common:

Look-up tables: With this technique, a table of temperatures versus all possible measured voltages is stored, and the appropriate temperature is obtained via an indexing operation. This is very fast, but requires large amounts of memory. Cold-junction compensation is also difficult to handle.

Polynomial compensation: Using this technique, polynomial approximations are used to obtain temperature from voltage. The number of polynomial terms used depends on the temperature range, and the type of thermocouple. For example, type J thermocouples can be approximated to 0.1° over 0 to 760°C with a fifth-order polynomial, but an F-type thermocouple requires a ninth-order equation for only 0.5° accuracy.

For wide temperature ranges, several lower-order polynomials over narrower ranges are often used. For example, there are thermocouple board drivers that use three eighth-order polynomials for voltage-to-temperature conversions. The range of each equation is optimized for each type of thermocouple. In addition, a second-order polynomial is used to convert the cold-junction temperature to a thermocouple voltage for compensation.

The use of a second-order polynomial is only possible because the terminal block temperature varies from 0° to 70°C.

2.6.4 Thermocouple types and standards

Thermocouple standards specify the voltage vs temperature characteristics, color codes, error limits and composition of standard thermocouples. There are five standards for thermocouples in general use, namely NBS/ANSI (American), BS (British), DIN (German), JIS (Japanese), and NF (French).

Eight main types of thermocouples are general used in industry. These are divided into two main groups: base metal thermocouples (types J, K, N, E & T) and noble metal thermocouples (types R, S & B). Their composition and operating temperature range according to the NBS standard is shown in Table 2.1.

In addition, there are several high temperature tungsten-based thermocouples (types G, C & D), which allow temperature measurements between 0°C and 2320°C. As these thermocouples do not follow any official standards, manufacturers' data sheets should be consulted to ensure correct use.

Type	Positive	Negative	Temperature range □C
B	Pt, 30% Rh	Pt, 6% Rh	+300 to 1700
C	W, 5% Re	W, 26% Rh	0 to 2320
D	W, 3% Re	W, 25% Re	0 to 2320
E	Ni, 10% Cr	Cu, 45% Ni	−200 to 900
G	W	W, 26% Re	0 to 2320
J	Fe	Cu, 45% Ni	−200 to 750
K	Ni, 10% Cr	Ni, 2% Mn, 2% Al	−200 to 1250
N	Ni, 14% Cr, 1% Si	Ni, 4% Si, 0.1% Mg	−200 to 1350
R	Pt, 13% Rh	Pt	0 to 1450
S	Pt, 10% Rh	Pt	0 to 1450
T	Cu	Cu, 45% Ni	−200 to 350

Table 2.1
Thermocouple specifications (NBS Standard)

2.6.5 Thermocouple construction

In addition to thermocouple type, thermocouple style is another important factor in performance. Three basic styles are available, as illustrated in Figure 2.11(a).

The exposed, or bead, junction thermocouple has its junction exposed to air. Thermocouples with exposed junctions (Figure 2.11(b)) are generally used to measure gas temperature, and they have an extremely fast response time.

In ungrounded-junction thermocouples (Figure 2.11(c)), a conductive sheath protects the thermocouple junction. This sheath is electrically isolated from the thermocouple itself. This con-struction is particularly useful where high levels of electrical noise are present. The ungrounded junction thermocouple has the disadvantage that response time is long, typi-cally of the order of several seconds. Problems can also arise from thermal shunting, re-sulting in the junction being at a different temperature to the sheath.

In grounded-junction thermocouples, a conductive sheath also protects the thermocouple junction, and the sheath is electrically connected to the thermocouple junction. This has the advantage that response time is faster than for the ungrounded-junction type, and thermal shunting effects are minimized, while still maintaining good noise immunity. A disadvantage is the susceptibility to ground loop problems, which are particularly difficult to solve in thermocouples, due to low voltages.

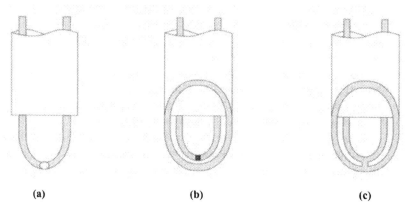

(a)   (b)   (c)

Figure 2.11
Thermocouple styles

2.6.6 Measurement errors

When making temperature measurements using thermocouples there are several possible sources of error, in addition to any errors that occur due to the accuracy of the measuring equipment.

These are:

- Reference junction isothermal characteristics and reference junction temperature sensor accuracy – the most significant sources of error. Temperature gradients between the temperature sensor and the terminals to which the thermocouples are connected result in errors of the magnitude of the temperature difference. Added to this is the magnitude of any inherent inaccuracies in the temperature sensor used to measure the ambient temperature.

- Induced electrical noise. Due to the low signal voltage levels from thermocouples, typically in the order of µV/°C, temperature measurements using thermocouples are susceptible to the effects of noise. This is especially true where long thermocouple cables are used in the measurement process. The effects of noise can be reduced by amplifying the low-level thermocouple voltages as close to the source as possible, and where this is not possible, by using twisted, shielded cables.

- Quality of the thermocouple wire. Where inhomogeneities occur in the thermocouple manufacturing process, the quality of thermocouple wire and its standard voltage temperature characteristics may vary.
- Linearization errors occur because polynomials are only approximations of the true thermocouple voltage output.

2.6.7   Wiring configurations

As the voltage levels from thermocouples are very small, typically in the order of μV/°C, temperature measurements using thermocouples are susceptible to the effects of noise. Three wiring configurations are shown in the following figures:

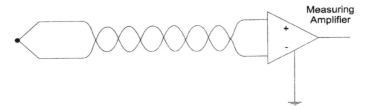

Figure 2.12
Thermocouple with no shielding

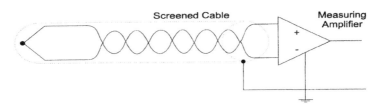

Figure 2.13
Thermocouple with thermocouple sheath and ungrounded junction

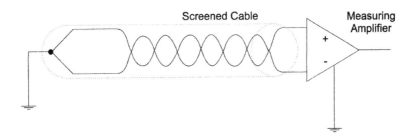

Figure 2.14
Thermocouple with thermocouple sheath and grounded junction

In addition to the wiring suggestions made above, it is important to consider isolation and over-voltage protection in the measurement circuitry, especially as a safeguard from charge buildup and other transient over-voltages on long thermocouple cables.

2.7 Strain gauges

Strain gauges are the most widely used devices for the measurement of force, or more particularly strain resulting from force. The most common type of strain gauge is the bonded resistance strain gauge, which consists of a resistive material, usually metal film a few micrometers thick, bonded to a polyester backing plate. A typical strain gauge is shown in Figure 2.15.

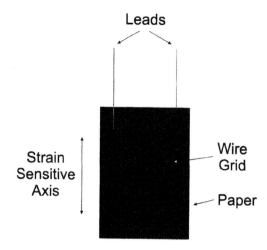

Figure 2.15
Typical bonded resistance strain gauge

The strain gauge operates on the principle that when strained, the length, cross-sectional area and resistivity of the metal film changes, thus changing the resistance of the conductor. When attached to a unit under test by an adhesive of some kind, the strain gauge experiences the same strain as the unit. The amount of strain can be measured by detecting changes in the resistance. Provided the change in length of the strain gauge is small, the relationship between resistance and strain is linear.

The ratio of the percentage change in resistance to the percentage change in length is known as the 'gauge factor' (GF) and is a measure of the sensitivity of the gauge.

$$GF = \frac{\Delta R / R_0}{\Delta L / L_0} = 1 + 2\sigma + \frac{\Delta \rho / \rho}{\Delta L / L_0}$$

Where:

R_0	=	resistance in ohms
ρ	=	resistivity in ohms per meter
L_0	=	length in meters
$\Delta R/R_0$	=	fractional resistance change
σ	=	Poisson's ratio
$\Delta L/L_0$	=	fractional change in length
$\Delta \rho/\rho$	=	fractional change in resistivity

The gauge factor, provided by manufacturers for a particular strain gauge, typically lies between 2 and 4 for commonly used metal foil gauges with nominal resistance of 120 Ω, 350 Ω and 1 kΩ. Thus, if a 350 Ω gauge with a gauge factor of 2.0 is stretched by 1%, then its resistance will change by 2% or 0.57 Ωs.

2.8 Wheatstone bridges

2.8.1 General characteristics

Due to its sensitivity, the Wheatstone bridge circuit is a commonly used circuit for the measurement of small changes in electrical resistance, particularly for strain gauges. It comprises four resistive elements and can be excited by either a voltage or current source. The standard Wheatstone bridge configuration is shown in Figure 2.16.

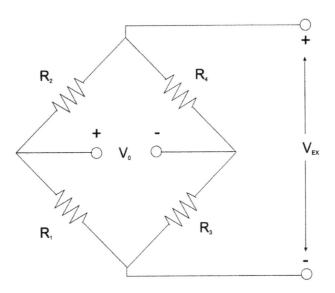

Figure 2.16
Standard Wheatstone bridge configuration

When excited by an input voltage V_{EX} it can be shown that the output voltage V_0 is given by the equation:

$$\frac{V_0}{V_{EX}} = \frac{R_1}{R_1 + R_2} - \frac{R_3}{R_3 + R_4}$$

When the ratio of resistances R_1 to R_2 is equal to the ratio of resistances R_3 to R_4, then the measured output voltage is 0 V, and the bridge is said to be *balanced*.

When a resistive element changes its resistance in response to the physical parameter being measured (e.g. a strain gauge) it is called the active element, while the remaining resistors are called completion resistors. If R_1 is an active element, then an increase in the resistance of the active element R_1 increases the output voltage. A decrease in this resistance will decrease the voltage appearing at the output. It is conversely true that if R_2 is an active element, then an increase in its resistance would result in a reduction of the voltage appearing at the output, while a decrease in this resistance would result in the output voltage increasing.

It can be shown that if any one of the bridge resistances is an active element whose nominal resistance (R_0) is precisely matched to each of the other completion resistors (i.e. $R_0 = R_2 = R_3 = R_4$), then for a small change in the active element resistance (ΔR), the ratio of the output voltage to the input voltage is given by:

$$\frac{V_0}{V_{EX}} = \frac{\Delta R}{4R_0}$$

This equation holds true irrespective of which arm of the bridge contains the active element.

Further to this, it can be shown that if there are (N) arms of the bridge which contain an active element, then for a small and equal change in the active element resistances ΔR, the ratio of the output voltage to the input voltage is given by:

$$\frac{V_0}{V_{EX}} = \frac{N}{4} \times \frac{\Delta R}{R_0}$$

This equation is true only if the sensitivity, of adjacent active elements of the bridge (i.e. R_1 & R_2, R_3 & R_4, R_1 & R_3 or R_2 & R_4) to changes in the physical parameter being measured, is of opposite polarity. This means that if R_1 and R_2 are active elements, then for an incremental change in the physical parameter being measured, the resistance of R_1 increases by ΔR and the resistance of R_2 decreases by ΔR. If the values in resistance of the active elements increase by the same amount, then the resistance in both arms would theoretically remain the same, the ratio of their resistances would remain the same, and their effects would cancel.

The above equation shows that the Wheatstone bridge is a ratiometric circuit whose output voltage sensitivity is proportional to the excitation voltage and the number of active elements in the bridge. The more closely matched the completion resistances are to the active resistive element(s), the smaller will be the unbalanced output voltage compared to the input excitation voltage. In addition, the output voltage polarity is dependent on where the active elements are positioned in the bridge, and whether these active elements increase or decrease resistance to an increase in the physical parameter being measured.

The quarter bridge, half bridge and full bridge configurations, in which strain gauges form the active elements, are discussed in the following sections.

2.8.2 Quarter bridge configuration

Where only one of the four resistors in the Wheatstone bridge is active, as shown in Figure 2.17, the circuit is known as a quarter bridge.

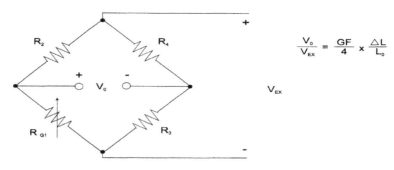

Figure 2.17
Quarter bridge circuit

In this configuration, an increase in the resistance of the active strain gauge resistance R_{G1} increases the output voltage, while a decrease in this resistance will decrease the voltage appearing at the output. Therefore, for the quarter bridge configuration, the polarity of the output voltage, and whether the voltage increases or decreases with increasing strain, depends on the position of the strain gauge in the bridge circuit and whether the strain gauge resistance increases or decreases with increasing strain.

Where the completion resistors are precisely matched ($R_2 = R_3 = R_4$) and the nominal strain gauge resistance is chosen to be equal to these values then it can be deduced from the previous equations that for a small change in the active resistance ΔR, the micro-strain ($\mu E = \Delta L / L_0 \times 10^6$) of the strain gauge is given by:

$$\mu E = \frac{4\,V_0}{V_{EX} \times GF} \times 10^6$$

Where:

μE	=	micro-strain ($\Delta L / L_0 \times 10^6$)
GF	=	gauge factor
V_0	=	unbalanced output voltage
V_{EX}	=	excitation voltage
ΔL	=	change in length
L_0	=	unstrained length

This equation assumes that the change in strain gauge resistance from its nominal value is very small, compared to the nominal resistance value.

2.8.3 Half-bridge configuration

As we have seen, it is possible to increase the sensitivity of a quarter bridge circuit by replacing one or more of the completion resistors with other active elements. Adding a second strain gauge, as shown in Figure 2.18, subjected to the same strain will double the output from the bridge. This is known as a half bridge circuit.

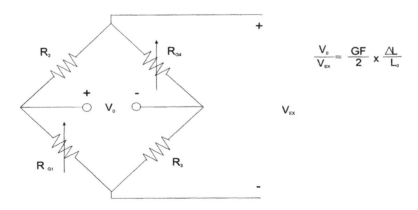

Figure 2.18
Half bridge circuit

Note: The placement of an identical strain gauge in the same side of the bridge would have no effect on the output voltage. Since the change in resistance in the adjacent arms would theoretically remain the same, the ratio of their resistances would remain the same and their effects would cancel.

2.8.4 Full bridge configuration

In circumstances where it is possible to place strain gauges, which have equal, and opposite strain (i.e. on opposite sides of a bending beam), it is possible to make all arms of the bridge active and get four times the sensitivity. This configuration, shown in Figure 2.19, is referred to as a full bridge.

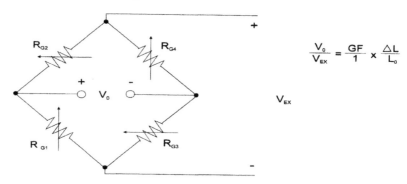

$$\frac{V_0}{V_{EX}} = \frac{GF}{1} \times \frac{\triangle L}{L_0}$$

Figure 2.19
Full bridge circuit

2.8.5 Wiring connections

As well as providing a choice of voltage or current excitation for the bridge circuit, signal conditioning equipment used to measure the output from a Wheatstone bridge, often provides two of the precision trimmed compensation resistors as part of its own circuitry. This provides flexibility in configuring quarter bridge, half bridge or full bridge circuits, but requires the user to add the active element(s) and any required matching compensation resistors. Any compensation resistors added by the user, and external to the signal conditioning equipment, should be precision-trimmed, with high accuracy and stability, especially with regard to temperature.

As the voltage output sensitivity from the Wheatstone bridge is proportional to the input excitation voltage, it is possible that cable and connector resistance voltage drops may reduce the excitation voltage seen at the bridge circuit and lead to inaccuracies in the measured output. Consider the three-wire half bridge configuration of Figure 2.20.

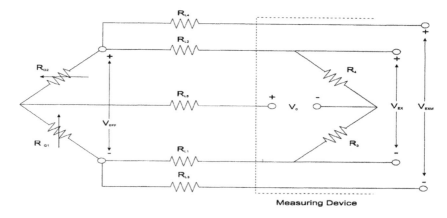

Measuring Device

Figure 2.20
Three-wire half bridge circuit wiring configuration

For the three-wire half bridge configuration shown, the wire lead resistances R_{L1} and R_{L2} appear in the opposite arms of the bridge and therefore have little effect on the bridge balance. However, they do affect the effective excitation voltage V_{EFF} by a small amount. If the nominal strain gauge resistance is 120 Ω and the lead resistance is 1 Ω then the effective excitation voltage V_{EFF} is given by the expression:

$$V_{EFF} = \frac{120}{121} \times V_{EX} = 0.992V_{EX}$$

The measured excitation voltage, V_{EXM}, would therefore be 0.8% higher than the effective excitation voltage. This 0.8% error should be seen in context with typical uncertainty in the gauge factor of ± 1%.

Where the lead wire resistance is more significant (especially for long cable runs) compared to the active element resistance, the five-wire configuration shown in Figure 2.21, should be used to eliminate this error. In this configuration, two leads are used to provide current or voltage excitation for the completed bridge circuit, while two separate leads are used to measure the effective excitation voltage. The voltage drops caused by (R_{LI} and R_{L2}) will still result in the effective voltage excitation being reduced by the same amount as the three-wire half bridge configuration. However, since negligible current flows in the lead resistances (R_{L3} and R_{L4}), the effective excitation voltage can be accurately measured (V_{EXM}).

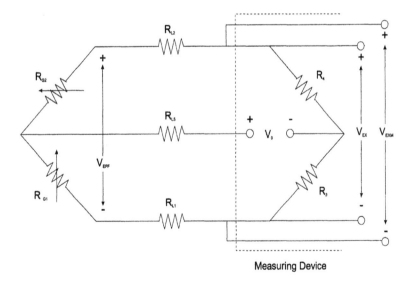

Measuring Device

Figure 2.21
Five-wire half bridge circuit wiring configuration

When using the three-wire quarter bridge configuration shown in Figure 2.22, both a single active element and matching completion resistor must be provided external to the signal conditioning equipment.

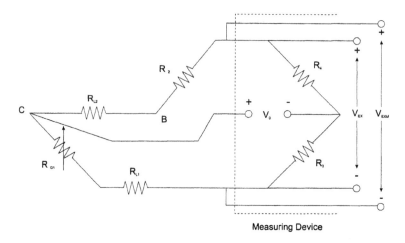

Figure 2.22
Three-wire quarter bridge circuit wiring configuration

In this configuration, the lead resistances (R_{L1} and R_{L2}) appear in opposite arms of the left-hand side of the bridge and therefore have little effect on the bridge balance. Assuming that the lead resistances (R_{L1} and R_{L2}) are insignificant compared to the strain gauge nominal resistance (i.e. $V_{EFF} = V_{EX}$) then the use of the third wire ensures that the unbalanced output voltage V_0 will be correctly measured between points A and C of the bridge. If only two wires are used, then V_0 will be measured as the voltage difference between points A and B and the lead resistances (R_{L1} and R_{L2}) will both be in series with the strain gauge. Changes in R_{L1} and R_{L2} due to temperature would therefore be indistinguishable from changes in R_{G1} due to strain, using the two-wire method.

Where the lead resistances (R_{L1} and R_{L2}) are significant compared to the strain gauge nominal resistance value, then the error in the effective excitation voltage is the same as for the three-wire half bridge configuration. In this case, a five-wire quarter bridge configuration should be used.

2.8.6 Temperature considerations

Changes in the resistance of a strain gauge can be caused by changes in the stress applied to the device, as well as variations in temperature.

Typical resistive changes for metal foil gauges due to temperature and strain are 0.015% /°C and 0.0002 %/μE. For a strain gauge with these specifications, a one-degree change in temperature would cause an effective strain error of approximately 75 μE.

Using a second strain gauge for temperature correction (unstressed), in the same arm of the bridge, gives significant reduction in the errors resulting from temperature changes. This is because the change in resistance due to temperature is the same for each of the strain gauges and therefore has a canceling effect.

2.8.7 Measurement errors

There are a number of error sources when measuring strain using the Wheatstone bridge. These are:

Gauge Factor uncertainty (typically 1%).

Bridge non-linearity. The equations derived in the sections above assumed that the change in strain gauge resistance is very small compared to the nominal gauge resistance. The error that would be introduced with an imbalance of 10000 μE is approximately 1%.

This can be reduced by modeling the non-linearity of the gauge in software with a suitable polynomial.

Matching of compensation resistors to the strain gauge. Where the compensation resistor in the same arm of the gauge is different by 1%, the error is 0.5%.

Measurement errors caused by accuracy; resolution of the measuring device and lead resistances

Temperature effects. The resistance of both the strain gauge and the compensation resistors vary with changes from the temperature at which a bridge is calibrated. This effect is greatly reduced by including an unstressed strain gauge in the same arm of the bridge.

Self-heating of gauges. This can be greatly reduced by energizing the bridge only while measurements are being made.

3

Signal conditioning

3.1   Introduction

PC based data acquisition (DAQ) systems and plug-in boards are used in a wide range of applications. Typically, general-purpose DAQ plug-in boards are used for measuring analog and digital input and output voltages.

As we have seen, many transducers' signals must be conditioned in some way before a DAQ board or measuring system can accurately acquire the desired signal. Signal conditioning is the term generally used to describe the front end pre-processing required to convert the electrical signals received from transducers into signals which DAQ plug-in boards or other forms of data acquisition hardware can accept.

In addition, many transducers require excitation currents or voltages, Wheatstone bridge completion and linearization to allow accurate measurement of the required signal. Therefore, most PC based DAQ systems include some form of signal conditioning equipment.

The fundamental functions that a signal conditioning equipment performs are:

- Amplification
- Isolation
- Filtering
- Excitation
- Linearization

The type of signal conditioning equipment required, and the manner in which this is interfaced within the DAQ system, is largely dependent on the number and type of transducers, their excitation and earthing requirements, and no less importantly, how far the transducers are located from the personal computer, which must acquire, analyze and store the transducer signal data.

The signal conditioning functions performed are implemented in different types of signal conditioning products, covering a range of price, performance, modularity and ease of use.

This chapter discusses several of the main hardware configurations used when integrating signal conditioning products into a DAQ system, as well as the general signal conditioning functions that must be performed.

3.2 Types of signal conditioning

3.2.1 Amplification

Amplification is one of the primary tasks carried out by signal conditioning equipment. It performs two important functions:

- Increases the resolution of the signal measurement.
- Increases the signal-to-noise ratio (SNR).

Amplification is primarily used to increase the resolution of the signal measurement. Consider a low-level signal of the order of a fraction of an mV, fed directly to a 12-bit A/D converter with full-scale voltage of 10 V. There will be a resultant loss of precision because the A/D converter has a resolution of only 2.44 mV (15.2 µV for 16 bit resolution). The highest possible resolution can be achieved by amplifying the input signal so that the maximum input voltage swing equals the maximum input range of the ADC.

Another important function of amplification is to increase the SNR. Where transducers are located a long way from the data acquisition board and the signal measurements are transmitted through an electrically noisy environment, then low-level voltage signals can be greatly affected by noise. Where the low-level signals are amplified at the data acquisition board after they have been transmitted through the noisy environment, then any noise superimposed on the signal will also be amplified by the same amount as the signal. If the noise is of the same order of magnitude as the signal itself (i.e. the SNR is low), then the signal measurement may be lost in noise, leading to inaccurate and meaningless measurements.

Amplifying the low-level signals before they are transmitted through the noisy environment increases the level of the required signal before they are affected by noise, thereby increasing the SNR of the signal for the same level of noise. Consider for example, a J-type thermocouple, which outputs a very low-level voltage signal that varies by about 50 µV/°C. If the thermocouple leads were to travel through a noisy electrical environment for say 10 m, then it is possible that the amount of noise coupled onto the thermocouple leads could be of the order of 200 µV. This noise-induced error corresponds to 4°C at the measuring device. Amplifying the signal with an amplifier gain of 500, close to the thermocouple, produces a thermocouple signal that varies by approximately 25 mV/°C. At this higher signal level, the 200 µV of induced noise coupled onto the 10m cable would result in a much smaller error, adding only a fraction of a degree Celsius of noise to the measured temperature.

3.2.2 Isolation

An isolated signal conditioner passes a signal from its source to the measurement device without a galvanic or physical connection. The most common methods of circuit isolation include opto-isolation, magnetic or capacitive isolation. Opto-isolation is primarily used for digital signals. Magnetic and capacitive isolations are used for analog signals, modulating the signal to convert it from a voltage to a frequency and transmitting the frequency signal across a transformer or capacitor without a direct physical connection before being converted back to a voltage.

Isolation performs several important functions. Firstly, isolation provides an important safety function by protecting expensive computer equipment and DAQ boards, as well as the equipment operators, from high voltage transients that could be caused by electrostatic discharge, lightning, or high voltage equipment failure. While isolated signal conditioning equipment provides an effective physical barrier and transient voltage protection for the computer and DAQ equipment, typically up to 1500 V, separate over-voltage protection is usually provided at the input(s) of the signal conditioning equipment to prevent internal damage to the signal conditioning equipment itself. In medical applications, isolation prevents the possibility of potentially fatal voltage or current signals from reaching sensors or transducers attached to or implanted in the human body.

Another important function of isolation is to ensure that ground loops or common-mode voltages do not affect the accuracy of measured signals. Ground loops, caused by a potential difference between the source ground and the ground reference of the measuring device, may cause inaccuracies in the measured signal, or if too large, may damage DAQ equipment. Using isolated signal conditioning modules will eliminate the ground loop, and ensure that the signals are accurately measured.

We shall see later that common-mode voltage signals are those that appear equally on each input of a measurement system. They can be caused by potential differences in the ground references of the source and the measurement system (i.e. ground loops) or are a necessary part of the measurement process (e.g. measuring the temperature of a device that is many volts above ground potential).

3.2.3 Filtering

Filtering removes unwanted noise from signal measurements before they are amplified and presented to the A/D converter. In intelligent signal conditioning modules, integrating A/D converters go a long way to averaging (filtering) out any cyclical noise appearing at the input. Alternatively, software averaging may also be used to digitally filter out periodic noise signals such as mains hum. This technique involves taking many more measurements than is necessary to acquire the wanted signal, then averaging them to produce a single measurement. If the samples are averaged over the period of the cyclical noise signal then this signal will be averaged to zero.

Where there is no other form of filtering, an analog hardware filter provides the cheapest option. There are two types of analog filter, namely passive filters that use only passive components (such as capacitors and resistors), and active filters that utilize operational amplifiers.

Ideally, filters should eliminate all data at frequencies outside the specified frequency range, providing a very sharp transition between the frequencies that are passed and those that are filtered out. Most practical filters are not ideal and do not usually eliminate all the undesirable amplitude components outside a specified frequency range.

Attributes common to filters are:

- Cut-off frequency
  This is the transition frequency at which the filter takes effect. It may be the high-pass cut-off or the low-pass cut-off frequency and is usually defined as the frequency at which the normalized gain drops 3 dB below unity.

- Roll-off
  This is the slope of the amplitude versus the frequency graph at the region of the cut-off frequency. This characteristic distinguishes an ideal filter from a practical (non-ideal) filter. The roll-off is usually measured on a logarithmic scale in units of decibels (dB).

- Quality factor 'Q'
    This variable is an adjustable characteristic of a tuned filter and determines the gain of the filter at its resonant frequency, as well as the roll-off of the transfer characteristic, on either side of the resonant frequency.

Active filters are more frequently used since they provide a sharper roll-off and better stability. Such filters are described below.

Low pass filter

Low pass filters pass low frequency components of the signal and filter out high frequency components above a specific high frequency. An active low pass filter is shown in Figure 3.1.

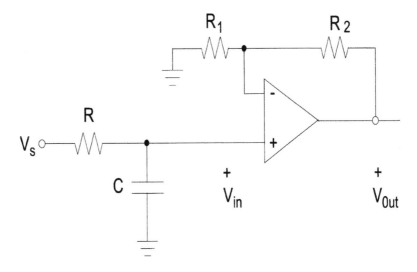

Figure 3.1
Active low pass filter

The transfer characteristic of an ideal low pass filter is shown in Figure 3.2.

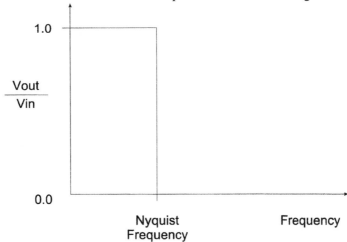

Figure 3.2
Ideal low pass filter transfer characteristics

The transfer characteristics of a practical filter for minimum 'Q' and maximum 'Q' are shown in Figure 3.3.

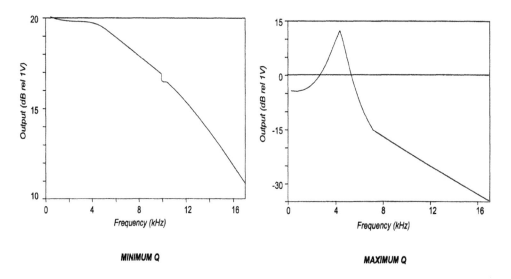

Figure 3.3
Practical active low pass filter transfer characteristics

High pass filter

High pass filters pass high frequencies and filter out low frequencies beginning at a specific low frequency. An active high pass filter is shown in Figure 3.4.

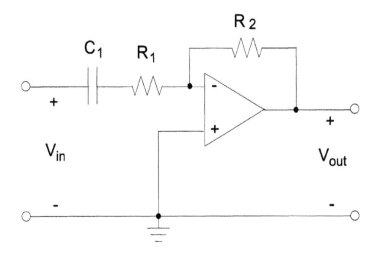

Figure 3.4
Active high pass filter

The transfer characteristic of an ideal high pass filter is shown in Figure 3.5.

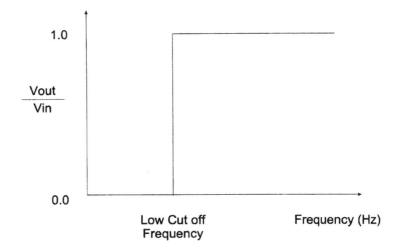

Figure 3.5
Ideal high pass filter transfer characteristics

The transfer characteristics of a practical filter for minimum 'Q' and maximum Q are shown in Figure 3.6.

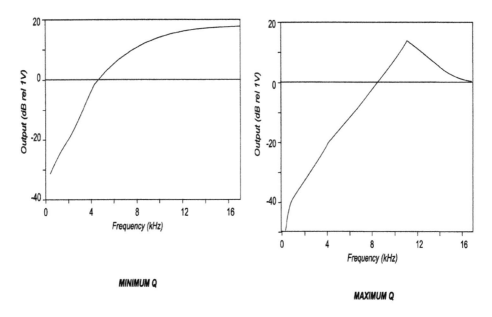

Figure 3.6
Practical active high pass filter transfer characteristics

Band pass (selective) filter

Band pass filters pass only those frequencies within a certain range specified by a low and high cut-off frequency.

This is also known as a selective filter and combines a low pass and high pass filter in series, each tuned to the low and high cut-off frequencies respectively. The ideal transfer characteristic of an active band pass filter is shown in Figure 3.7.

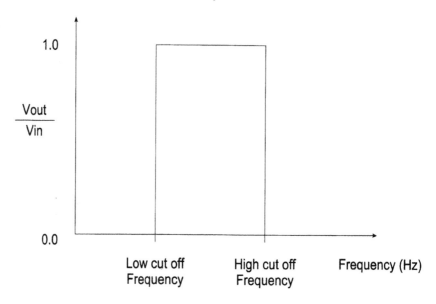

Figure 3.7
Ideal band pass filter transfer characteristics

The transfer characteristics of a practical filter for minimum 'Q' and maximum 'Q' are shown in Figure 3.8.

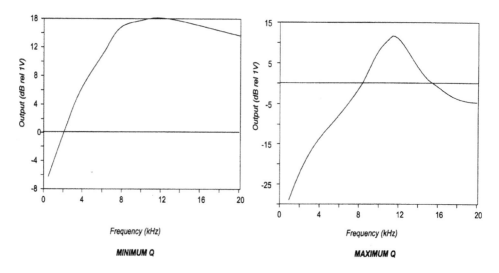

Figure 3.8
Practical active band pass filter transfer characteristics

Band stop (notch) filters

Notch filters filter out a certain range of frequencies specified by a start and stop frequency, and pass all others. These filters combine a high pass and a low pass in parallel, each tuned to the low and high cut-off frequencies respectively. The ideal transfer characteristic of an active band stop filter is shown in Figure 3.9.

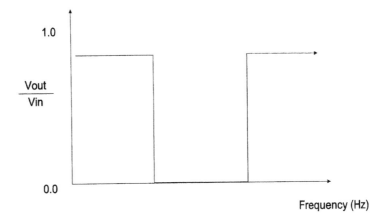

Figure 3.9
Ideal band stop filter transfer characteristics

The transfer characteristics of a practical filter for minimum 'Q' and maximum 'Q' are shown in Figure 3.10.

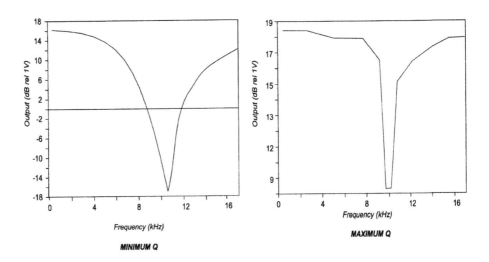

Figure 3.10
Practical active notch transfer characteristics

Butterworth filter

Butterworth filters provide a higher level of low pass filtering, containing two or more low pass filter stages. The number of stages 'n' of the filter determines how sharp the roll-off is at the cut-off frequency. A two-stage filter of this type is known as a second order Butterworth filter as shown in Figure 3.11.

A fourth order Butterworth filter would have two of the filter sections shown in Figure 3.11 cascaded together.

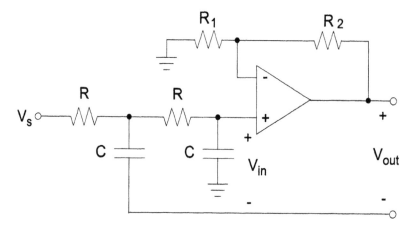

Figure 3.11
Two-stage Butterworth filter

3.2.4 Linearization

As we have seen, the output signals from transducers such as thermocouples exhibit a non-linear relationship to the phenomena being measured over a given input range. The data acquisition software typically performs linearization of these signals. However, where the non-linear relationship is predictable and repeatable this task can be performed by intelligent signal conditioning hardware. This typically requires the signal conditioning equipment to be programmed for a particular type of transducer, but once completed, the measurements returned to the host PC or stored as part of the measurement process are directly related to the phenomena (e.g. temperature) being measured.

3.3 Classes of signal conditioning

Signal conditioning products, available from many different equipment manufacturers, are provided in many different forms covering a range of price, performance, modularity and ease of use. The type of signal conditioning hardware should be matched to the specific application. The main forms are discussed below.

3.3.1 Plug-in board signal conditioning

This range of signal conditioning hardware typically covers specialty plug-in data acquisition boards where the signal conditioning hardware is contained on the board itself. This is shown in Figure 3.12.

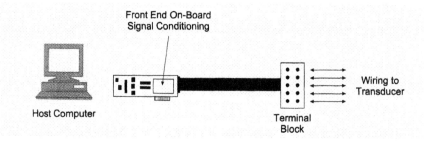

Figure 3.12
Plug-in DAQ board signal conditioning

Each board specializes in one type of transducer; thermocouple boards for interfacing to thermocouples, strain gauge boards for strain gauges, etc. These boards are typically used for small, specialized data acquisition systems that have a limited number of transducers located near the host computer.

3.3.2 Direct connect modular – two-wire transmitters

Two-wire transmitters are two-port modular signal conditioning modules that input an unconditioned signal on the input port and output a conditioned signal on the output port. A single module is required for each type of transducer (or actuator). These signal conditioning modules are not intelligent devices and do not perform on-board A/D conversion. Instead, the conditioned analog signal is transmitted over two lines to the data acquisition board in the host PC, either as a voltage, or converted into a standard current loop signal (4–20 mA) to the data acquisition board, hence the name two-wire transmitters. The simplified functional block diagram of a typical two-wire transmitter signal conditioning module is shown in Figure 3.13.

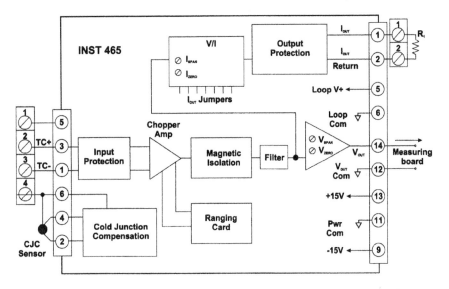

Figure 3.13
Functional block diagram of a two-wire transmitter signal conditioning module

Voltage outputs (±10 V or 0–10 V), compatible with the single ended inputs of most data acquisition boards allow easy interfacing to the latest data acquisition board technology. However, due to voltage drops, which may occur on the signal lines and the effects of noise that is proportional to the length of the transmission lines, voltage outputs should only be used for short transmission lines.

Current signals have much greater immunity to noise and can be transmitted over hundreds of meters (up to 1000 meters), to a receiver that converts the currents back into a voltage, for A/D conversion at the PC. The receiver is principally a resistor, nominally in the order of 500 Ω for full-scale deviation of 10 V (500 Ω X 20 mA). A separate pair of wires is used for the current loop of each individual sensor, resulting in many cable pairs to the PC. A power supply (between 15–40 V), capable of driving as many current loops as there are modules, is required.

As individual signal conditioning modules require external power, they are typically designed to plug into a mounting board with on-board power supply as shown in Figure 3.14.

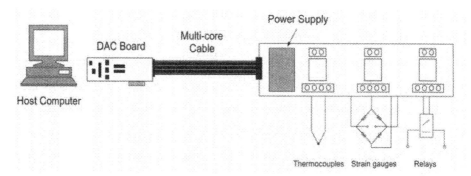

Figure 3.14
Board mounted modular signal conditioning

A single connector on the mounting board is used for easy cable connection between the mounting board and the I/O of the plug-in data acquisition board. Cables are typically a multi-core twisted-pair. This allows many different types of transducers to be interfaced to the latest plug-in data acquisition boards, but does not facilitate distributed I/O.

3.3.3   Distributed I/O – digital transmitters

Often sensors must be remotely located from the personal computer in which the processing and A/D conversion of the analog data takes place. This is especially true in industrial environments where sensors such as thermocouples and strain gauges are located in hostile environments over a wide area, possibly hundreds of meters away. In noisy environments, it is very difficult for the very small signals received from sensors, such as thermocouples and strain gauges (in the order of mV), to survive transmission over such long distances, especially in their raw form, without the quality of the sensor data being compromised.

An alternative to running long (and possibly expensive) wires from the transducers directly, or from two-wire transmitter modules, is the use of distributed I/O. Distributed I/O is available in the form of signal conditioning modules that are remotely located from the host PC, near the sensors to which they are interfaced. One module is required for each sensor used, allowing for high levels of modularity (single point up to hundreds per location). While this can add a reasonable expense to systems with large point counts, the benefits in terms of signal quality and accuracy may be worth it.

One of the most commonly implemented forms of distributed I/O is the digital transmitter. These intelligent devices perform all the functions of simple signal conditioning modules (two-wire transmitters) but also contain a micro-controller and A/D converter to perform the digital conversion of the signal within the module itself. Converted data is transmitted to the computer via an RS-232 or RS-485 communications interface. The simplified functional block diagram of a typical digital transmitter is shown in Figure 3.15.

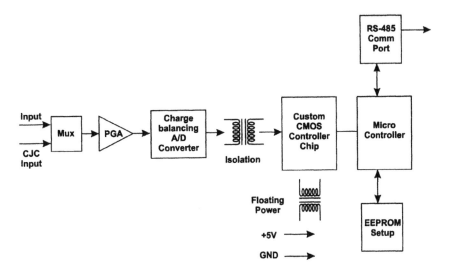

Figure 3.15
Functional block diagram of a digital transmitter signal conditioning module

The use of RS-485 multi-drop networks greatly reduces the amount of cabling required because each signal conditioning module shares the same cable pair. It does however require an RS-232 to RS-485 converter to allow communications between the computer and the remote signal conditioning modules.

Digital transmitters are available that provide two alternatives for configuring the distributed I/O system. In the first system configuration, shown in Figure 3.16, the digital transmitter modules are designed to plug into a mounting board with facilities to accept an external power supply.

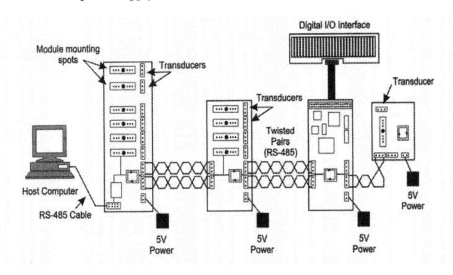

Figure 3.16
Distributed I/O signal conditioning network using board mounted digital transmitter modules

The second distributed I/O system configuration, shown in Figure 3.17 makes use of individual digital transmitter modules. Individual modules can be easily stacked together where many transducers are located in close proximity or can be positioned individually where they are required.

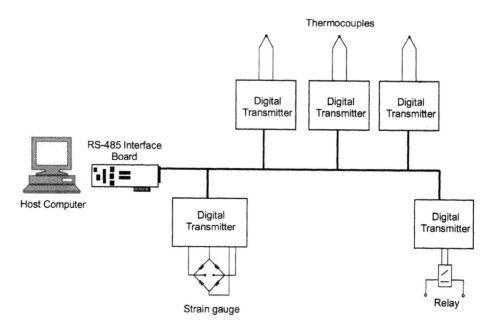

Figure 3.17
Distributed I/O signal conditioning network using individual digital transmitter modules

Like other signal conditioning modules, these devices require an external power supply. The power supply should be located to supply as many signal-conditioning modules as its rating will allow.

3.4   Field wiring and signal measurement

When measuring analog input signals from transducers, it is unfortunately not just a simple matter of wiring the transducer leads to the signal conditioning equipment or data acquisition board, or connecting the signal conditioning equipment to the data acquisition board itself.

Signal conditioning equipment and data acquisition boards typically provide a variety of methods for taking measurements of input signals. When determining the wiring connections and analog input configuration that will produce accurate and noise free measurements, careful consideration must be given not only to the type of signal produced by the transducer but also to the nature of the signal source.

The most common electrical signal output by transducers or signal conditioning equipment is in the form of voltage. In certain situations, where the output signal from signal conditioning equipment must be transmitted over long distances or is particularly susceptible to noise, it may be converted to a current or frequency signal. In most cases however, the signal is converted back to a voltage signal before a measurement is taken. It is therefore necessary to understand the voltage signal source and the various methods of taking measurements of voltage signals.

Two categories of voltage signal source are defined:

- Grounded signal source
- Floating (ungrounded) signal source

Three types of measurement are available on most signal conditioning equipment and data acquisition boards:

- Single-ended
- Differential
- Pseudo-differential

Since an understanding of the types of signal sources and measurement systems is necessary to determine the best methods of taking analog signal measurements, these topics are discussed in the following sections.

3.4.1 Grounded signal sources

By definition, voltage is a measurement of the potential difference between two points. Grounded signal sources have one of their signal leads connected to the system ground as shown in Figure 3.18. This is theoretically shown as earth potential, although the system ground is not necessarily at earth potential. The voltage output from the signal source is the potential difference between the system ground and the positive signal lead of the signal source.

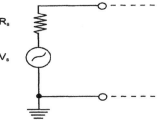

Figure 3.18
Grounded signal source

A common example of a grounded signal source is an instrument that is earthed via its AC plug to the building ground.

3.4.2 Floating signal sources

Floating or ungrounded signal sources, as shown in Figure 3.19, do not have either of the signal source leads connected to the system ground. This means that the signal source is not referenced to any absolute reference. The potential difference, that each of the signal lines may have, with respect to the system ground or earth potential between the signal lines, is not indicated in anyway by the voltage potential.

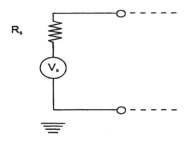

Figure 3.19
Ungrounded signal source

Examples of floating signal sources are transformers, isolation amplifiers, batteries, and battery powered instruments.

3.4.3   Single-ended measurement

A ground-referenced measurement system, as shown in Figure 3.20, is one in which the voltage measurement is taken with respect to 'ground'. It is known as a single-ended measurement because only one signal line is required to determine the signal voltage, provided it is ground-referenced.

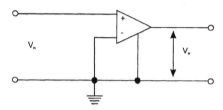

Figure 3.20
Single-ended measurement

3.4.4   Differential measurement

A differential measurement system, as shown in Figure 3.21, has neither of its inputs tied to a fixed reference, such as earth or system ground.

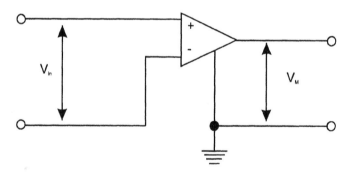

Figure 3.21
Differential measurement

Differential measurement is beneficial because, noise induced equally into each of the signal lines appears as a common mode voltage at the input and is largely rejected (see Common mode voltages and CMRR, below).

3.4.5   Common mode voltages and CMRR

Common mode voltages

Ideally a differential measurement system measures only the potential difference between its positive and negative terminals. Where a signal source is measured using differential inputs, and there is a voltage measured with respect to the measurement ground that is present on both input lines, then this voltage is referred to as a common mode voltage. This is shown in Figure 3.22

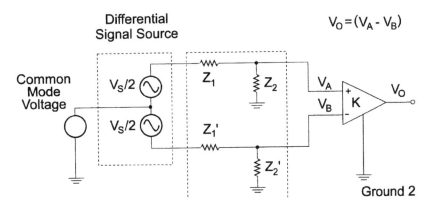

Figure 3.22
Common mode voltages

The common mode voltage V_{cm} can be calculated from the following:

$$V_{cm} = \frac{(V_A + V_B)}{2}$$

Where:

V_A = Voltage at the non-inverting terminal of the measurement system with respect to the instrumentation amplifier ground.

V_B = Voltage at the inverting terminal of the measurement system with respect to the instrumentation amplifier ground.

An example of a common mode voltage is the output from a bridge circuit, in which the small differential signal is superimposed over a much larger common mode voltage introduced by the excitation of the bridge circuit.

Common mode rejection ratio (CMRR)

Ideally, a differential amplifier would completely reject any common mode voltages present on its input signal lines and only amplify the potential difference between them. Practically, however, these devices do not totally reject common mode voltages. The common mode rejection ratio (CMRR) measures the ability of a differential input amplifier to reject signals that are common to both signal inputs.

The CMRR is defined as the ratio between the common mode signal present at the input to the amplifier and the signal produced by this voltage at the output of the amplifier, as defined by the following equation:

$$CMRR = 20 \log 10(\frac{V_{cm}}{V_{out}})$$

This ratio, normally expressed in dB, can be used to calculate the output voltage error, which would occur due to a common mode voltage appearing at the input. The higher the CMRR, the better the rejection of common mode signals, and the more accurate the output due to the differential signal being measured. Typically, a CMRR of 60 dB–80 dB could be expected for a well-designed system.

Common mode input voltage limits

Practically, measurement systems also have another limitation, and this is that there is a maximum and minimum common mode input voltage allowable on each input, with respect to the measurement system ground. Applying common mode voltages to either input beyond this input range will result in measurement errors, or, in the worst case, possible damage to the measurement circuitry.

3.4.6   Measuring grounded signal sources

Differential measurement of grounded signal source

A grounded signal source is best measured with a differential or pseudo-differential measurement system as shown in Figure 3.23. In this configuration, any potential difference (ΔV_g) between the ground references of the source and the measurement system appears as a common-mode voltage to the measurement system. The measured differential voltage is defined as:

$$V_m = (V_s. + \Delta V_g) - \Delta V_g = V_s$$

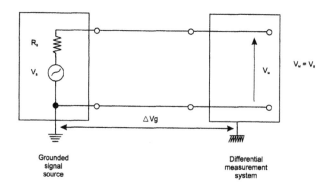

Figure 3.23
Differential measurement of a grounded signal source

Single-ended measurement of a grounded signal source

When a single-ended measurement system is used to measure a grounded signal source, as shown in Figure 3.24, measurement problems may occur. In this configuration, any potential difference (ΔV_g) between the signal source ground and the measuring system ground is added to the signal source voltage as part of the measurement. The measured voltage is defined as:

$$V_m = V_s. + \Delta V_g$$

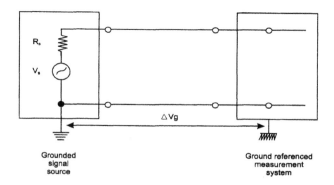

Figure 3.24
Single-ended measurement of a grounded signal source

If the signal voltage levels are quite high compared to the reference ground potential difference, and the wiring between the source and the measurement system has low impedance, then the inaccuracies in the signal voltage measurement may be acceptable.

3.4.7 Ground loops

The classic ground-loop problem arises because true earth ground is not necessarily the same potential at different locations. Where the ends of a wire are earth grounded at different locations, the potential difference between them (which may vary from micro-volts to many volts) can cause significant currents, referred to as ground-loop currents, to flow through the wire. In addition, this potential difference is not necessarily a DC level. As well as introducing DC offset errors, ground-loop currents contain AC components, such as AC mains hum (50–60 Hz), and are a continual source of noise. This is especially true when multiple ground points in a system separated by large distances are connected to AC power ground, or when the magnitude of signal levels in analog circuits is low compared to the noise voltage levels.

Where signal lines are used to connect grounds then ground currents will flow with unpredictable results. A possibly more serious result of ground loops is the undefined current loop area, which may couple magnetic fields and induce other unwanted noise voltages in the signal conductors.

3.4.8 Signal circuit isolation

Where a signal conductor is required to be earthed at both ends and additional noise immunity is required, the ground loop should be broken by isolating the signal source from the measuring equipment. Isolation by the use of transformers, opto-couplers and common mode chokes, is shown in Figures 3.25, 3.26 and 3.27 respectively.

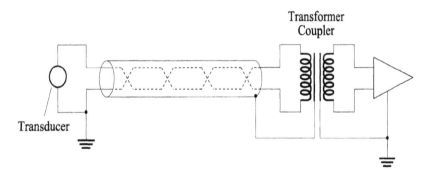

Figure 3.25
Transformer isolation of ground loop

When a transformer is used to isolate the signal source from the measurement system the common mode voltage appears between the windings of the transformer and not at the input to the measurement circuit. Noise coupling between the circuits is very small and dependent on any stray capacitance between the transformer windings. Disadvantages with using transformers are that they are quite large and costly, especially where several signal circuits have to be isolated. In addition, transformers have limited frequency response and provide no DC continuity from the signal source to the measurement system.

The opto-isolated circuit, shown in Figure 3.26, is more typically used for digital signals because of the non-linearity of the opto-coupler to analog signals.

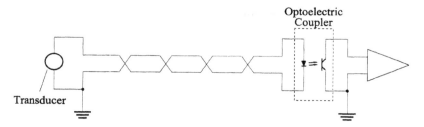

Figure 3.26
Opto-coupler isolation of ground loop

When a transformer is connected as a common mode choke, as shown in Figure 3.27, DC and differential analog signals are transmitted while common mode AC signals are rejected. The common mode noise voltage appears across the windings of the choke. One big advantage with this type of isolation circuit is that multiple signal circuits can be wound on a common core without coupling.

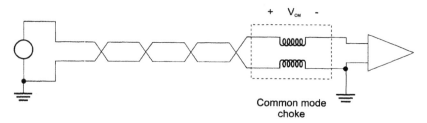

Figure 3.27
Common mode choke isolation of ground loop

3.4.9   Measuring ungrounded signal sources

Ungrounded or floating signal sources can be measured using the single-ended, pseudo-differential or differential measurement methods.

Differential measurement of ungrounded signal sources

When using the differential measurement system to measure the voltage signal from an ungrounded source, care should be taken to ensure that the common mode voltage level of the signal with respect to the measurement ground does not exceed the common mode input voltage limits of the measurement device.

In addition, where there is no return path to the measurement system earth for the instrumentation amplifier input bias currents, then the flow of these currents through the source impedance, as well as charging stray capacitances, can cause the voltage level of the source to float beyond the valid range of the input stage of the measurement system. This is especially true where the source impedance is high. The degree to which the source voltage will float depends on the magnitude of the input bias currents and the system imbalance.

A balanced measurement system meets the following criteria:

- The input impedances to ground of each terminal of the instrumentation amplifier are equal.
- The impedances of each signal cable to ground are equal.
- The impedances to ground of each terminal of the source are equal.

Increased noise immunity is also achieved using a balanced system, since induced noise voltages appearing on the signal wires, are equal and should be cancelled out by the differential amplifier measurement.

Bias resistors, connected between each input lead and the ground reference of the measurement system, as shown in Figure 3.28, provide a DC return path for bias currents from the inputs of the instrumentation amplifier to the reference ground.

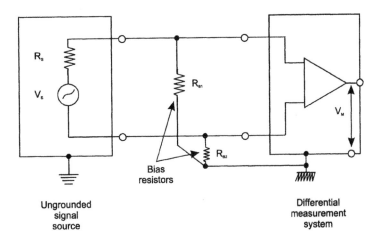

Figure 3.28
Differential measurement of an ungrounded signal source

Where the signal contains both AC and DC components (i.e. DC coupled) and the signal source has low impedance, only one bias resistor is required to be connected between the negative input and the ground reference. If the source impedance is relatively high compared to the input impedance of the instrumentation amplifier, then the imbalance caused by using a single bias resistor could lead to erroneous results. Therefore, for high source impedances both input bias resistors should be used.

For input signals, which contain no DC component (i.e. AC coupled), both bias resistors are required.

The bias resistors should be large enough to allow the source to float with respect to the measurement system ground and not to load the signal source (i.e. much greater than the source impedance), but small enough to keep the voltage at each input terminal within the input stage common mode voltage range of the measurement system. Bias resistors between 10 kΩ and 100 kΩ are typically used for low impedance sources such as thermocouples or when connecting the outputs of signal conditioning modules to data acquisition boards.

3.4.10　System isolation

To allow the measurement of signals that contain large common mode voltages, special hardware and measurement techniques are used. This typically involves isolating the measurement system from the ground reference so that signal lines, such as amplifiers, commonly used as a measurement reference, become a floating reference point.

System isolation can be carried out in the following ways:

- Using isolation transformers to reject the common mode voltage appearing on the signal lines.

- Using isolation amplifiers to isolate the input signals from the measurement system ground reference.
- Permanently isolating the measurement system ground using isolation transformers.
- Temporarily isolating the measurement system ground reference with a digital switch whilst an input signal measurement is taken.

3.5   Noise and interference

3.5.1   Definition of noise and interference

Noise, by definition, is the presence of an unwanted electrical signal in a circuit. Interference is the undesirable effect of noise. Where a noise voltage causes improper operation of a circuit, or its relative magnitude is of the same order as the desired electrical signal, then it is interference.

Noise itself cannot be totally eliminated but only reduced in magnitude until it no longer causes interference. This is especially true in data acquisition systems where the analog signal levels from transducers measuring a physical quantity can be very small. Compounding this in many instances is the physical cable distance over which these signals must be transmitted and the effect that noise may have on this extended circuitry.

3.5.2   Sources and types of noise

Before considering the cabling and shielding requirements of data acquisition systems, it is important to understand the nature and source of interference caused by the coupling of noise into data acquisition systems.

Figure 3.29 illustrates that there are three components involved in any noise-induced problem:

- A noise source (AC power cables, high voltage or high current AC or switching circuitry)
- A coupling channel (common impedance, capacitance, mutual inductance)
- A receiver (the circuitry that is susceptible to the induced noise)

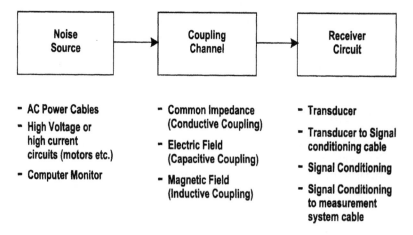

Figure 3.29
Noise coupling between a noise source and a receiver

The mechanisms for coupling noise most common to data acquisition and control applications are as follows:

- Conductive coupling
- Capacitive coupling
- Inductive coupling

Conductive coupling

Conductive coupling occurs where two or more circuits share a common signal return. In such cases, return current from one circuit, flowing through the finite impedance of the common signal return, results in variations in the ground potential seen by the other circuits. A series ground connection scheme resulting in conductive coupling is shown in Figure 3.30. If the resistance of the common return lead is 0.1 Ω and the return current from all other circuits is 1 A, then the voltage measured from the temperature sensor, (V_T), would vary by 0.1 $\Omega \times 1$ A $= 100$ mV, corresponding to 10 degrees error in the temperature measured.

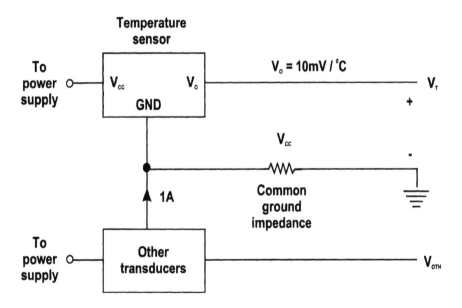

Figure 3.30
Series ground connections resulting in conductive coupling

Capacitive coupling

Electrical fields occur in the vicinity of voltage-varying sources. Capacitive coupling is the transmission of external noise through mutual and stray capacitances between a noise source and receiving circuit. This is sometimes referred to as electrostatic coupling, although this is a misnomer, since the electrical fields are not static. Since cables tend to be the longest circuit elements, capacitive coupling is best demonstrated by considering a signal circuit connecting a signal source to a measurement system by a pair of long signal-carrying conductors.

The physical representation of electric field coupling between a noise source and such a signal circuit is shown in Figure 3.31.

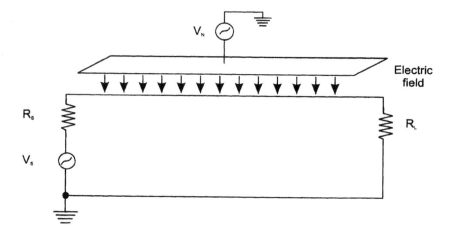

Figure 3.31
Physical representation of an electrical field coupling into a signal circuit

The equivalent circuit representation for this system is shown in Figure 3.32.

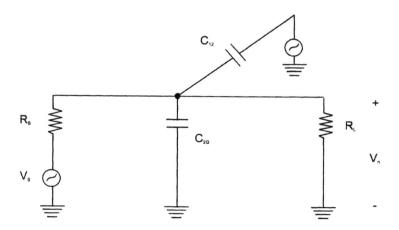

Figure 3.32
Equivalent circuit representation of an electric field coupling into a signal circuit

Where the source resistance (R_S) is much less than the load resistance (R_L) and also much lower than the impedance of the stray capacitances (C_{12} and C_{2G}) (i.e. $R_S \ll 1/j\omega[C_{12} + C_{2G}]$), then $V_n = j\omega R_S C_{12} V_N$

The preceding equation clearly shows that the capacitively coupled noise voltage is directly proportional to the frequency and amplitude of the external noise source, the resistance to ground of the signal circuit, which in this case is R_S, and the mutual capacitance between them.

Where the signal source resistance is comparable in magnitude to the load resistance, and their combined resistances to ground are much larger than the impedance of the stray capacitances (C_{12} and C_{2G}) (i.e. $R_S \; 1/j\omega[C_{12} + C_{2G}]$), then it can be shown that the capacitively-coupled noise voltage, is independent of the frequency of the noise source, and is much greater than in the case where the same resistance is relatively small.

This equation shows that the capacitively-coupled noise voltage is independent of the frequency of the noise source and is much greater in magnitude than in the case where the source resistance is relatively small.

Where the amplitude and the frequency of the noise source cannot be altered, the only means for reducing capacitive coupling into the signal circuit is to reduce the equivalent signal circuit resistance to ground or reduce the mutual stray capacitance. The mutual stray capacitance can be reduced by increasing the relative distance of the signal wires from the noise source, correct orientation of the conductors, or by shielding.

Magnetic field coupling

Magnetic field coupling or inductive coupling is the mechanism by which time-varying magnetic fields produced by changing currents in a noise source, link with current loops of receiving circuits. The physical representation of magnetic field coupling between a noise source and a signal circuit is shown in Figure 3.33.

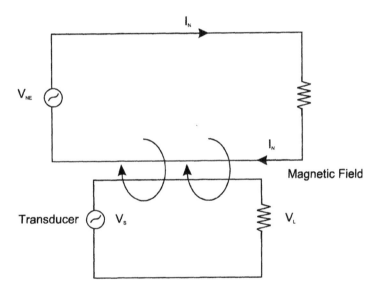

Figure 3.33
Physical representation of magnetic field coupling between a noise source and a signal circuit

Lenz's law states that the voltage, V_n induced into a closed loop signal circuit of area A is proportional to the rate of change of the magnetic field coupling the circuit loop, the flux density (B) of the magnetic field and the area of the loop. This is represented by the formula:

$$V_n = 2 f B A \cos\phi \ (10^{-4})$$

Where:

f	=	the frequency of the sinusoidal varying flux density
B	=	the rms value of the flux density (gauss)
A	=	the area of the signal circuit loop (m²)
ϕ	=	the angle between the flux density (B) and the area (A).

This equation indicates that the noise voltage can be reduced by reducing B, A, or $\cos\phi$. The flux density (B) can be reduced by increasing the distance from the source of the field or if the field is caused by currents flowing through nearby pairs of wires, twist-ing those wires to reduce the net magnetic field effect to zero and or by alternating its direction.

The signal circuit loop area (A) can be reduced by placing the signal wires of the receiving circuit current loop closer together. For example, consider a signal circuit whose current carrying wires are 1 meter long and 1 centimeter apart, lying within a 10 gauss 60 Hz magnetic field, typical of fans, power wiring and transformers. The maximum voltage induced in the wires occurs for $\phi = 0°$.

$$V_n = (2\pi \times 60)(1)(1 \times 10^{-2})(10^{-4}) = 3.7 \text{ mV}.$$

If the distance between the wires is reduced to 1 mm the noise voltage is reduced tenfold to 0.37 mV.

The $\cos\phi$, term can be reduced by correctly orienting the wires of the signal circuit in the magnetic field. For example, if the signal wires were perpendicular to the magnetic field ($\phi = 90°$) the induced voltage could be reduced to zero, although practically this would not be possible. Running the signal wires together in the same cable as the wires carrying the noise current source would maximize the induced noise voltage.

The equivalent circuit model of magnetic coupling between a noise source and a signal circuit is shown in Figure 3.34. In terms of the mutual inductance (M), V_n is given by:

$$V_n = 2 \pi f M I_N$$

Where:

I_N is the rms value of the sinusoidal current in the noise circuit and f is its frequency. The mutual inductance (M) is directly proportional to the area (A) of the signal circuit current loop and the flux density, (B).

The physical geometry of the current loop of the receiving signal circuit, specifically its area, is the key to why it is susceptible to magnetic fields and how to minimize the effect. Cables provide the longest and largest current loop. The effect of magnetic coupling is best demonstrated by considering the circuit of Figure 3.34, in which the signal cable current loop is coupled by a sinusoidal changing magnetic field with a peak flux density of $B\phi$.

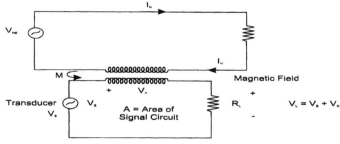

Figure 3.34
Equivalent circuit model of magnetic coupling between a noise source and a signal circuit

Ideally, the only voltage appearing across the load should be V_S – the source signal voltage. However, the magnetic flux induces a voltage in the loop that appears in series with the receiver signal circuit. The voltage appearing across the load is the sum of the source voltage and the unwanted magnetic field induced voltage (V_N).

Twisting the insulated conductors of the loop together, as shown in Figure 3.35, can greatly reduce the amount of magnetic coupling into the signal lines.

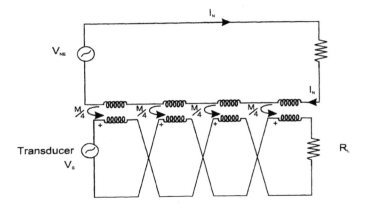

Figure 3.35
Reducing magnetic noise coupling by twisting of wires

The voltage induced in each section of the loop now alternates phases; its magnitude reduced by the reduction in area of each twisted loop (i.e. 1/4). Provided there is an even number of twists in the signal conductors, the voltages due to the magnetic field cancel out and only the desired signal voltage appears across the load.

3.6 Minimizing noise

3.6.1 Cable shielding and shield earthing

The effects of noise due to capacitive coupling can be greatly reduced by the use of a cylindrical metal shield placed around the signal-carrying conductor. Consider the equivalent circuit shown in Figure 3.36, in which the signal conductors are completely enclosed by the ungrounded shield.

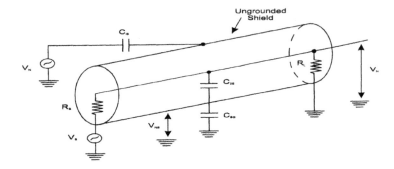

Figure 3.36
Equivalent representation of a signal circuit, completely surrounded by a capacitive shield

Note that as the signal conductors are completely enclosed, there is no stray capacitance between the signal conductors and ground.

Where the source resistance (R_S) is much less than the load resistance (R_L) and also much lower than the impedance of the shield to signal conductor stray capacitance (C_{2S}), (i.e. $R_S \ll 1 / j\omega C_{2S}$), then the noise voltage capacitively coupled onto the signal line can be shown to be:

$$V_n = j\omega \, R_S \, C_{2S} \, V_{NS}$$

Where the shield is grounded (i.e. $V_{NS} = 0$), then the noise voltage induced in signal conductor is also zero.

Completely surrounding the signal carrying conductors is not practical in most instances, since conductors will extend beyond their shield. Also, in the case of a braided shield there is a small stray capacitance due to the holes in the braiding.

Where signal conductors extend beyond the shield, coupling capacitance between the signal conductor and the noise source (C_{12}) and between the conductors and ground (C_{2G}) will still exist, although they will be much smaller. This is shown in Figure 3.37.

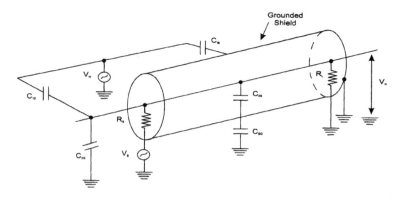

Figure 3.37
Equivalent representation of a practical circuit in which the capacitive shield does not completely surround the signal circuit

Where the source resistance (R_S) is much less than the load resistance (R_L) and also much lower than the impedance of the stray capacitances (C_{12} and C_{2G}) (i.e. $R_S \ll 1/j\omega[C_{12} + C_{2G} + C_{2S}]$), then the noise voltage induced by external noise source onto the signal conductor is given by:

$$V_n = j\omega \, R_S \, C_{12} \, V_N$$

This is the same as for the unshielded conductor, however the mutual stray capacitance (C_{12}) will be much less because of the shield.

The value of C_{12} depends on the length of the signal conductor extending beyond the shield.

Capacitive shielding works by bypassing or providing another path for induced noise currents to flow, so that they are not carried in the signal circuits.

The rules of shielding are as follows:

- For a shield to be effective it should be well grounded and the length of conductors extending beyond the end of the shield minimized. The screen continuity should be maintained at each termination point.
- The screens of individually screened cores in the same cable should be electrically isolated from each other, but continuous for each line through terminal junctions.

3.6.2 Grounding cable shields

To be fully effective, capacitive shielding also requires attention to the number and location of shield earths. In the way that the grounding of signal lines at both ends of a circuit may cause significant ground currents to flow, the same is also true for cable shields. For example, a potential difference of only 1 V between the grounds at either end of a circuit will drive a current of 2 A around the current loop if its resistance is 0.5 Ω. Where the current flow is significant, and the ground loop created by earthing of the shield has a large area, shield currents may inductively couple unequal voltages into the signal cables and be a source of interference. Where possible, shields should be earthed at one end only.

The placement of shield earths depends on the grounding of the signal source and the type of measurement system used. Figure 3.38 shows the preferred shield grounding when measuring an ungrounded signal source, using a measurement system where the signal lines are referenced to the amplifier common. It is assumed that amplifier common, although normally connected to ground may have a potential ($\Delta Vg1$) relative to ground potential. $\Delta Vg2$ represents the difference in ground potential. The circuit equivalent for this system shows that in this configuration neither of the noise voltages ($\Delta Vg1$ or $\Delta Vg2$) appears across the input terminals of the amplifier. Instead, if the shield was earthed at point B, then the noise voltage across the input terminals of the amplifier would be the voltage across the impedance of C_2 as part of the voltage divider formed with C_1.

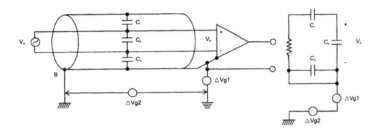

Figure 3.38
Shield grounding when measuring an ungrounded source with a grounded measurement system

When an ungrounded (differential) measurement system is used to measure a grounded source the preferred cable shielding is shown in Figure 3.39. The voltage $\Delta Vg1$ represents the potential of the source common above earth ground potential.

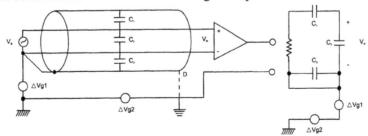

Figure 3.39
Shield grounding when measuring a grounded source with an ungrounded measurement system

The equivalent circuit for this measurement system again shows that the noise voltage appearing across the input terminals of the amplifier, is zero. If the shield was grounded at the other end of the cable at point D, then the noise voltage across the input terminals of the amplifier would be the voltage across the impedance of C_2 as part of the voltage divider formed with C_1.

Where the signal circuit is required to be grounded at both ends, the difference in ground potential and the susceptibility of the ground loop to inductive coupling determines the amount of noise in the circuit. The preferred shield grounding configuration when there is no other alternative is shown in Figure 3.40, in which a portion of the ground loop current is bypassed through the lower impedance shield.

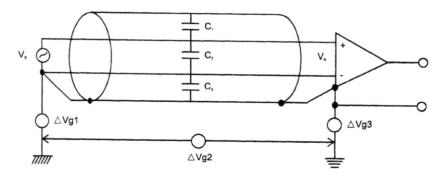

Figure 3.40
Preferred shield grounding when measuring a grounded source with a grounded measurement system

Breaking the ground loop on the signal lines using transformers or optical couplers can provide additional noise reduction.

The rules of shield grounding are as follows:

- Where possible, cable shields should be earthed at one end only.
- Where the source is ungrounded and the signal amplifier is grounded, the input shield should always be connected to the amplifier common terminal, even if this point is not at earth ground.
- Where the source is grounded and the signal amplifier is ungrounded, the input shield should be connected to the source common terminal, even if this point is not at earth ground.

Grounding the shield has additional benefits such as providing a path for RF currents and preventing the build-up of static charge by providing a discharge path to ground.

3.7   Shielded and twisted-pair cable

Cables with copper conductors and plastic insulation are still the most common and reliable solution. This is not surprising as they combine the important elements of good electrical characteristics, low cost, mechanical flexibility, ease of installation and ease of termination. Aluminum conductors are seldom used for data communication cables because of the higher resistance and other physical limitations.

The cable resistance depends on the cross-sectional area of the conductor (usually expressed in mm^2) and the length of the cable. The thicker the conductor, the lower the resistance, the lower the signal volt drop, and the higher the current it can carry without excessive heating.

The signal voltage drop, $V_{drop} = I (R + (2 \pi f L - 1/2 \pi f C))$, depends on the:

- Frequency of signal
- Line current, which is dependent on the receiver input impedance, and
- Conductor resistance, which is dependent on wire size and length.

For DC voltages and low-frequency signals the resistance of the conductor is the only major concern. The voltage drop along the cable affects the magnitude of the signal voltage at the receiving end. In the presence of noise, this affects the signal-to-noise (S/N) ratio and thus the quality of the signal received.

As the frequency (or data transfer rate) increases, the other characteristics of the cable, such as capacitance and series inductance, become important. Inductance and capacitance are factors that depend on the construction of the cable and on the type of insulation material.

The resistance, inductance and capacitance are distributed along the length of the cable. At high frequencies they combine to present the effects of a low pass filter. The simplified electrical single-line diagram of a cable shows these electrical parameters distributed along the length of the cable and can be seen in Figure 3.41. Note, however, that a more complex model would also need to include a minor conductance factor (the inverse of resistance) in parallel across the cable.

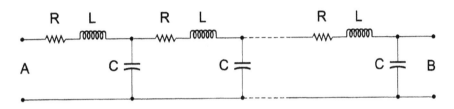

Figure 3.41
Main parameters of a cable

To derive the optimum performance from a cable, the correct type and size must be used. The following general rules apply to most applications:

- **Low data transfer rates:** use low-frequency cables (for example, twisted-pair cables)
- **High data transfer rates:** use high-frequency cables (for example, coaxial cables, optical fiber; though there are some new types of twisted-pair cables that give very good high-frequency performance)
- **High noise environment:** use shielded copper or optic fiber cables

3.7.1 Twisted-pair cables

Twisted-pair cables are the most economical solution for data transmission (differential circuit). They allow for transmission rates of up to 100 Mbps on communication links of up to 300 m (or even longer distances but with lower data transfer rates). Some new types of twisted-pair cables (e.g., 'Twistlan') are suitable for up to 100 Mbps. Twisted-pair cables can be STP (shielded twisted-pair) or UTP (unshielded twisted-pair).

Twisted-pair cables are made from two identical insulated conductors that are twisted together along their length a specified number of times per meter, typically 40 twists per meter (12 twists per foot). The wires are twisted to reduce the effect of electromagnetic and electrostatic induction. An earth screen, or shield, is often placed around them as well to reduce the capacitance-induced noise, and an insulating PVC sheath usually provides mechanical protection. As the cross-sectional area of the conductor affects IR loss, heavier conductor sizes are recommended for long distances. The capacitance of a twisted-pair is low at about 15 to 50 pF/m, allowing a reasonable bandwidth and an achievable slew rate.

For full-duplex systems using balanced differential transmission, two sets of screened twisted-pair conductors are required in one cable, with both individual and overall screens. The entire cable is covered with a protective PVC sheath.

3.7.2   Coaxial cables

Coaxial cables are used in applications that require high data transfer rates of up to 10 Mbps or high-frequency analog signals over long distances.

Coaxial cables are more expensive than twisted-pair cables. They consist of a central conductor running through an enclosing cylinder on the same axis. This enclosing cylinder is made of a conducting material and is braided for flexibility. The insulating material separating the two conductors affects the cable capacitance, and hence the rate of signal propagation. The cable is usually covered with a protective PVC sheath, and sometimes with an additional shield as well.

Several types of standard coaxial cables are manufactured; each has a different combination of electrical and mechanical characteristics to suit different applications. The main variables are:

- Cable characteristic impedance
- DC resistance
- Power capacity
- Bandwidth
- Type of shielding
- Mechanical characteristics (such as flexibility)

Coaxial cables are more difficult to terminate than multi-core or twisted-pair cables. They are also more difficult to splice and connect to tee-offs. Special tools and connectors are required for good coaxial cable terminations. The ends of the cable should be terminated in a *dead-end* terminator to prevent signals reflecting from the ends of the cable. Coaxial cables can sometimes be ordered to specified lengths and with terminators already in place.

4

The PC for real time work

Introduction

The key to the effective use of PCs in data acquisition and process control is the careful matching of the specific requirements of a particular application to the appropriate hardware and software available.

The personal computer consists of the following main components:

- System unit (CPU, memory, interrupt controller, DMA controller, power supply)
- I/O devices (hard disk, floppy disk, keyboard, mouse, display, COM port(s), CD
- Hardware BIOS (Basic input/output system)
- Operating system (WIN 95, 98, 2000, NT)

This chapter examines some of the important features of the PC as they relate to the data acquisition techniques studied in other sections of this course. The topics covered are:

- Operation of interrupts
- Operation of direct memory access (DMA)
- Data transfer speeds (polled I/O, interrupt, DMA, repeat inst)
- Memory (base memory, expanded memory, extended memory)
- PCI, Compact PCI, ISA bus, EISA bus
- Interfacing techniques to the PC bus
- Compact PCI

4.1   Operating systems

An operating system is the software responsible for managing the computer's resources (including hardware and software), processing commands, and controlling program execution. It provides an interface between the application software and the hardware of a particular system. Operating systems manage communications with the disk drive, display, printer, and usually consist of a small machine dependent section of code, accompanied by a standard command interpreter.

Programming languages, such as Basic or C, interact with the computer hardware through the operating system. The operating system provides a platform at a level less dependent of the computer hardware, providing a simpler and more uniform approach to software development. Two popular operating systems available today are DOS and UNIX, each with their own advantages and disadvantages.

4.1.1 DOS

DOS is a 16 bit operating system that was originally developed in 1980, for the Intel 8088 microprocessor in the IBM personal computer. When IBM developed the IBM PC, they used the new Intel 16 bit microprocessor, and therefore needed a new operating system. IBM contracted the development of the new operating system to Bill Gates at Microsoft, and his team bought a program called QDOS (Quick and dirty operating system) from Seattle Computer Products. Using this program as a starting point, they developed an operating system called MS-DOS.

DOS structure

DOS comprises several major components, each with a certain task within the system. The three most important components are the DOS-BIOS, the DOS kernel, and the command processor.

DOS-BIOS

DOS-BIOS is stored in a system file and appears under various file names such as IBMBIO.COM, IBMIO.SYS or IO.SYS. The DOS-BIOS contains the device drivers for the keyboard, display, printer, serial interfaces, real time clock, and floppy and hard disk drives.

If DOS wants to communicate with one of these hardware devices, then it accesses the specific DOS-BIOS device driver. The DOS-BIOS is the most hardware dependent component of the operating system and varies from one computer to another.

DOS kernel

The DOS kernel in the IBMDOS.COM or MSDOS.SYS file is normally invisible to the user. It contains file access routines, character input and output, and more. The routines operate independently of the hardware and use the device drivers of DOS-BIOS for keyboard, display, and disk access.

Application programs can access the kernel functions in the same manner as the ROM-BIOS functions. The functions are accessed via the software interrupt mechanism, and microprocessor registers are used to pass the function number and any applicable parameters.

Command processor

Unlike the DOS-BIOS and kernel, the command processor is contained in the DOS file COMMAND.COM. The command processor displays the command prompt (i.e. C:\>) on the screen, accepts input from the user and controls input execution. Many users incorrectly think that the command processor is actually the operating system. In reality, it is only a spe-cial program that executes under DOS control.

The command processor, also called a shell in programming terminology, actually consists of three modules. These modules are the resident portion, a transient portion, and the initialization portion.

The resident portion, or the part that remains in memory, contains various routines called critical error handlers. These routines allow the computer to react to different events, such as when the user presses the <CTRL> <C> or <CTRL> <BREAK> key sequences, or when errors occur during communication with external devices (e.g. disk drives).

The transient portion contains code for displaying the prompt, reading user input from the keyboard and executing input. The name of this module is derived from the fact that the memory where it is located is unprotected, and can be overwritten in certain circumstances.

When program execution ends, control returns to the resident portion of the command processor. It executes a checksum routine to determine whether the transient portion was overwritten by the application program, and reloads the transient portion if necessary.

The initialization portion loads during the booting process and initializes DOS. When the initialization process is complete, the memory it occupies can be over written by another program. The commands accepted by the transient portion of the command processor can be divided into three groups – internal commands, external commands, and batch files.

Internal Commands are contained in the resident portion of the command processor. DIR, COPY, and RENAME are examples of internal commands.

External Commands must be loaded into memory from diskette or hard disk as needed. FORMAT and CHKDSK are examples of external commands.

Batch Files are text files containing a series of DOS commands. When a batch file is started a special interpreter in the transient portion of the command processor, executes the batch file commands. Execution of batch file commands is the same as if the user entered them from the keyboard. An important example of a batch file is the AUTOEXEC.BAT file, which is executed immediately after DOS is first loaded.

DOS device drivers

Device drivers are software modules that are responsible for controlling and communicating with the hardware. They represent the lowest level of an operating system and permit all other levels to work independently of hardware. When adapting an operating system to various computers this is an advantage, as only the device drivers need to be modified.

In earlier operating systems, device drivers resided in the operating system code. This meant that changes or upgrades of these routines to match new hardware were very difficult, if not impossible. DOS version 2.0 introduced the flexible concept of device drivers, making it possible for the user to adapt even the most exotic hardware to DOS.

Since communication between DOS and a device driver is based on relatively simple function calls and data structures, the assembly language programmer can develop a device driver to adapt DOS to any hardware device.

During the DOS boot process, standard display, printer and drive device drivers are installed sequentially, in memory. If the user wants to install his own driver, he has to inform DOS using the CONFIG.SYS file. This text file contains the information that DOS requires for configuring the system. The contents of the CONFIG.SYS file are read and evaluated during the boot process after linking the standard drivers. If DOS finds the DEVICE command in the CONFIG.SYS file, it knows that a new driver should be included.

4.1.2   Microsoft Windows 3.1, 95, 98, 2000 and NT

Microsoft Windows 3.1 was an extension for DOS that supported both multitasking and a graphical user interface. Its graphical user interface (GUI) was a symbolic interface that attempted to improve the speed of communication between people and computers. Windows' ability to run multiple applications simultaneously (multitasking), and transfer information between applications provided further advantages for any application.

With the advent of Windows as a graphical interface we have gone from Windows 3.11 to 95 and NT, to 98 and now Windows 2000. It has been the direction of the Microsoft Corporation to move the PC operating system away from DOS. This has been done for many very good reasons. DOS had many problems, the worst of which was the memory allocation system. Microsoft used 95 and 98 as small steps to move the PC operating system further and further away from DOS. 2000 and NT do not use DOS at all. These operating systems have one major problem when it comes to data acquisition. It is in the timing. The graphical interfaces from 95 and up do not consider the peripherals as high priority. Windows itself considers the operating system as the most important thing in the computer. The 95 and higher computers do true multitasking with every window opened as a virtual computer within the computer. Even applications that are not viewable by the user i.e. running in the background, are handled like virtual computers. This means that if the operating system, background program, or running application, need service, the operating system decides who is serviced. Just because a data acquisition program is running, it does not mean that it will have high priority. Timing therefore cannot be guaranteed when using these operating systems.

Due to this problem, data acquisition users and designers have often looked elsewhere for an operating system that will guarantee direct access to the data acquisition system. Some have locked themselves into DOS while others have tried UNIX. The problem with these operating systems is that they have their own problems. DOS is old and not supported by anyone, including Microsoft. UNIX is limited by the software available and a good graphical interface. This may change with Linux. The best one can do is not use Windows products when timing is critical, especially when the computer is used to control a time critical process, or alternatively, to strip everything off the computer except the data acquisition program. However, even this will not guarantee correct timing.

Some of the advantages of Microsoft Windows are:

- Standard graphical user interface for all applications
- Natural user interface that is easy to learn and use
- Application programs that are independent of system devices
- Multitasking support
- Support for dynamic data exchange
- Virtual memory management

Multitasking

Multitasking is the ability of the operating system to run multiple application programs simultaneously. With Microsoft Windows, you can execute two or more application programs at the same time. Each application has its own window allowing the user to switch between applications.

Graphical user interface

The graphical user interface allows the user to execute application programs, by simply selecting a graphic symbol or icon representing that application. Icons are small symbols that are used to represent files, programs, or program tools.

Graphical user interfaces are important for computer systems because they provide symbolic interfaces, giving the user visual or graphic control over program execution. It also provides a common interface with the user for all application programs, making it easier to learn new application programs. At the same time, graphical user interfaces provide software developers with visual tools for developing programs that take advantage of the user interface.

Graphical user interfaces offer several advantages to an application program:

- Communication between the application program and the user is in a natural symbolic form that more closely resembles the human thinking process
- Communication between the user and the computer is faster, as the user is not required to enter program names
- The learning process for programs is faster, since the communication is more natural and symbol oriented
- Programs can be more powerful. In a conventional program, intricate operations that are generally difficult for the user and often purposely omitted by the program designer, are now more feasible.

Because of these advantages, most developers believe that graphical user interfaces will be the primary interfaces used by application programs in the future. However, the graphical user interface does have some disadvantages with respect to current hardware and software:

The amount of memory and disk space needed to support the graphic environment is much larger than that needed for conventional (or text oriented) environments

Higher processing speeds are required to support the graphic environment, as larger amounts of data must be moved around, particularly from the hard disk to the display

Since hardware must be faster and contain more memory, the hardware costs are consequently higher

The amount of programming support needed for a graphic environment is more extensive.

Virtual tools

The Microsoft Windows graphical user interface, provides an ideal environment for virtual tools. In conjunction with digital and analog acquisition and control boards, virtual tools allow users to set up experiments, acquire data, and graphically display the results on the screen in real time.

Virtual tools can be developed by professional software engineers, or by less experienced users, by using packaged software. Specific software packages are available that allow the user to edit, control and logic strategies. The user can move function block icons from on-screen libraries onto the desktop and connect them into a flowchart.

Display editors allow the user to set up color graphic displays, such as instrument panels, control panels and charts. When the display panel is complete, the user simply connects function icons to it for real-time graphical output. A typical example of a virtual tool would be an oscilloscope with the CRT display and front panel controls emulated on the screen, and signal input provided by an analog to digital conversion board.

Virtual memory

Microsoft Windows allows an application program to apparently access more memory than is physically available. This virtual memory is achieved by using reserved space on the hard disk drive to emulate physical memory. Windows can also manage memory for the application programs on a dynamic basis, through the use of swap files.

4.1.3   UNIX

The UNIX operating system, originally developed by AT&T Laboratories in the late 1960s, is a powerful multi-user and multitasking operating system. UNIX is available for nearly any computer that contains sufficient real memory and a fast hard disk.

UNIX shell

Similar to the DOS command processor, the UNIX shell software processes commands entered by the user into instructions for the system's internal syntax. The name shell really describes the function (i.e. hard material that stands between the core of the system and the outside world) providing a robust user interface for the operating system.

Commands are usually separate executable programs that the shell finds and executes in response to typed instructions. But the shell is actually much more powerful and useful than just a means of passing commands to the system for execution.

UNIX file system

The UNIX file system is based on a hierarchical structure, where files are arranged within directories that are, themselves, files inside other directories. A tree-like structure emerges, with all branches eventually tracing back to the root directory.

Unlike DOS, in the UNIX operating system all hardware and software objects are treated as files and behave like files. All input and output is done by reading or writing files, because all peripheral devices, even the user terminal, are files in the file system. This philosophy means that a single, homogeneous interface handles all communication between a program and peripheral devices.

4.2   Operation of interrupts

Interrupts are the mechanism by which the CPU of a computer can attend to important events such as keystrokes or characters arriving at the COM port only when they occur. This allows the CPU to execute a program and only service such I/O devices as needed without requiring constant attention.

An interrupt is not an expansion bus cycle, but a cycle on the CPU board.

There are three groups of interrupts that may occur in a PC:

Hardware interrupts

These are generated electrically by I/O devices that require attention from the CPU.

Software interrupts

There are 256 possible interrupt types that can be generated by software.

Processor exceptions

Exceptions are generated when an illegal operation is performed in software (for example divide by zero).

The following sections will examine hardware interrupts in detail, as these are mechanisms by which expansion cards transfer data to the PC's memory.

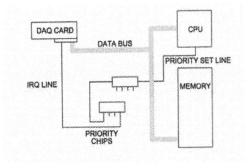

Figure 4.1
IRQ

4.2.1   Hardware interrupts

Two types of hardware interrupt are available, non-maskable interrupts (NMI) and the mask-able interrupt (INTR), connected respectively to the NMI and INTR input pins of the CPU.

4.2.2   Non-maskable interrupts

Since NMIs are not maskable internally in the CPU, logic on the system board uses an I/O register port (A0h) bit to mask and unmask the NMI.

The CPU's NMI interrupt input is transition-sensitive or edge-sensitive with a low to high transition triggering the interrupt. It is used to indicate to the CPU serious conditions, such as system RAM parity errors or impending power failure.

NMIs can also occur when the bus expansion signal /IOCHK (I/O channel check) is asserted low, bit 3 of Port 61h is cleared and NMIs are enabled with port A0h. The /IOCHK signal is used by I/O devices to indicate that a serious error such as a memory parity error or an uncorrectable hardware fault has occurred on their expansion board.

When the CPU receives an NMI, it automatically begins executing the code pointed to by interrupt vector 2 (i.e. a type 2 interrupt is generated).

4.2.3   Maskable interrupts

The CPU recognizes a hardware maskable interrupt when its INTR input goes from low to high with the interrupt enabled. INTR interrupts are enabled/disabled by setting/clearing the interrupt enable flag bit IF in the CPU FLAGS register.

4.2.4   Programmable interrupt controller(s)

As there is only one maskable hardware interrupt line (INTR) and many I/O devices which must inform the CPU that they require servicing (by generating interrupts), there needs to be a method of prioritizing the PC's interrupt structure for more than just one interrupt. This task is performed by the 8259A programmable interrupt controller (PIC), which accepts interrupt requests from I/O devices, prioritizes and stores them, and generates the interrupt request signal (INTR) to the CPU as required.

The PC/XT has only one 8259A PIC with interrupts IRQ0 to IRQ7, whilst the 80286/80386 and 80486-based PCs have two 8259As. The output of the second (or slave PIC) 8259A is connected directly to the IRQ2 channel of the master PIC.

Hardware interrupt requests by I/O devices are made on the interrupt lines IRQ (15..14), IRQ (12..9), IRQ (7..3). IRQ 0,1,2, 8 and 13 are used by the system board and are not available to the expansion bus. A list of standard and common interrupt allocations is provided in Table 4.1 below.

Interrupt level	Interrupt type number	Standard device	Available on ISA bus
IRQ0	8 (8h)	System timer	No
IRQ1	9 (9h)	Keyboard	No
IRQ2	10 (Ah)	Redirected to IRQ9	Same as IRQ9
IRQ8	112 (70h)	Real time clock	No
IRQ9	113 (71h)	Display adapter (VGA) network card	Yes
IRQ10	114 (72h)	Free	Yes
IRQ11	115 (73h)	Free	Yes
IRQ12	116 (74h)	Free	Yes
IRQ13	117 (75h)	Math coprocessor	No
IRQ14	118 (76h)	Hard drive controller	Yes
IRQ15	119 (77h)	Free	Yes
IRQ3	11 (Bh)	Serial port 2 (and / or 4)	Yes
IRQ4	12 (Ch)	Serial port 1 (and / or 3)	Yes
IRQ5	13 (Dh)	Parallel port 2	Yes
IRQ6	14 (Eh)	Floppy drive controller	Yes
IRQ7	15 (Fh)	Parallel port 1	Yes

Table 4.1

The 8259A PIC has the following features:

- IRQ inputs are prioritized with the lower numbered inputs having the higher priority. When the slave PIC is cascaded into IRQ2 of the master PIC on the PC/AT type systems, IRQ0 and IRQ1 have a higher priority than the slave PIC IRQs (IRQS–IRQ15). However IRQ3–IRQ7 have a lower priority than the slave PICs IRQ8–IRQ15
- Since IRQ2 is not available for expansion boards, any interrupt requests on IRQ2 (from expansion boards originally designed for use in XT type boards) are transparently routed to IRQ9 by the system board
- IRQs can be individually masked (enabled or disabled). This is performed by writing to the PIC's operation command register (OCWl) – a single byte describing the interrupt status for the eight IRQ inputs
- It automatically issues the interrupt type bytes to the CPU during the INTA cycles (see I/O devices requesting interrupt service p. 74). The master PIC can generate interrupt types 08h to 00h, whilst the slave PIC (when available) is programmed to generate interrupt types 70h to 77h, see Table 4.1
- It automatically tracks which interrupts are being serviced by the CPU, to prevent multiple occurrences of the same interrupt

- IRQ inputs can be configured as edge-sensitive (normal) or level-sensitive via the initialization command word (ICW) register

4.2.5 Initialization required for Interrupts

Before interrupts can be handled correctly the following functions must be performed:

Initialize the 'interrupt vector table' located in the first 1024 (1 k) bytes of system memory to contain the addresses of the interrupt service routines of each of the 256 possible interrupts. Each four-byte address consists of an instruction pointer (IP) and code segment (CS) value. A large number of these are initialized by the BIOS and DOS as part of the system boot and operating system startup procedures. Initialize the 8259A PIC(s). This is largely initialized by the BIOS as part of the system boot.

Enable the system interrupt INTR by setting the interrupt enable flag bit IF in the FLAGS register.

4.2.6 I/O devices requesting interrupt service

When an I/O device asserts an interrupt request, an ordered sequence of events occurs, to direct the CPU to the interrupt service routine (ISR) that will service the specific request.

We will assume that any system and remote I/O device initialization that is required to allow the interrupt request to be handled correctly has already occurred.

The sequence of events is as follows:

- The I/O device hardware activates an interrupt request by asserting its IRQx line from 'low' to 'high'. This signal is usually 'latched' high by an interrupt request latch on the I/O device and remains high until the latch is reset and the interrupt is acknowledged (thus allowing further interrupts). These last tasks are performed by the ISR (see Interrupt service routines, p. 75).
- The interrupt controller receives this IRQx interrupt request and prioritizes it with other requests that may be coming in or pending. The interrupt controller will then send an interrupt request to the CPU on the INTR signal line under the following conditions:
  - This is the only interrupt request.
  - A lower priority interrupt is in progress.
  - Several interrupts are pending but this interrupt has the highest priority.
- If the CPU has interrupts enabled, it acknowledges the interrupt request by sending two INTA pulses to the interrupt controller. The first freezes the priority levels in the interrupt controller, while the second requests an 8-bit pointer value, called the 'interrupt type'.
- The interrupt controller places the 8-bit interrupt type onto the CPU data bus. This 'interrupt type' byte is the means by which the CPU knows where to look for the address of the interrupt service routine that will service the I/O device.

  It is used to index the 'interrupt vector table' located in the lowest 1024 (1 k) bytes of system memory. Each of the table's 256 entries (four bytes each) contains the segmented memory addresses of all the ISRs.

  The CPU multiplies the interrupt type by four (4) to get the offset to the interrupt vector table where the address of the appropriate ISR will be located. Therefore, for interrupt type S, the address of the initial service routine (ISR) to be executed will be address 20h (32nd byte).
- The CPU saves the information necessary to allow the program currently being executed to resume execution at its next instruction upon completion of the ISR.

It does this by saving the current code segment instruction pointer (CS:IP) and system FLAGS registers onto the stack.

It is important to note that at this time the interrupt enable flag bit IF of the CPU FLAGS register is also cleared, disabling further interrupts.

The CPU finally fetches the instruction pointer (IP) and code segment (CS) value from the interrupt vector table at the correct location for the interrupt being serviced, and branches to this address.

It should be noted that the CPU assumes that the address retrieved is in fact the starting address of a valid ISR. If it is not, then the CPU begins executing at this address anyway and will probably never return, causing the program being executed when the interrupt occurred to appear 'locked up' or, even worse, to malfunction, with potentially dangerous consequences. Due to the possible severity of this error, it is important to correctly initialize the interrupt vector routine for the interrupt service required (see Initialization required for interrupts, p. 74).

4.2.7   Interrupt service routines

Apart from the function(s) that are expected to be performed by the ISR and the reason it was requested in the first place, there are several other considerations and certainly some important tasks that must be performed within the ISR.

These are listed below:

- Any CPU registers that might be used by the ISR, (registers that the interrupted program could have been using), should be saved by pushing their current values onto the stack. Only those registers whose values will be altered need be saved, although it is safe programming practice to save all registers (remember that the CS, IP and FLAGS registers are automatically saved by the CPU).

- Since the interrupt enable flag bit IF in the FLAGS register has been cleared and interrupts to the CPU are disabled, a decision must be made whether to set this flag bit and re-enable INTR interrupts to the CPU. If this is the case higher level interrupts may interrupt the current ISR. Although this may be required, the consequences of allowing this must be considered carefully.

  Note that even when interrupt requests from the master PIC to the CPU are disabled, interrupts may still be received by the PIC(s) and will remain pending until serviced.

  The interrupt request latch on the I/O device requesting service must be reset so that further interrupts can be received from the same device.

  It is usual practice (but strictly dependent on the hardware of the expansion board) that resetting the interrupt request latch will leave the IRQ line disabled or in a high impedance tri-state.

  Pull up resistors on the system board are used to take the IRQx signal lines to a 5 V logic level when not in use and, when floated, are guaranteed to be high after 500 ns. This means that the ISR must reset the I/O devices interrupt latch at least 500 ns before issuing an end of interrupt (EOI) command to the PIC(s) (see below).

  This must be strictly observed to avoid the possibility of an unknown transition of the IRQ line from low to high in the high impedance state, inadvertently triggering an interrupt on the IRQ signal line.

- An EOI command must be sent to the master PIC to re-enable interrupts on the same IRQ line. Where an interrupt occurred on one of the interrupt lines IRQ8–IRQ15 via the slave PIC an EOI command must be sent to this PIC as well.

- Upon completion of the ISR all registers saved onto the stack at the start of the ISR are retrieved and restored to their original values.
- The very last instruction executed within the ISR is the IRET (return from interrupt) instruction. This signals the CPU that the ISR is complete. Upon executing this command the CPU retrieves the original CS, IP and FLAGS register values from the stack and begins executing the interrupted program at the segmented memory location CS:IP.

   At this point the CPU is in exactly the same state as it was when the interrupt was first acknowledged. This allows the program (or possibly a lower priority interrupt), which was being executed when the ISR was called, to continue execution unaffected at its next instruction.

   Restoration of the FLAGS register automatically re-enables interrupts to the CPU.

4.2.8   Sharing interrupts

As we have seen an I/O device requesting an interrupt latches its IRQx line high until reset by the ISR. This precludes more than one I/O device from using the same interrupt line reliably at the same time.

It is possible however for more than one expansion device to use the same interrupt request line IRQx if each device is guaranteed not to make interrupt requests when another device might be using the line. This is achieved by using IRQ line drivers with three state (tri-state) outputs. When an expansion board's IRQ line is disabled, the output is in a high impedance state. Pull-up resistors on the system board are used to take the IRQx signal lines to a 5 V logic level when not in use.

Where I/O devices use the same IRQx line, the interrupt service routine must be able to differentiate which device is the source of the interrupt.

4.3   Operation of direct memory access (DMA)

In many I/O interfacing applications and certainly in data acquisition systems, it is often necessary to transfer data to or from an interface at data rates higher than those possible using simple programmed I/O loops.

Microprocessor controlled data transfers within the PC (using the IN(port) and OUT(port) instructions) require a significant amount of CPU time and are performed at a significantly reduced data rate. Further to this, the CPU cannot perform any other processing during program controlled I/O operations.

While the use of interrupts might allow the CPU to perform some concurrent tasks, certain applications exist where the amount of data to be transferred and the data rate required is too high.

Two such applications are as follows:

- Transferring screen information to the 'video card adapter' on board memory
- Transferring data from a remote I/O device (data acquisition board) to the PC's memory

Direct memory access (DMA) facilitates the maximum data transfer rate and microprocessor concurrence. Unlike programmed or interrupt controlled I/O, where data is transferred via the microprocessor and its internal registers, DMA (as its name implies) transfers data directly between an I/O device and memory (memory to memory DMA transfers are also possible).

Whichever CPU is being used, it must have a DMA feature to determine when DMA is required, so that it can relinquish control of the address and data buses, as well as the control lines required to read and write to memory. In addition, the CPU must inform the I/O device that requires the DMA data transfer when it again requires control of the address and data buses and I/O control lines.

Further to this, a separate DMA controller is required to actually perform the DMA I/O operations.

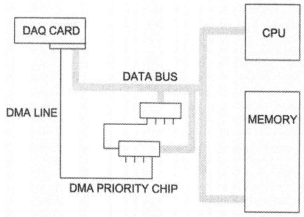

Figure 4.2
DMA

4.3.1   DMA controllers

In the PC/XT a single Intel 8237-5 DMA controller, with four channels (Ch0–Ch3), was used. The PC/AT has two DMA controller chips, cascaded together in a similar manner to the 8259A PIC. The supplementary DMA channels (Ch5–Ch7) are 16-bit channels. Standard DMA channel usage is shown in Table 4.2.

DMA channel	Standard device	Page register I/O address
DRQ0	Free	87h
DRQ1	SDLC	83h
DRQ2	Floppy disk	81h
DRQ3	Free	82h
DRQ4	Used for cascading controller 1	–
DRQ5	Free	8Bh
DRQ6	Free	89h
DRQ7	Free	8Ah

Table 4.2

The 8237-5 DMA controller contains:

- A two-byte address register containing the starting memory address from which data must be read.

- A two-byte address register containing the current address to which data must be written.
- A two-byte count register, which holds the number of bytes/words to be written in total.
- A two-byte count register containing the current bytes/word count of each channel.
- Control lines that allow the reading/writing of data from/to memory.

As each device only supports 16-bit addresses (limiting access to 64 kB of memory), each DMA channel has an associated 'page register' in system I/O memory to provide the added upper 4 address lines required to address the complete 20-bit (1 MB) system base address space.

4.3.2   Initialization required for DMA control

Before any DMA operation can occur, the DMA controller must be initialized.
Items requiring initialization are as follows:

- Select whether the DMA controller will read or write to memory.
- Configure the type of DMA data transfer. Four modes of DMA data transfer are available:
  - Single transfer mode
    The DRQx signal must be asserted for every byte/word transferred.
  - Block transfer mode
    A single DRQx signal DMA request initiates the transfer of an entire block of data.
  - Demand transfer mode
    Data is transferred as long as the DRQx signal DMA request is asserted and the terminal count has not been reached.
  - Cascade mode
    All DMA channels are programmed for single transfer mode.
- The total number of bytes to be transferred is loaded into the appropriate total byte/word count register. The current byte/word count register is then automatically initialized.
- The memory address to which the first data byte will be read/written is loaded into the start memory address register. The current memory address register is automatically initialized.
- The 4-bit page register corresponding to the upper four bits of the 20-bit address is written using the I/O port addresses of the PC.
- The DMA channel priorities should be set. When the PC is booted up, the ROM BIOS sets the priorities so that the lowest numbered channel has the highest priority.
- The DMA controller(s) channels that are to be used should be enabled. Channels to be enabled have the channel mask register bits cleared.

4.3.3   I/O devices requesting DMA

Assuming that the necessary DMA controller has been initialized, the standard operation, when an I/O device requests DMA data transfer on one of the channels of the DMA controller, is as follows:

- An I/O device requests a DMA transfer on a specified channel by asserting its DRQx (DMA request) signal from 'low' to 'high'. The requesting device must hold its DRQx line 'high' until the DMA controller responds by asserting the corresponding /DAK line.
- The DMA request is prioritized by the 8237 DMA controller and if it is the highest priority request, the controller asserts the HOLD signal to the CPU, requesting that the CPU relinquish control of the bus and float all its address, data, and control outputs.
- After floating the address, data and control outputs, the CPU asserts the HLDA signal to the DMA controller.
- When the DMA controller detects that the HLDA signal has been asserted, it asserts AEN, acknowledges the DMA request by asserting the corresponding DAKx 'low' to the requesting I/O device. This signal is usually used as a chip select for the I/O device, enabling it onto the bus.

  The DMA controller drives the lower 16 ISA bus address lines SA[15..0], with the address contained in the corresponding channel's current address register. The upper address lines are driven by the corresponding DMA channel's page register.

  For a write cycle, this address represents the destination in memory for the data supplied by the I/O device. The DMA controller first asserts the /IORC bus signal to instruct the I/O device to drive the data onto the bus. The /MWTC signal is then asserted low, instructing the memory to latch the data at its trailing edge.

  Throughout this cycle the bus command signals /IOWC and /MRDC remain inactive (high).
- The DMA controller then performs the following functions:
    - Decrements the corresponding channel's byte/word transfer count register.
    - Increments the same channel's current address register.
    - Asserts the I/O device's corresponding DACKx line.
- Once an I/O device's /DACKx line has been asserted by the DMA controller, it will release the DRQx line to an inactive state (low).
- Upon detecting the DRQx line 'low', the DMA controller now drives the HOLD signal 'low'. The CPU responds by dropping the HLDA signal to the DMA controller, thus indicating it is ready to again take control of the bus.

In much the same way a DMA 'memory read' cycle is performed, the difference being that /IORC and /MWTC are inactive, while asserting /MRDC instructs the memory to drive the data onto the bus and asserting /IOWC instructs the I/O device to latch the data at its trailing edge.

4.3.4   Terminal count signal

The terminal count (T/C) is a bi-directional signal from the system-board, which can act in one of two modes, depending on the programming of the DMA controllers.

In output mode, the system board asserts T/C to indicate that a DMA channel's current byte/word count has rolled over from 0h to FFFFh when last decremented as part of a 'memory read' or 'memory write' DMA cycle.

The T/C signal is only asserted while the corresponding channel's /DACKx signal line is also asserted. In this manner, hardware on the I/O device can be used to determine that the T/C has been asserted because of completion of a DMA transfer it initiated, and not one on another channel by a different I/O device.

In input mode, an I/O device to stop a DMA transfer can use T/C. T/C is sampled by the system board when /IORC or /IOWC are asserted.

If T/C is found asserted, the DMA transfer is terminated.

If the channel is programmed for auto-initialize then:

- The transfer starts at the address found in the starting address register (this is transferred to the current address register)
- The current word/byte count register is set to the value contained in the total word/byte count register

The auto-initialize feature allows the DMA controller to be automatically set up to accept more DMA requests after an earlier DMA transfer has been completed.

4.3.5 DMA modes

DMA cycles operate in single mode, because a DMA request initiates one DMA cycle in which one data transfer occurs. DMA allows the direct transfer of data from I/O devices to memory devices and vice versa (and from memory to memory) without involving the CPU. This makes it possible to transfer large amounts of data to and from memory in the background, at high speed.

The DMA system is based on two 8237-type DMA controllers. Controller 2 provides DMA channels 5, 6 and 7 as well as the cascade input for controller 1.

The 8237 device only supports 16-bit addresses (limiting access to 64 kB of memory). Each DMA channel has an associated *page register* on the main board to provide the additional addresses, so that up to 16 MB of memory may be accessed via DMA. This means that if more than 64 kB is to be transferred via DMA, the page register must be repro-grammed after each 64 kB block and a new block of DMA transfer started. This can lead to time gaps in the DMA-transferred data if the data is arriving at high speed from a real-time data acquisition expansion board.

To overcome this limitation, the following techniques may be used:

Dual channel gap-free DMA

At the end of the first 64 kB block of data, the board immediately switches to a second DMA channel and begins transferring the second block of data on it. The host software reprograms the first DMA channel to point the following (third) 64 kB block. When the second block is complete, the board immediately reverts back to the first DMA channel and continues the transfer, which allows the software to reprogram the second DMA channel. This ping-pong approach continues, filling up memory, until all the data has been transferred. This method is called *dual channel gap-free DMA*. Its advantages are that it can access virtually all memory up to 16 MB. It can operate without program intervention if the DMA device provides an interrupt-on-terminal count signal, but it uses up two of the six available DMA channels.

Circular buffer DMA

A buffer (up to 64 kB) may be set up in memory, and the DMA controller programmed to transfer data into the buffer. The controller is programmed for auto-initialize so that when it reaches the end of the buffer, it immediately continues transfer to the beginning of the buffer. The host computer software must process or transfer the data out of the buffer before the next cycle overwrites the previous cycle's data. This is called *circular buffer DMA*. The advantages are that it is simpler to implement, uses less memory, and allows processing while transferring the data. However, it generally requires much program intervention.

Normal DMA using on-board FIFO

The DMA device uses a single DMA channel but it has an on-board cache large enough to store buffer data while the DMA controller is reprogrammed to point to the starting address of the next 64 kB memory block. This is normal DMA as far as the host PC is concerned. It allows gap-free DMA using one channel, with the additional hardware on the expansion board. The host DMA system must operate fast enough to be able to transfer not only the data that has built up in the cache while the DMA channel was being reprogrammed, but also subsequent data arriving at the cache. The process may also operate completely in the background if the board has an interrupt-on-terminal count signal.

DMA operational timing is as follows:

- The DMA device requests a DMA transfer by asserting its DRQ (DMA request) signal high. This causes the DMA controller to begin the process of requesting the bus. It sends a HOLD signal to the CPU, which causes the CPU to float all its address, data, and status outputs and assert its HLDA signal. When the DMA controller finds the CPU's HLDA signal asserted, it asserts AEN, acknowledges the DMA request by asserting the channel's /DAK line low and drives the ISA bus address, read, and write lines according to the DMA timing charts below.

- For a memory-write / I/O-read DMA cycle, as shown in Figure 4.15, the lower 16 address lines, SA[15..0] are driven by the DMA controller with the DMA channel's current address register. The upper address lines are driven by the DMA channel's page register. This address is the destination address in memory for the data. After asserting /DAK (which the DMA device normally uses as a chip select signal), the controller asserts /IORC and then /MWTC (/SMWTC). The /IORC instructs the I/O device to drive the data bus with its data, while the /MWTC instructs the memory to latch the data at its trailing edge. Throughout the cycle the other two command signals /IOWC and /MRDC remain inactive high.

- The same timing and principle applies to an I/O-write / memory-read DMA cycle, shown in Figure 4.3. The difference is that /IORC and /MWTC are held inactive while /MRDC instructs the memory device to drive the data lines, and /IOWC instructs the I/O device to latch the data.

On some computers the bus clock BCLK has different timings when DMA cycles are run and may be called S clocks instead of T clocks. Some computers may not allow wait states to be inserted into the cycle, while on others the wait state mechanism operates in the same way as CPU initiated bus cycles. A DMA cycle is normally completed in seven BCLK periods, but this may vary too. The first four DMA channels may only support 8-bit transfers on some machines.

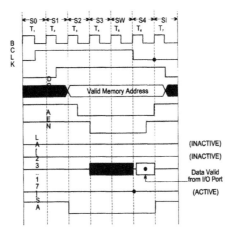

Figure 4.3
Timing chart of a memory-write / I/O-read DMA cycle

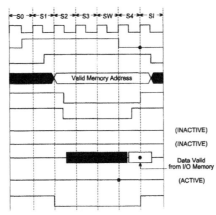

Figure 4.4
Timing chart of an I/O-write / memory-read DMA cycle

4.4 Repeat string instructions (REP INSW)

You may know that the XT used DMA to transfer data from the hard disk, while the AT does not. The AT's disk transfer rates are much higher (over 500 kB/sec for a standard AT bus interface) than the XT's. The question arises: how can the AT achieve such high data transfer rates without using DMA?

The answer is found in new instructions added to the Intel 80286 (and higher) processors. These are the repeat input and output string instructions. On the 8088/8086, repeat instructions are available for moving and processing strings in memory, but not to memory add-resses from I/O addresses (strings are just continuous sequences of bytes). Subsequent processors extended this to include I/O locations, from where a data acquisition board's samples originate.

To obtain a series of data samples from a board, the program initializes a counter register, sets up a destination memory address register and executes the repeat input string word instruction (REP INSW), giving the I/O address of the board's data register. The instruction automatically reads a sample value, stores it at the memory address pointed to by the destination address register, increments this register, decrements the counter and continues transferring samples until the counter reaches zero. This method has distinct advantages, but the data acquisition board must meet certain requirements.

The advantages are:

- Very high speed (slightly higher than DMA because there is no overhead associated with the single-sample nature of PC type DMA)
- Very simple to program
- Does not need to use hardware methods at all

The requirements are:

- Since the PC just reads a block of converted samples, the board must have an on-board or external pacer clock, convert all the readings without program intervention, as well as have a fair-sized sample buffer to pre-store the samples in readiness for the REP INSW instruction.
- The board must have a status flag to indicate the 'buffer half full' or 'nearly full' condition, so that the program can execute the REP INSW loop. If this flag is capable of causing an interrupt request, collection may be done in the background, otherwise the program must poll the status flag at regular intervals to check if the data is ready for transfer.
- As high-level language compilers do not support these instructions, the code must be written in assembly language.

There is a 32-bit version of the repeat string instruction, but as the ISA I/O bus data width is 16 bits, this does not offer significant advantages over the 16-bit instruction. 32-bit buses (EISA, VL and PCI) take advantage of this.

4.5   Polled data transfer

The term 'polled data transfer' describes transfers of data to or from the CPU that are initiated by a CPU instruction. These are memory and I/O reads and writes.

There are two sizes of data transfer: 8-bit and 16-bit, each with its own default timing. For backward compatibility with 8-bit devices, a 16-bit instruction, if executed by the CPU, and the expansion board does not indicate that it is a 16-bit device (with either of the /M16 or /IO16 signals), then the system board performs *data bus translations*. The 16-bit operation is converted into two 8-bit operations, and two 8-bit cycles are run instead of a single 16-bit cycle.

The 80286, 80386 and 80486 processors have a machine cycle consisting of two clock periods or states. These are called T_S – send status, and T_C – perform command. The processor machine cycle may be extended by additional command (T_C) states, when the processor is in the command state, by driving its /READY input. This is achieved on the ISA bus with the CHRDY signal, and the additional T_C states are called wait states (annotated T_W and T_{AW} in the timing charts).

The system board, to ensure compatible timing, adds wait states and they may also be added and reduced by expansion boards. As BCLK, the I/O clock is generally slower than the CPU clock, the system board lengthens the periods of the machine states in the machine cycles that are to be run on the I/O bus. For example, if the CPU clock is 400 MHz and the I/O clock is 100 MHz, each T state in an I/O cycle will be lengthened by a factor of four over that of the main CPU. In this example, the bus cycle looks to the I/O device like a system with a CPU running at 100 MHz.

The next few sections describe the actual timing relationships of the signals on the bus for the various types of bus cycles.

These are:

- Memory-read: data transfer from a memory device to the CPU
- Memory-write: data transfer from the CPU to a memory device
- I/O-read: data transfer from an I/O device to the CPU
- I/O-write: data transfer from the CPU to an I/O device
- DMA Write I/O: data transfer from a memory device to an I/O device
- DMA Read I/O: data transfer from an I/O device to a memory device

Memory-read and memory-write cycles are essentially similar and are discussed together, as are I/O-read and I/O-write cycles and the two different directions of DMA cycles.

A further type of cycle run on the system board is the interrupt acknowledge cycle. It is not present on the expansion bus and therefore not relevant to expansion board design.

Lastly, there may be cycles on the bus, run by bus master boards.

The timing of these cycles obviously depends on the bus master board, but the initiation and conclusion (i.e. the bus arbitration) of bus master cycles have fixed timing specifications. This is not discussed here but guidelines were given in the previous section on bus signal descriptions.

In the timing charts that follow, dots are used to indicate sampling points and shaded areas indicate a 'don't care' state.

Memory-read and memory-write cycles

The following three timing charts show instances of 8-bit memory access:

- A standard 6-BCLK cycle
- A cycle which the expansion board extends (to seven BCLKs) with CHRDY
- A cycle shortened by the expansion board to three BCLKs with the /NOWS signal

As all the I/O bus cycles described here are fairly similar in mechanism, the first description below applies to all of them; the individual cycle descriptions that follow focus on additional details as well as on points of difference with the general description.

Standard 6-BCLK 8-bit memory access

Figure 4.5 shows the XT-compatible cycle, which is the default for I/O memory access in the absence of intervention by an expansion board. It consists of a T_S state, the send status state, followed by four wait states inserted by the system board, followed by the perform command state, (T_C).

In the standard memory cycle, BALE goes active in the last half of T_S. It indicates a valid address on the latchable address lines LA[23..17], and its trailing edge may be used to latch these addresses if they are required by the expansion board for the remainder of the cycle.

Note that LA[23..17] are valid before the machine cycle begins and go invalid in the latter half of the state following T_S. This allows them to be setup before the next machine cycle begins, and is called *address pipelining*. (On the EISA bus, all the address lines from A2 to A32 are available in latchable form, which allows more flexible shortening of the bus cycle.) Compare this to the address lines SA[19..0], which go valid just before the second machine state begins and remain valid throughout the remainder of the machine cycle (and a little into the next). The address lines are used to decode the address(es) of the device(s) on the expansion board.

The read or write command signals (/SMRDC, /SMWTC, /MRDC or /MWTC) go active (low) just after the second half of the second machine state, T_2. This indicates to the expansion board that it may begin latching the data on the data lines in a write cycle (note that the write data from the CPU is valid in the first half of T_2, before the command signal goes active), or it may begin driving the data lines in a read cycle. The CPU latches the read data on the trailing edge of the read (/SMRDC, /MRDC) signal, while the adapter-board must latch the CPU data on the trailing edge of the write (/SMWTC, /MWTC) signal. This occurs at the end of the T_S state.

/M16 is sampled at the end of T_S. Since it is inactive (high), an 8-bit, 6-wait state cycle is run.

/NOWS is sampled on the trailing edge of BCLK, which is halfway through the wait states, T_2 to T_5 and is acted on if a command signal (a read or write line) is active. Since it is found inactive (high) in each wait state, the next wait state is inserted by the system board until the default number (four) of wait states has been run. If /NOWS is active at any of the sampling points, the remaining wait states will be discarded and the T_C state is run, completing the cycle.

The sampling of CHRDY starts at the end of the last default wait state, T_5. Since it is inactive (high), no more wait states are added and the cycle is completed with T_C.

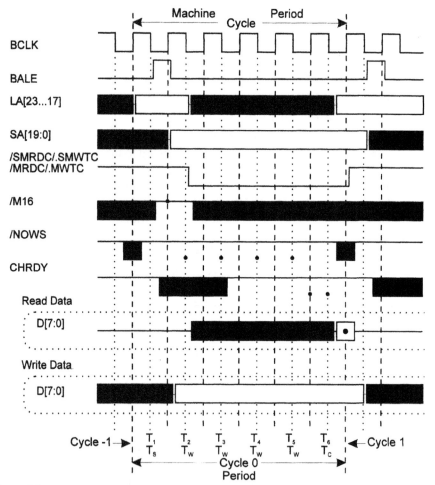

Figure 4.5
Timing chart of a standard 6-BCLK 8-bit memory access cycle

Extended 7-BCLK 8-bit memory access

The 8-bit machine cycle with additional wait states is almost exactly the same as the standard cycle shown above in Figure 4.5. CHRDY is driven and held by the expansion board somewhere during the cycle before T_C begins. The system board begins sampling CHRDY twice every machine state, beginning in the last default wait state (T_5 in Figure 4.6). Since it finds it active (low), it runs another wait state (T_6), during which it continues to sample /NOWS and CHRDY. Since CHRDY is found to be inactive at the end of T_6, the machine cycle is completed with a T_C state.

The additional wait state has the effect of lengthening the bus cycle by one bus clock, which may allow slower devices to interface to the bus. If more time is required by the I/O bus device, CHRDY may be held low until the required cycle time is completed. 2.5 µs is a recommended maximum length of time for which CHRDY may be asserted.

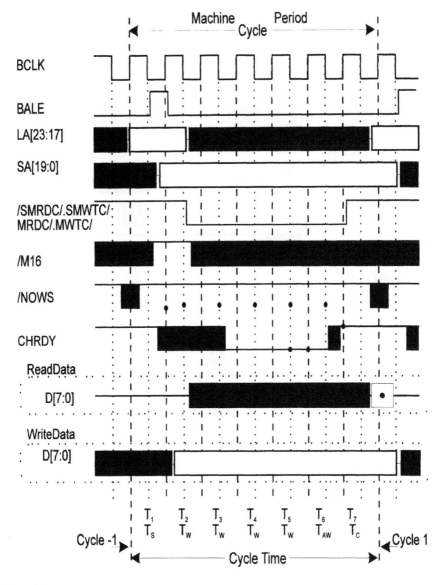

Figure 4.6
Timing chart of an extended 7-BCLK 8-bit memory access cycle

Shortened 3-BCLK 8-bit memory access

The shortened 8-bit machine cycle is also almost exactly the same as the standard cycle. The /NOWS line is sampled in the middle of the machine states, on the trailing edge of BCLK. In Figure 4.7, the system board finds /NOWS asserted (low) at the middle T_3, and a command signal is active. T_3 would have been a wait state, but because /NOWS is asserted, the system board immediately completes the machine cycle by converting it to a T_C state. /NOWS may be asserted anywhere in the bus cycle to indicate that no further wait states are required. If the expansion board had asserted /NOWS before the middle of T_2, a 2-BCLK cycle would not have been generated because in 8-bit cycles, the command signal is not active until after the middle of T_2.

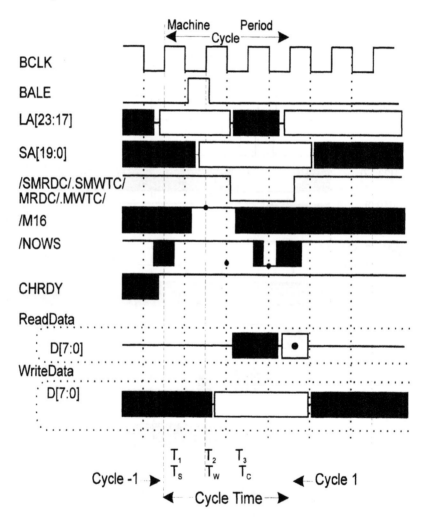

Figure 4.7
Timing chart of a shortened 3-BCLK 8-bit memory access cycle

The next three timing charts show instances of 16-bit memory access:

- A standard 3-BCLK cycle (Figure 4.8)
- A cycle which the expansion board extends (to six BCLKs) with CHRDY (Figure 4.9)

- A cycle shortened by the expansion board to two BCLKs with the /NOWS signal (Figure 4.10)

If the expansion board indicates that it is a 16-bit memory device by asserting /M16 (which is sampled at the beginning of T_2), the system board runs a standard 3-BCLK 16-bit memory cycle. The timing is virtually the same as for the standard 8-bit memory cycle. The differences are: /SBHE is of interest and goes active at the end of T_1, the read or write command goes active just after the beginning of T_2 instead of halfway through T_2, and now the data is transferred on all 16 of the data lines.

/NOWS and CHRDY are sampled in the same places as the 8-bit cycle, and as they are found inactive here, they do not influence the bus cycle.

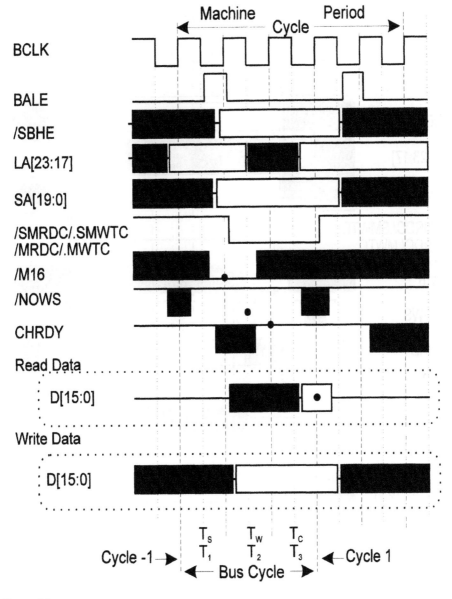

Figure 4.8
Timing chart of a standard 3-BCLK 16-bit memory access cycle

Extended 6-BCLK 16-bit memory access

If the access and setup times of the 16-bit attached bus memory device are longer than provided for by the default 3-BCLK bus period, it may, after asserting /M16, assert CHRDY to extend the bus cycle.

The system board finds /M16 asserted at the end of T_S and begins to run a standard 3-BCLK 16-bit memory access. However, at the end of T_2, it finds CHRDY asserted. It therefore runs another wait state (T_{AW1} in the chart). It expects to find CHRDY disasserted at the end of this additional wait state but, since it is sampled still active, another wait state is added. When CHRDY is found inactive (here, at the end of T_{AW3}), the bus cycle is completed with a T_C state.

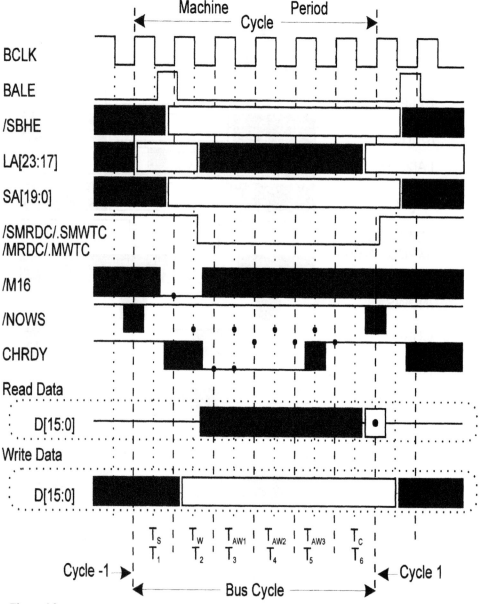

Figure 4.9
Timing chart of an extended 6-BCLK 16-bit memory access cycle

Shortened 2-BCLK 16-bit memory access

If the memory device is capable of transferring data in a shorter time than the default 3-BCLK period, it may, after asserting /M16, assert /NOWS to execute the cycle in two BCLKs.

As before, the system board finds /M16 asserted at the end of T_S, so it begins to run a standard 3-BCLK 16-bit memory access. However, halfway through T_2 (which would have been a wait state), it finds /NOWS asserted. It therefore converts the wait state into a T_C state and completes the bus cycle.

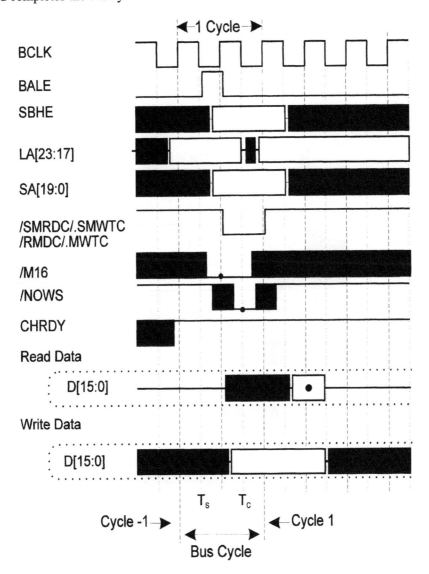

Figure 4.10
Timing chart of a shortened 2-BCLK 16-bit memory access cycle

I/O-read and -write cycles

The following three timing charts show instances of 8-bit I/O access:

- A standard 6-BCLK cycle (Figure 4.11)

- A cycle which the expansion board extends (to seven BCLKs) with CHRDY (Figure 4.12)
- A cycle shortened by the expansion board to three BCLKs with the /NOWS signal (Figure 4.13)

8-bit I/O cycles are initiated when the processor executes an 8-bit IN or OUT instruction, while 16-bit cycles are run when the processor executes a 16-bit IN or OUT instruction and the expansion board responds by asserting /IO16. The command signals in I/O cycles go active ½ BCLK later than the command signals in memory cycles.

Standard 6-BCLK 8-bit I/O access

The 8-bit I/O access cycle is almost exactly the same as the 8-bit memory access. The differences are: the AEN signal goes low at the start of the cycle, indicating that an I/O cycle is in progress and not a DMA cycle; only the SA[15..0] address lines are used, and the read and write command signals are /IORC and /IOWC respectively which become active ½ BCLK later than the corresponding memory-read and memory-write signals.

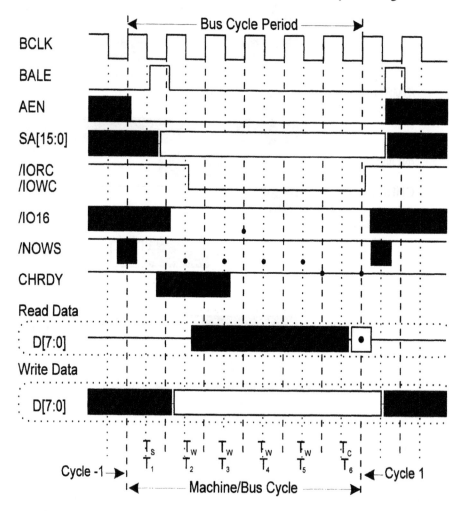

Figure 4.11
Timing chart of a standard 6-BCLK 8-bit I/O access cycle

Extended BCLK 8-bit I/O access

If the I/O device needs to extend the cycle, it does so by asserting CHRDY in exactly the same manner as an 8-bit memory device. When the system board finds CHRDY active, it runs additional wait states at the end of the default number (four in this case) of wait states.

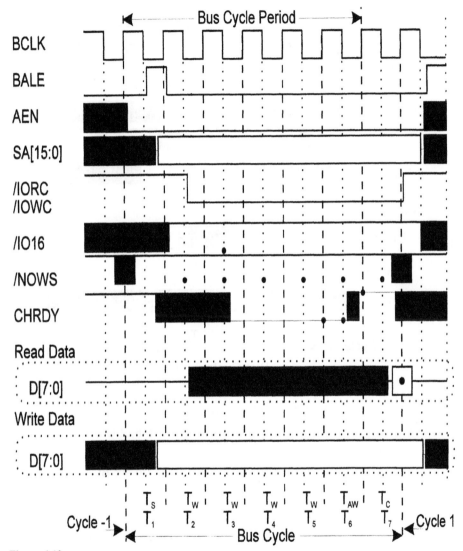

Figure 4.12
Timing chart of an extended BCLK 8-bit I/O access cycle

Shortened 3-BCLK 8-bit I/O access

The shortened I/O access operates in a similar manner to the shortened memory access. The system board finds /NOWS asserted in the T_3 state so it completes the bus cycle at the end of the T_3 state.

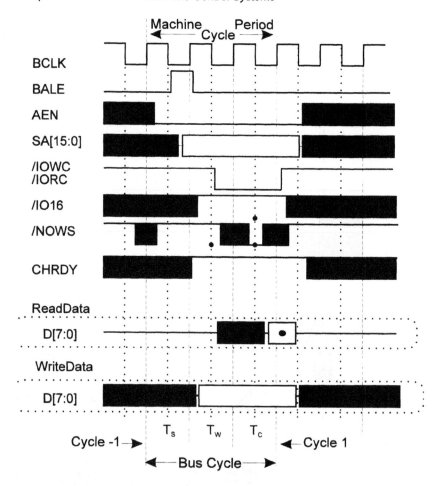

Figure 4.13
Timing chart of a shortened 3-BCLK 8-bit I/O access cycle

The following two timing charts show instances of 16-bit I/O access:

- A standard 3-BCLK cycle (Figure 4.14)
- A cycle that the expansion board extends (to six BCLKs) with CHRDY (Figure 4.15)

Note that it is not possible for I/O devices to shorten cycles to run with no wait states (that is, to two BCLKs) because, for I/O cycles, the command signal goes active only after the trailing edge of BCLK in T_2 (when /NOWS is sampled). Also /IO16 is sampled halfway through T_3 and not at the end of T_1 as in memory cycles.

Standard 3-BCLK 16-bit I/O access

If an I/O device indicates that it is a 16-bit device with the /IO16 signal, the system board runs a 3-BCLK I/O cycle. This cycle is almost exactly the same as the shortened 3-BCLK 8-bit I/O cycle, except that the expansion board does not have to drive /NOWS and /SBHE is of interest in this cycle.

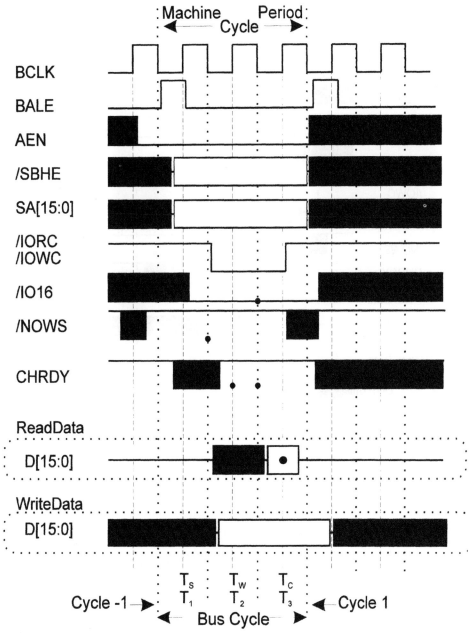

Figure 4.14
Timing chart of a standard 3-BCLK 16-bit I/O access cycle

Extended 6-BCLK 16-bit I/O access

Again, if a 16-bit I/O device cannot meet the setup and hold times of the standard 16-bit I/O cycle, it may drive CHRDY to cause the system board to insert more wait states, until CHRDY is disasserted by the expansion board.

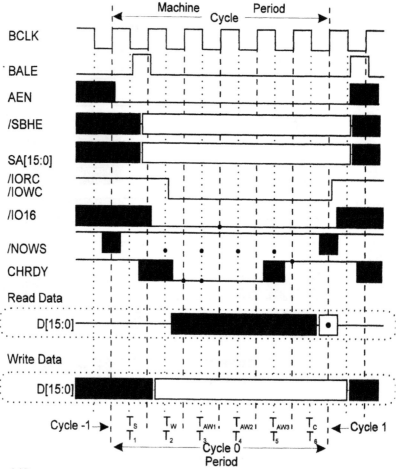

Figure 4.15
Timing chart of an extended 6-BCLK 16-bit I/O access cycle

4.6   Data transfer speed (polled I/O, interrupt I/O, DMA)

One of the key considerations for data acquisition systems using I/O expansion cards is that of the speed of data transfer between the I/O device and the PC's memory.

There are only two ways to transfer data:

- Under software control (simple polled I/O, interrupts)
- Under hardware control (DMA)

The answer of relative speed can be found simply by looking at what is required to perform the transfer of one word from an I/O mapped I/O device to system base memory.

Polled I/O is the simplest and most common method of data transfer between I/O devices and memory.

Assuming that the readiness of data in the I/O device is indicated by an addressable I/O status register, then the minimum functions which the software must perform are as follows:

- I/O read – data transfer from I/O device to the CPU (status)
- I/O read – data transfer from I/O device to the CPU (data)
- Memory write – data transfer from CPU to memory (data)

If the CPU is not doing anything else while waiting for the data to be ready, then data transfer speeds of 40 kHz are possible. In this case, the CPU is tied up completely (in a single task operating system) while executing the polling loop and is doing nothing useful while waiting for the data to become ready.

Should any additional processing be required, this must be inserted in the polling loop, further reducing the sampling rate.

Interrupt I/O assumes that the I/O device is capable of notifying the CPU that data is ready by driving its corresponding interrupt line active (see Operation of interrupts, p. 71).

The CPU no longer wastes time by waiting for data to become ready and is free to execute other programs in the foreground.

However, when interrupted, the minimum functions the CPU is required to perform (within the interrupt service routine) are as follows:

- I/O read – data transfer from I/O device to the CPU (data)
- Memory write – data transfer from CPU to memory (data)

It would appear that the interrupt driven I/O approach should be faster as this requires fewer I/O bus cycles to perform the data transfer. This, however, does not take into account the hardware delays in signaling and acknowledging an interrupt (see Operation of interrupts, p. 71) and the overheads in servicing an interrupt (saving registers, etc), which can be quite significant. Further to this, is the fact that even if an interrupt occurs at the start of the CPU instruction cycle, it must wait until the instruction has been executed before being serviced.

In fact, interrupt driven I/O is significantly slower (about half the speed) than polled I/O for these very reasons.

Direct memory access (DMA) represents the greatest improvement in both the speed of data transfer (100 kHz) and the amount of data that can be transferred without CPU intervention (32 k word samples). In much the same way, that an I/O device's interrupt line informs the CPU that data is ready, the I/O device informs the DMA controllers that data is ready by asserting its DRQx line (see Operation of DMA, p. 76).

Once the CPU releases the bus, the transfer process is totally transparent to the CPU. In fact, the CPU may still operate at full throughput, if it has local high-speed cache memory, from which it can execute, and none of the instructions reference anything not already in the cache. If this happens, then the CPU must wait until the DMA cycle is finished. Caches are common on 386DX and 486 machines.

4.7   Memory

There are three main classifications of memory used in PC systems, called base memory, expanded memory and extended memory. They are explained below.

4.7.1   Base memory

The memory from address 0 up to either the amount of memory installed in the computer or address FFFFFh (that is, up to a total of 1 MB) is called base memory. The first 640 kB of this is RAM and is usually used by the operating system and application programs. The latter 384 kB of address space is reserved; it is used for the BIOS ROM and other adapter ROMs, display adapter memory, other adapter memory, and expanded memory (see below). This, and the first 64 kB of extended memory (see below), is sometimes called *high memory*. A system may have the ability to map physical memory into the spaces in high memory not used by any device in the computer. This memory may then be used by some applications.

4.7.2   Expanded memory system (EMS)

Early processors (the 8086/8088) and all PC processors running in real mode are limited to a memory space of 1 MB because only the first 20 address lines are available. (The same holds for DOS, being a 16-bit operating system.) To make more memory available for applications, a scheme was developed by Lotus, Intel, and Microsoft called expanded memory (LIM EMS 4.0 is a common version). In hardware, a second linear array of memory, called the *logical expanded memory* is designed into a system. This may be up to 32 MB in size. A block of memory space is then set aside in the high memory area (normally 64 kB) and divided into four separate pages of 16 kB each. This acts as a window into the expanded memory. Thus, four pages of the actual expanded memory are accessible at any one time through the window in high memory. These windows are called *page frames*. The required portion of expanded memory is mapped into the page frame through registers in the computer's I/O space.

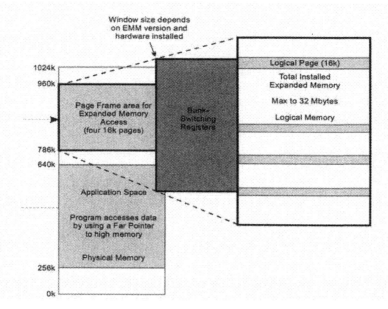

Figure 4.16
Organization of expanded memory

The management of the memory is handled by an operating system extension called the expanded memory manager (EMM), normally installed at system startup. Application programs use the expanded memory for data. It is not usually possible to place program code in EMS. The application program communicates with the EMM via software interrupt 67h and accesses the memory via a far pointer into the page frame.

The EMM has two main sections:

- The driver, which has some of the characteristics of a real device driver, and includes initialization and output status functions.
- The manager, which acts as an interface between application software and the expanded memory hardware.

The EMM provides the following services:

- Hardware and software module status
- Allocation and de-allocation of EMS pages
- Mapping of logical pages into the physical frame

- Support for multitasking operating systems
- Diagnostic routines

For compatibility with different types of hardware (and EMM software revisions), applications should communicate with the EMM via the assigned software interrupt. If memory is acquired, used, and released via the EMM, other programs will also be able to make use of the memory.

The advantage of EMS is that it provides additional data memory to DOS and 8088-type computers.

The limitations of EMS are:

- It requires special hardware and software drivers to operate.
- It cannot execute programs from EMS; it is used for data storage only. For example, in a DOS system with 32 kB of expanded memory and over 1 MB of base and extended memory, the maximum executable program size (without using disk overlays) is the DOS 640 kB limit, less the amount of memory used by DOS and other installed drivers.
- Much slower access time than linear base and extended memory.
- Programs cannot access EMS directly; they must go via the memory manager and this results in more complex code.

4.7.3   Extended memory (XMS)

Extended memory is the physical linear memory found above the 1 MB mark. 80286 and 80386SX processors can address up to 16 MB of base and XMS while 80386DX and 80486 processors can address up to 4 GB of this type of memory. XMS is memory addressed directly by the processor (and hence the application program) and is therefore simpler, quicker, and more efficient. Extended memory is only available as normal application memory when the processor is in protected mode; it follows that only 32-bit protected mode operating systems and extensions - such as OS/2, UNIX and MS-Windows, but not DOS – can make this memory available to programs.

4.7.4   Expansion memory hardware

All systems support a certain amount of memory on the main board. On most PCs, eight SIMM (single in-line memory module) sockets are standard. Each socket usually accepts 256 kB, 1 MB, 4 MB SIMMs and higher. Additional memory may be added on using an expansion card in the I/O channel. Some system boards provide access to the high-speed local processor bus with additional proprietary memory board slots. On 80386 systems and higher, memory may be configured dynamically in software as XMS or EMS, as is needed by applications. Main board memory has a shorter access time than memory on add-on boards; these in turn have much shorter time than memory attached to the I/O channel.

4.8   Expansion bus standards (ISA, EISA, PCI and PXI bus)

4.8.1   ISA bus

One of the primary reasons for the success of the first IBM personal computer (PC) was its 'open system' design. IBM encouraged the development of compatible add-on products by third party manufacturers by releasing details of its 8-bit expansion bus. This bus formed the basis of the industry standard architecture (ISA) bus.

The IBM PC and early versions of the IBM XT model of personal computer were based on the 8-bit, 20 address-line 8088 CPU running at 4.77 MHz. In later versions of the XT, some manufacturers also changed the microprocessor to the 16-bit 8086. In all these machines, the expansion bus remained the same, operating at the CPU clock frequency of 4.77 MHz.

With the introduction of the Intel 80286 ('286') 16-bit microprocessor, IBM released the PC AT. To accommodate this 16-bit bus architecture, the 8-bit ISA slot was extended to the 16-bit ISA slot with the addition of eight extra data lines, four additional address lines, extra interrupt and DMA channels and various other control signals.

To allow backward compatibility with existing expansion cards, IBM kept the original 62-pin connector intact (the one closest to the rear of the expansion card), and added an extra 36-pin connector to accommodate the new signal, data and address lines. (For connector pin-outs etc, refer to the reference text 'PC Instrumentation for the 90s.')

With the release of the PC AT also came an increase in CPU speed and subsequent increases in ISA bus speed, originally to 6 MHz and finally to 8 MHz.

When IBM decided to define the maximum ISA bus speed at 8 MHz, leading manufacturers of expansion boards began utilizing only the ICs with access times required to meet the timing specifications of an 8 MHz ISA bus. Before long, however, AT clone manufacturers were soon producing 286 systems with CPU clock speeds of 10, 12, 16, 20, and 25 MHz. This led to backward compatibility problems between the faster CPUs and the many slower expansion boards already in existence.

At this time the ISA expansion bus changed from a local bus to a translated (or split bus) bus slot, whereby the CPU or local bus signals were buffered and the memory, I/O and DMA bus cycles slowed down to meet the 8 MHz ISA bus limit. This 'slowing down' was achieved by adding 'wait states' to the normal bus cycles, a 'wait state' being the condition where all bus lines remain in their current state for another full bus clock cycle.

The 16-bit ISA bus is an extension of the 8-bit I/O bus of the original PC and PC XT. They had between five and eight slots each, with a 62-way connector for plug-in cards. ISA extended the slots with a second connector, adding 32 more signal lines. (While XT cards are generally compatible with ISA computers, XT systems have become less common nowadays and will not be discussed.)

The descriptions of the ISA signals are based on the EISA specification, revision 3.12, which includes the ISA specification. Unless specifically noted, the acronym ISA refers to the part of the EISA specification that deals with the ISA bus. Not all PC systems comply totally with the specification but they all comply, to some degree. Any adapter board that complies with the specification has a very good chance of working correctly even in systems that do not comply exhaustively with the specification.

A common misunderstanding regarding PC specifications concerns the term *clock speed*. In modern PCs the CPU (or system) clock speed, or to use the more correct term *clock frequency*, differs from that of the *I/O bus clock*. The CPU clock speed is generally 30 to 50 MHz while the bus clock speed is 8 to 12 MHz. The compatibility of expansion cards is not dependent on the CPU clock but on the I/O clock (together with DMA clock and I/O cycle timing). ISA specifies the bus clock (BCLK) to have a frequency between 8.333 MHz and 4 MHz with a duty cycle of 50%, but many PCs have bus clocks with a higher frequency. 10 MHz and 12 MHz are common while I/O clocks can be as high as 13.7 MHz.

A problem that sometimes occurs when connecting an interface to the PC bus concerns the matching of the PC bus cycle with that of the interface design. For example, the interface board may operate at a lower speed than that provided for by the PC bus cycle. Lowering the PC bus cycle to that of the hardware interface can solve the problem. If the bus cycle is four clock cycles in length, a READY bus signal (derived from the interface card) tells the CPU to add an additional one or more WAIT states to the bus cycle, thus ensuring a match of the data rates between the PC bus and the hardware interface.

Many systems also allow the bus and DMA clock (and possibly other system parameters such as default number of I/O wait states) to be programmed within the CMOS setup. These can then be set for the highest speed/lowest number of wait states, etc, that is still supported by the slowest board in the system.

ISA signal descriptions

Some of the terms and conventions used in the following sections are described in the following table.

Bus master	An expansion board that is capable of taking over the bus and running its own bus cycles on it. Bus masters use a DMA channel to gain access to the bus.
Command cycle/bus cycle	The sequence of events (or bus clock periods) that make up a complete cycle on the bus. Usually one piece of data is transferred per bus cycle.
Command signals	The bus signals that provide instructions to the expansion board, for example, the Read and Write signals.
DMA cycle	A bus cycle controlled by the DMA controller (not involving the CPU) in which data is transferred directly to and from memory.
Slave	Used to refer to an ISA expansion board that responds to bus signals directed at it.
Standard cycle	For each type of cycle (e.g. 8-bit I/O-read of 16-bit memory-write etc), the default type of cycle the system board runs, is called a standard cycle. Adapter boards may modify standard cycles.
/	Active Low. The symbol P preceding any signal indicates that the signal is active when it is in the low state.
D[...]	Bits on the data bus.
LA[...]	Latchable addresses.
SA[...]	Address lines.

Table 4.3
Some ISA bus terminology

Figure 4.17 tabulates the ISA signal mnemonics, pin locations, names and types. Each signal is then described fully.

Facing rear panel with mounting brackets

I/O PIN	SIGNAL NAME	I/O
B1	GND	GROUND
B2	RESDRV	O
B3	+5 Vdc	POWER
B4	IRQ 9	I
B5	-5 Vdc	POWER
B6	DRQ2	I
B7	-12 Vdc	POWER
B8	NOWS	I
B9	+12 Vdc	POWER
B10	GND	GROUND
B11	/SMWTC	O
B12	/SMRDC	O
B13	/IOWC	I/O
B14	/IORC	I/O
B15	/DAK3	O
B16	DRQ3	I
B17	/DAK1	O
B18	DRQ1	I
B19	/REFRESH	I/O
B20	BCLK	O
B21	IRQ7	I
B22	IRQ6	I
B23	IRQ5	I
B24	IRQ4	I
B25	IRQ3	I
B26	/DAK2	O
B27	T-C	O
B28	BALE	O
B29	+5 Vdc	POWER
B30	OSC	O
B31	GND	GROUND

I/O PIN	SIGNAL NAME	I/O
A1	/10CHK	I
A2	D7	I/O
A3	D6	I/O
A4	D5	I/O
A5	D4	I/O
A6	D3	I/O
A7	D2	I/O
A8	D1	I/O
A9	D0	I/O
A10	CHRDY	I
A11	AEN	O
A12	SA19	I/O
A13	SA18	I/O
A14	SA17	I/O
A15	SA16	I/O
A16	SA15	I/O
A17	SA14	I/O
A18	SA13	I/O
A19	SA12	I/O
A20	SA11	I/O
A21	SA10	I/O
A22	SA9	I/O
A23	SA8	I/O
A24	SA7	I/O
A25	SA6	I/O
A26	SA5	I/O
A27	SA4	I/O
A28	SA3	I/O
A29	SA2	I/O
A30	SA1	I/O
A31	SA0	I/O

I/O PIN	SIGNAL NAME	I/O
D1	/M16	I
D2	/IO16	I
D3	IRQ 10	I
D4	IRQ 11	I
D5	IRQ 12	I
D6	IRQ 15	I
D7	IRQ 14	I
D8	/DAKO	O
D9	DRQO	I
D10	/DAK5	O
D11	DRQ5	I
D12	/DAK6	O
D13	DRQ6	I
D14	/DAK7	O
D15	DRQ7	I
D16	+5 Vdc	POWER
D17	/MASTERIG	I
D18	GND	GROUND

I/O PIN	SIGNAL NAME	I/O
C1	/SBHE	I/O
C2	LA23	I/O
C3	LA22	I/O
C4	LA21	I/O
C5	LA20	I/O
C6	LA19	I/O
C7	LA18	I/O
C8	LA17	I/O
C9	/MRDC	I/O
C10	/MWTC	I/O
C11	D8	I/O
C12	D9	I/O
C13	D10	I/O
C14	D11	I/O
C15	D12	I/O
C16	D13	I/O
C17	D14	I/O
C18	D15	I/O

Facing front of computer

Figure 4.17
ISA signal mnemonics, signal directions and pin locations

The ISA signals are divided into four groups according to their function:

- Address and data bus signal group
- Data transfer control signal group
- Bus arbitration signal group
- Utility signal group

Address and data bus signal group

This group contains the signal lines that are used to address memory and I/O devices and the signal lines used to transfer the actual data.

D[7..0]

D[7..0] are the low eight bits of the 16-bit bi-directional data bus, used to transmit data between the microprocessor, memory and I/O ports. During CPU-initiated write bus cycles, the CPU data is valid on these lines at the trailing edge of /MWTC, /SMWTC and /IOWC and may be latched using this edge. During CPU read cycles, the expansion device must drive these data lines with its data before the rising edge of /MRTC, /SMRTC and /IORC. During

DMA cycles, data is transferred directly from the I/O device to memory (or vice versa) on the data lines while the processor is disconnected from them. The DMA controller drives the control lines in this case. Bus masters may also take control of these lines.

D[15..8]

D[15..8] are the high eight bits of the 16-bit bi-directional data bus. They are similar to the lower eight data lines, D[7..0]. 8-bit wide transfers must use D[7..0]. If the currently running software requests a 16-bit transfer from an 8-bit device, the system board automatically converts it into two 8-bit cycles on D[7..0]. Adapters capable of 16-bit transfers must indicate this using /M16 or /IO16 during cycles addressed to them or the system board will convert the instruction into two 8-bit instructions. /SBHE (explained later), is asserted by the system board during 16-bit cycles.

LA[23..17]

The LA17 to LA23 (latchable address) lines form part of the latchable address bus. (The remaining lines of the latchable address bus, LA[16..2] and LA[31..24] are wired to the EISA connector and are not available in ISA systems. SA[19..0] must be used instead.)

LA16 to LA23 are unlatched and, if required for the whole bus cycle, must be latched by the addressed slave. During standard cycles, they are valid during the active time of the BALE signal (explained later) and remain valid for at least ½ BCLK period after the command signals are asserted.

During DMA or ISA bus master cycles, LA[23..17] are valid at least one BCLK before the command signals are asserted. They may be driven by an expansion board acting as a bus master. These lines may be latched with the trailing edge of BALE.

These address lines are provided in this way because they are pipelined from one cycle to the next, and to reduce address delay when they are used to decode a block of bus-attached memory.

SA[19..0]

Address lines SA0 through SA19 are used to address system bus I/O and memory devices. They form the low-order 20 bits of the 32-bit address bus. (However, only 24 of the 32 address lines are normally available in ISA systems.)

On normal cycles SA0 to SA 19 are driven onto the bus while BALE is high and they are latched by the system board on the trailing edge of BALE and are therefore valid throughout the bus command cycle.

During DMA and 16-bit ISA bus master cycles, they are driven by the DMA logic and bus master respectively. They should be valid one BCLK before the command signals and normally stay valid one BCLK after the command signals end.

With 20 address lines it is possible to address 1 MB of memory, but not all address locations are available. Base system memory, system ROMs and display memory all use addresses in this range.

The processor, using the IN and OUT instructions, addresses I/O devices with lines SA0 through SA15, while SA16 through SA19 are not used and are held inactive. Most PC I/O devices only decode the first ten address lines, (SA0 to SA9, which correspond to I/O addresses 0h to 3FFh) so care must be taken when addressing I/O devices with SA10 to SA15.

/SBHE

/SBHE (system bus high enable) is an output-only signal. When low, it indicates to the expansion board that the present cycle expects to transfer data on the high half of the D[15..0] data bus. An example of this happens when the expansion board had previously indicated to the system board that it is capable of transferring 16-bit data with the /IO16 or /M16 signals.

The type of bus cycle can be decoded from /SBHE and SA0 as follows:

/SBHE	SAO	BUS CYCLE
0	0	Full 16-bit transfer
0	1	Upper byte transfer on D[15..8]
1	0	Lower byte transfer on D[7..0]
1	1	Invalid combination

Table 4.4
Decoding bus cycle type from /SBHE and SA0

AEN

When low, AEN (address enable) indicates that an I/O slave may respond to addresses and I/O commands on the bus. It is an output signal issued by the DMA control logic during DMA cycles which, when asserted (high), is used to prevent I/O slaves from misinterpreting DMA cycles as I/O cycles. The system board also uses this signal to disable the processor's address, data, and control lines from the I/O bus during DMA cycles.

Data transfer control signal group

This group contains signals that are used to control data transfer cycles on the bus.

BCLK

BCLK (bus clock) is provided to synchronize events with the main system clock. According to EISA specifications, BCLK should operate at a frequency between 8.333 MHz and 4 MHz, with a normal duty cycle of 50%. However, most ISA systems have a BCLK frequency of 8 MHz to 12 MHz. (The original XT had a high time of 66⅔% and a low time of 33⅓% and a frequency of 4.77 MHz.) BCLK is driven by the system board. Its period is sometimes extended for synchronization to the main CPU or other system board devices. During bus master cycles, the system board extends BCLK only when required to synchronize with main memory. Events must be synchronized to BCLK edges without regard to frequency or duty cycle. This signal can be used to generate system bus wait states.

BALE

When high, BALE (address latch enable) indicates that a valid address is present on the latchable address lines LA17 to LA23. It goes high before the addresses are valid and falls low after they have become valid. If the addresses are needed for the whole cycle, the expansion board should latch them with the trailing edge of BALE. This is a good synchronization signal when looking at normal bus cycles because it starts at the beginning of each bus cycle. It is high (and does not fall low) during DMA or bus master cycles.

/MRDC

This signal is asserted by the system board or ISA bus master to indicate that the addressed memory slave should drive its data onto the system data bus. This should be done before the rising edge of the /MRDC signal to ensure that the receiving device obtains valid data.

During DMA cycles, /MRDC is asserted for read accesses from memory addresses between 0h and 00FFFFFFh, regardless of the type of memory responding. This allows the /DAK selected I/O port to receive the data. (The I/O device should not use /MRDC to decode its I/O address.) /MRDC may be driven by expansion boards acting as ISA bus masters.

/SMRDC

This memory-read signal is derived from /MRDC and has similar timing; the difference between the two is that /SMRDC is only active for addresses between 0h and 000FFFFFh (that is, in the first megabyte of memory).

/MWTC

This signal is asserted by the system board or ISA bus master to indicate that the addressed memory slave may latch data from the system data bus. The data is valid at the rising edge of the /MWTC signal and maybe latched at this time. During DMA cycles, /MWTC is asserted for write accesses to memory addresses between 0h and 00FFFFFFh, regardless of the type of memory responding. This allows the /DAK selected I/O port to drive the data bus with its data. (The I/O device should not use /MWTC to decode its I/O address.) /MWTC may be driven by expansion boards acting as ISA bus masters.

/SMWTC

This memory-write signal is derived from /MWTC and has similar timing; the difference between the two is that /SMWTC is only active for addresses between 0h and 000FFFFFh (that is, in the first megabyte of memory).

/IORC

The I/O-read signal is asserted by the system board or ISA bus master to indicate that the addressed I/O slave should drive its data onto the system data bus. This should be done after /IORC goes low, and the data must be held valid until after the rising edge of the /IORC signal to ensure the receiving device obtains valid data. During DMA cycles, the address bus does not contain an I/O port address; it contains the memory address to which the I/O port data will be transferred. The I/O port is selected, not by an address decode, but by a /DAK signal.

/IOWC

The I/O-write signal is asserted by the system board, or ISA bus master, to indicate that the addressed I/O slave may latch data from the system data bus. This should be done at the rising edge of /IOWC to ensure the receiving device obtains valid data. The system board, DMA device or bus master must drive the data bus before asserting /IOWC. During DMA cycles, the address bus does not contain an I/O port address; it contains the memory address from which the I/O port will latch data. The I/O port is selected, not by an address decode, but by a /DAK signal.

CHRDY

An expansion device may use CHRDY (CHannel ReaDY) to lengthen a bus cycle from the default time. This allows devices with slow access times also to be attached to the system. The slave drives CHRDY low after decoding a valid address and finding a command signal (any of the six I/O or memory-read or write signals) asserted. This lengthens bus cycles by an integral number of bus cycles. If CHRDY is low, the command signals remain active at least one BCLK period after it goes inactive. CHRDY should be driven low by an open collector or tri-state driver and it should *never* be driven high. It should not be held low for more than 2.5 µs (or about 10 BCLK periods, whichever is less).

/NOWS

The /NOWS (NO wait state) signal may be driven by a memory device after it has decoded its address and command to indicate that the remaining BCLK periods in the present cycle are not required. This must happen before the falling (back) edge of BCLK to be recognized in that BCLK period. /NOWS should be driven low by an open collector or tri-state device capable of sinking 20 mA and never be driven high. A slave should not assert /NOWS and CHRDY at the same time.

/M16

If the addressed memory is capable of transferring 16-bits of data at once on the D[15..0] data lines, it may assert /M16, after decoding a valid address. This causes the system board to run a 3-BCLK memory cycle (that is, with only one wait state). /M16 should be driven low by an open collector or tri-state device capable of sinking 20 mA and never be driven high.

/IO16

If the addressed I/O port is capable of transferring 16-bits of data at once on the D[15..0] data lines, it may assert /IO16, after decoding a valid address. This causes the system board to run a 3-BCLK I/O cycle (that is, with only one wait state). /IO16 should be driven low by an open collector or tri-state device capable of sinking 20 mA and never be driven high.

Bus arbitration signal group

These signals are used to arbitrate between devices and the system board for control of the bus.

DRQ[7..5] and DRQ[3..0]

The DRQ (DMA request) lines are used to request a DMA service from the DMA subsystem, or for a 16-bit ISA bus master to request access to the system bus. The request is made when the DRQ line is driven high and may be asserted asynchronously.

The requesting device must hold its DRQ line active until the system board responds by asserting the corresponding /DAK line. For demand mode DMA memory-read I/O-write cycles, DRQx is sampled on the rising edge of BCLK, one BCLK from the end of the present cycle (the rising edge of /IOWC).

For demand mode memory-write I/O-read cycles, DRQx is sampled on the rising edge of BCLK, 1½ BCLKs from the end of the cycle (the rising edge of /IORC).

For 16-bit ISA bus masters, DRQx is sampled on the rising edge of BCLK, two BCLKs before the system board asserts DAKx. The trailing edge of DRQx must meet the setup and hold time to the sampling point for proper system operation. The ROM BIOS initializes the DMA controller so that DRQ0 has the highest priority and DRQ7 the lowest. Care must be taken to deactivate the DRQ line without delay, otherwise more than one cycle may be granted. The corresponding /DAK is typically used to reset the DRQ line.

/DAK[7..5] and /DAK[3..0]

The system board asserts a DMA channel's /DAK (DMA acknowledge) signal low to indicate that the channel has been granted the bus. The DMA device is selected if it finds its /DAK signal, together with either /IORC or /IOWC, asserted. The DMA controller then takes control of the bus and proceeds with the DMA cycle. /DAKx is also asserted to acknowledge granting the bus to a 16-bit ISA bus master. The bus master must then assert /MASTER16 if it finds its /DAK asserted and proceed with its cycle. Afterwards, the bus master must float the address and control lines and make /MASTER16 inactive before the system board disasserts the /DAK line.

T-C

T-C (terminal count) is a bi-directional signal acting in one of two modes, depending on the programming of the DMA channel.

In output mode, the system board asserts T-C to indicate that a DMA channel's word count has reached its terminal value. This happens when the decrementing count rolls over from 0 to FFFFFFh. T-C is only asserted when the corresponding channel's /DAK line is asserted, so that the DMA device can condition T-C with its /DAK signal to determine if the DMA transfer has been completed.

In input mode, a DMA device to stop a DMA transfer can use T-C. T-C is sampled by the system board when /IORC or /IOWC is asserted. If T-C is found asserted, the transfer is terminated, and if the channel is programmed for auto-initialize, the transfer starts at the beginning.

/MASTER16

This signal allows bus master cards to take over the system bus. A master asserts /MASTER16 when it receives a /DAK signal from a DRQ on its DMA channel. Asserting /MASTER cancels the DMA operation and tri-states the system address, data and control signals. This allows the bus master card to control and drive devices attached to the system bus as well as memory. It must obey all the timing requirements of the bus devices and memory, and return control to the processor (that is, release the bus) within 64 BCLK periods (nominally 8 μs).

/REFRESH B19

When low, /REFRESH indicates that a refresh cycle is in progress. /REFRESH causes SA[15..0] (or LA[15..2]) to drive the row address inputs of all DRAM banks so that when /MRDC is asserted, the entire system memory is refreshed at one time.

Utility signal group

OSC

OSC is a clock signal for use in general timing applications. Its frequency is 14.31818 MHz (roughly 70 ns) with a duty cycle of 50%.

RESDRV

RESDRV (reset driver) is an output signal, which, when asserted, produces a hardware-reset for devices attached to the bus. It is also asserted during power-up. All devices that can prevent operation of the CPU, memory or system board I/O must use RESDRV for hardware reset. When RESDRV is asserted, these devices must float all outputs that drive the bus. Examples of expansion board devices that must sample and use RESDRV are slaves that insert wait states, devices that require software initialization and DMA devices.

IRQ[15..14], IRQ[12..9], IRQ[7..3]

The input-only interrupt lines are used by expansion boards to interrupt the CPU to request some service. An interrupt is recognized when the IRQ line goes from low to high and stays high until the corresponding interrupt is acknowledged. This implies that it is not possible for more than one device to use an interrupt line at the same time, reliably. The interrupt lines are pulled high by the system board and, when floated, are guaranteed to be high after 500 ns. Interrupt routines must reset the device's interrupt latch at least 500 ns before issuing an end-of-interrupt command to the interrupt controller, to re-enable interrupts on that line. Missing interrupt lines (1RQ13, 8, 2, 1 and 0) are used by the system board and are not available in the bus.

/IOCHK

An expansion board can assert /IOCHK (I/O channel check) to indicate that a serious error has occurred. Assertion of /IOCHK causes an NMI (non-maskable interrupt) to the CPU if Port 61h bit 3 is 0 and if NMIs are enabled with Port A0h. Parity errors and uncorrectable hardware errors are examples of where expansion boards might assert /IOCHK.

4.8.2   Microchannel bus

In 1986 IBM included the newly released Intel 32-bit 80386 ('386') microprocessor into a new family of PC's called the PS/2 systems. At the same time, and rather unexpectedly, IBM introduced with this family of machines a new and proprietary expansion bus called the microchannel architecture (MCA) bus.

This bus provided many enhancements, which included:

- Bus mastering.
- Burst mode data transfers (where data is transferred a predefined block size at a time).
- Bus arbitration, which permitted up to eight processors and eight other devices, such as DMA controllers, to share the single data bus without interfering with each other.

However, the MCA bus was not compatible at all with ISA bus expansion boards, the most obvious difference being the change in connector size, layout and pin spacing (the MCA connector is much smaller, pin configurations were rearranged and pin spacing decreased to 0.050"). Moreover, there were two versions of the MCA bus slot (a 16-bit version and 32-bit version), each with different pinouts.

IBM's refusal to co-operate with the industry to create just one standard (development of the EISA standard bus was already in progress), has meant that this bus was largely shunned from the beginning. MCA systems have achieved limited success in the data acquisition environment, due mainly to the incompatible expansion bus and the availability of high performance, low cost, easily expandable ISA machines.

4.8.3   EISA bus

The introduction of the Intel 32-bit 80386 ('386') microprocessor marked the first departure by the industry from following developments in the IBM PC system architectures. While IBM produced the PS/2 family of machines incorporating the MCA bus, the rest of the industry produced 386-based ISA machines operating at 16, 20, 25, 33 and 40 MHz CPU speeds. Initially these were AT clones with 386 CPUs, the split bus architecture allowing the expansion bus to continue to operate at around 8 MHz, thus maintaining ISA compatibility, and access to the increasing range of ISA compatible expansion cards.

In 1988, and after almost two years of meetings, nine manufacturers collaborated to produce the extended industry standard architecture (EISA) bus specification, in direct opposition to IBM's MCA bus. This published standard encompassed all of the ISA features, almost all of the MCA enhancements and added some new features while maintaining the backward compatibility with existing expansion boards. The EISA bus is a full 32-bit data and address bus.

Some of its features include:

- True bus master capabilities.
- Additional DMA transfer modes, such as block demand or block burst mode.
- Ability to share interrupt lines between devices.
- Automatic expansion board configuration so as to achieve a conflict free system.

While the speed of the EISA slot is still limited to 8 MHz because of the ISA compatibility restriction, the increased bus-width allows a much higher data transfer rate (33 MHz). In burst mode, speeds up to 40 MHz can be achieved. Unfortunately, benchmarks show that EISA expansion cards are no faster than their ISA equivalents, largely due to the fact that few EISA peripherals make use of the extra speed features such as bus mastering.

4.8.4   The PCI, compactPCI, and PXI bus

In the spring of 1991 Intel Corporation began working on the PCI bus as an internal project. Intel engineers were concerned that the existing input / output (I/O) bus bandwidths were not keeping up with current CPU speeds and were falling even farther behind as the new generations of CPU's (486 / Pentiums) were becoming available.

The ISA bus was introduced by IBM in 1981 and was upgraded to 16 bit with the introduction of the AT in 1984. The ISA originally ran at 4.77 MHz. The top speed was 8 MHz. In 1987, the MCA (micro channel for the PS/2) was introduced. This bus didn't last long (neither did the PS/2). The MCA was a 16- or 32-bit bus that ran at 10 MHz. The EISA bus was presented in 1989. It was a 32-bit, 8MHz bus with the big advantage of more pin outs for larger cards. In 1992 the CPU VL bus was developed. It was a 32-bit bus running at 33 MHz. In 1993 the PCI was introduced. At first, it ran at the same 33 MHz, but soon the speeds increased to 100 MHz on 3.3 volt CPUs. The PCI bus can use either 32- or 64-bit bus lines.

Year	Bus	Bits	Frequency
1981	IBM PC ISA	8	4.77
1984	IBM AT ISA	16	6/8
1987	MCA (PS/2)	16/32	10
1989	EISA	32	8
1992	VL	32	33
1993	PCI	32	33
1995	PCI	32/64	33/66
1996	PCI	32/64	100
1998	PXI	same as PCI	

Table 4.5
Growth table of the PC bus system

The PCI or (Personal Computer Interface bus) is a relatively new addition to the PC motherboard. As time goes on it seems that personal computers have more and more PCI slots and less and less ISA or EISA SLOTS. Very soon all computers will have PCI slots but no ISA or EISA slots. Is this one of those transitional things that we have to put up with in an ever-changing industry, or is the PCI that much better? The answer to that question is that the PCI bus is that much better. Legacy bus systems such as the ISA and EISA were good in the days when only one application was running at a time. Now that the PC can run more than one application at a time, it is necessary to have a multi-card bus system. The PCI bus does this by running at faster clock rates then the legacy buses and by temporarily releasing the bus, therefore allowing other cards to have a chance to transmit or receive data from the CPU.

Card size is also a big advantage of the PCI bus. The fingers on the PCI bus are smaller and more numerous. This makes the average PCI card the same size as a legacy ISA card. With most electronic PCBs going to surface mount components, smaller PCI cards can have more functions than huge EISA cards. This will be a benefit for the manufacture (reduction in cost) and for the consumer (more functions for the same or smaller price). Another spin off of the PCI bus is better plug and play.

The PCI local bus is a high-performance bus that provides a processor-independent data path between the CPU and high-speed peripherals. PCI is a robust interconnect mechanism designed specifically to accommodate multiple high performance peripherals for graphics, full motion video, SCSI, LAN, etc.

The PCI bus is not a local bus but an intermediate bus. It talks to the CPU bus (local bus) by way of a CPU to PCI Bridge. This allows the CPU to talk to host memory, or the cache, over a very fast and short local bus, while the PCI bus connected devices (and even ISA bus devices) do their own thing. A bridge is used as a buffer / transmitter / receiver between the CPU bus (local bus) and the PCI bus. The PCI bus is really a multi-bus system with devices being allocated resources on each PCI bus. If legacy bus systems are present, and as of this writing they usually are, the legacy devices run on their own legacy bus not on a PCI bus.

The legacy bus is allocated resources by IRQ and DMA control chips in conjunction with the PCI bus system. The PCI devices still can use IRQs and DMA but in a different way than legacy devices. The PCI bus has its own internal interrupts called INT#...

Plug and play (or is that plug and pray) is the ability of the computer to recognize a new card has been plugged into the bus. On legacy cards (ISA and EISA), recognition for PnP (plug and play) was difficult. The ISA system cannot produce specific information about a card, so the BIOS has to isolate each one and give it a temporary handle so its requirements can be read. Resources are allocated only when all of the cards have been dealt with. All PnP cards are isolated and checked, but only those needed to boot the machine are activated. Another way is for the information to be stored in the registry and read on boot up. All PnP devices are configured and then activated on boot up.

It must be remembered that not all PCI cards are PnP, but because they have the resources to give the BIOS everything it needs, PCI cards are the perfect hardware vehicles for PnP. To do plug and play correctly it is necessary to have PnP compliant BIOS, PnP cards, and a PnP operating system. This will give the best PnP performance. Many people have had problems because one of these PnP systems was not truly compliant. Another problem was that the PnP devices interacted with each other and caused strange results. This was because the devices were not truly multi-functional.

Multi-functional cards are able run at the same time (apparently) without interfering with each other; sometimes this is accomplished by using a PCI bus master. The PCI bus master can allocate time for each card to access the bus. The bus master can define how long the PCI bus will delay a transaction between the given PCI slot and the ISA bus. This delay is accomplished by using a latency timer command.

The PCI also uses an internal interrupt system such as INT#1, INT#2, INT#3 etc. This interrupt system is an internal (to the PCI bus) interrupt system. It has nothing directly to do with the IRQ system, but can be mapped to if the card concerned needs it. Any available IRQ can be mapped to an INT#. The use of this INT# system means the PCI cards can be accessed without the use of IRQs. The card designated as the master can interrupt the other PCI (slaves) as needed. The INT# PCI cards are usually configured as edge triggered as opposed to ISA which is usually level triggered IRQ.

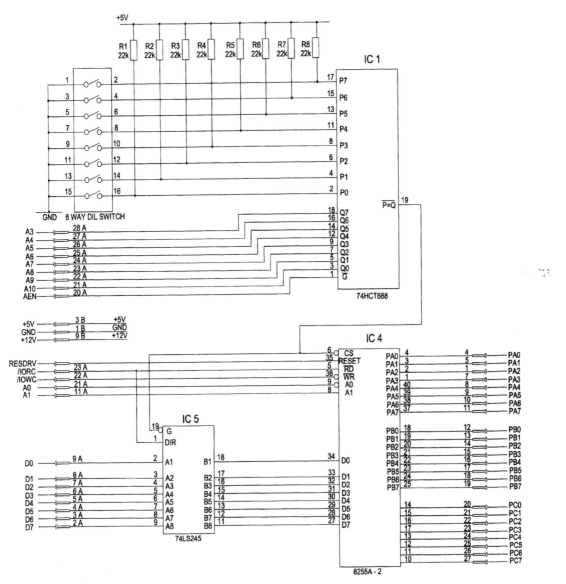

Figure 4.18
PCI structure

In 1998, the compactPCI and PXI bus were developed to combine the attributes of the PC's PCI bus with the industrial card connection system of the VME. The VME bus had been used in various forms for many years as an industrial bus rack for mounting electronic PCBs. The PCI bus is a very fast and easy way of connecting the computer to the outside world. By combining the PCI with the VME the best of both worlds was accomplished. Instead of having to open the PC and insert a PCI card, the user can just slide the card into the front of the chassis. It uses a 5-row 2 mm-pitch connector with impedance-matched pins and sockets. The integrated shielding system doubles the bus card's capacity from four to eight. The PXI version of the compactPCI has extra features that the basic compactPCI does not. These include timing and triggering functions so multiple boards can perform synchronous data acquisition, and the ability of one card to trigger another without the intervention of the system.

4.9 Serial communications

The need for PCs to exchange data with remote instruments and distributed or stand-alone data acquisition systems over long distances over the existing telephone network and to function as intelligent terminals, has furthered the development of various serial communications adapters. A standard PC has two asynchronous serial communications ports called COM1 and COM2. DOS supports up to four such ports, and some PCs may also have a COM3 and COM4.

The electrical characteristics of the ports conform to published Standard RS232 and they operate at speeds from 50 bps to over 115.2 kps. IBM compatible PCs are typically fitted with an RS-232 serial communications card with two serial ports.

4.9.1 Standard settings

I/O address of the serial ports:

> Serial port 1 (COM1) 03F8 Hex
> Serial port 2 (COM2) 02F8 Hex

Where additional serial port hardware is fitted these ports have the following settings:

> Serial port 3 (COM1) O3E8 Hex
> Serial port 4 (COM2) 02E8 Hex

Hardware interrupts:

> Serial port 1 IRQ 4
> Serial port 2 IRQ 3

For communication over an analog telephone line, a modem is connected to a serial port, or it may be incorporated in an add-on board. A modem converts the digital data generated by the PC into a signal that can be transmitted over an analog line. Some modems incorporate data compression and error correction techniques to achieve much higher transmission rates than the bandwidth of a telephone line would normally support.

The COM ports are also used to attach terminals to multitasking, multi-user environments such as UNIX; this often requires multiple COM ports. Add-on boards are available which support two or four COM ports; further COM ports may be added by using multiple 8-port COM boards. DOS will not support COM ports higher than 4, but the hardware itself is exactly the same as that for ports 1 to 4. The additional COM ports appear at different I/O channel addresses and may be accessed directly by the application software.

4.9.2 Intelligent serial ports

An intelligent serial port is designed to relieve the load on the main CPU of the host PC. It typically contains an Intel 80186 microprocessor with read only memory (ROM), containing an operating system and some RAM. When the board is initialized, the operating system is loaded into the RAM together with any other user programs. There is also dual ported RAM on the board for transferring information between the PC CPU and the on-board RAM. The 80186 then polls each of the serial ports on the board to read for any incoming data or to transfer any outgoing data. The user programs contained in the RAM can also pre-process the data before it is transferred to the PC system memory. The transfer is accomplished either by DMA, which does not use much memory, or by memory-mapping the dual ported RAM into PC memory, 64 kBytes of PC system memory.

4.10   Interfacing techniques to the IBM PC

The problems that may need to be considered when interfacing expansion cards to the IBM PC vary, depending on whether you are designing a board for a specific purpose or simply plugging in an off-the-shelf board that requires configuring.

While the basic considerations do not change, their complexity varies depending on the functions that the expansion board must perform.

The primary considerations are:

- Hardware compatibility

     These are the physical requirements such as connector configurations, board size, bus loading etc.
- Addressing considerations

     Whether the expansion board is addressed as part of the PC's memory or I/O map, it is obviously important that each address is unique and does not conflict with any other used addresses in the PC's memory.
- Timing requirements

     The timing of accesses to and from the memory and I/O as well as the system board timing for interrupts and DMA are strictly controlled within pre-defined limits (this is what makes it a 'standard'). Consideration of this timing is very important since it greatly affects the efficiency of functions that the board must perform, and can be affected by the physical hardware factors of the bus itself, the speed of the CPU and the speed of the integrated circuits (ICs) used on the expansion board.

We will consider these issues here as they relate to an 8-bit 24 line programmable I/O control board interfacing to a standard ISA compatible bus. This is shown in Figure 4.18.

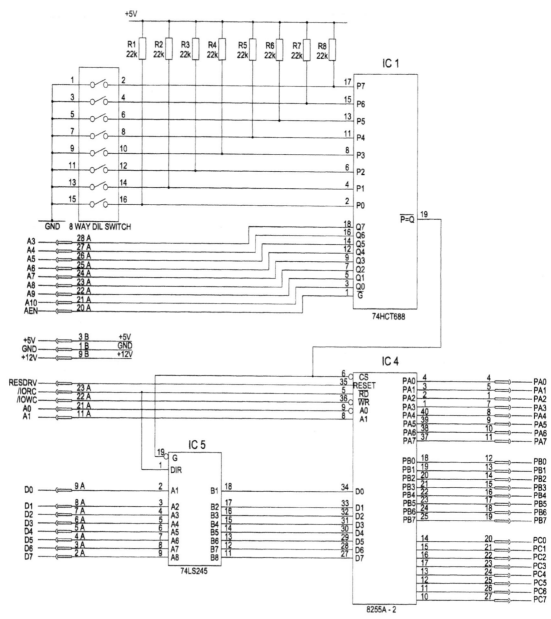

Figure 4.19
8-bit 21-line I/O board

4.10.1 Hardware considerations

Physical size

The physical characteristics of the expansion board that must be considered are its physical size, and connector size and configuration. There are two variations of this: the original ISA 8-bit add on card configuration with a single 62-pin connector (principally designed for the PC/XT) and an 8/16 bit version with an additional 36-pin connector (for the PC/AT).

Capacitive loading

Capacitive loading of the output bus signals is the capacitance that the signal line driver circuit sees, and is related to the length of the bus signal lines. As the capacitance is increased (with the addition of further expansion cards), the signal can become distorted and delayed, affecting critical timing conditions for bus signals.

Where the capacitive coupling between signal lines is significant, and especially when the signal lines are being driven at high frequency, then noise or signal chatter between signal lines can also affect the timing of bus signal lines and performance of the bus.

This is especially true when expansion boxes with a long connection to the PC are fitted and there are many cards installed.

A common engineering practice by manufacturers is to deliberately limit the length of these bus signal lines on expansion cards.

Bus bandwidth is the maximum frequency at which data can be reliably transmitted on the bus. This is related to the transmission line characteristics of each of the bus signal lines, the capacitive loading on each signal line and the capacitive coupling between them.

Although the manufacture of a multitude of IBM clones has meant that bus characteristics can vary between manufacturers, a common maximum transmission rate (bus bandwidth) is 100 MHz.

Bus loading

The simple and safe rule here is that the maximum loading presented to any bus signal line should not exceed more than 2 LS TTL devices.

Power supply (VCC) noise

The frequency of noise that may appear on the power supply is directly related to the frequency that an IC changes state, since the power that ICs draw from the power supply changes as they change state.

By decoupling each IC device on the board from its power supply with a 0.1 uF ceramic capacitor, the noise being applied back to the supply is greatly reduced. If this power supply noise is not decoupled, then the high frequency power supply changes can be induced into other ICs and also capacitively coupled to other important bus signal lines, possibly affecting the correct operation of the expansion bus.

4.10.2   Address decoding

I/O devices addressed within the memory or I/O address map of the PC must be uniquely addressable and must not conflict with other memory or I/O addresses in the system.

The base address of an I/O device determines where in the computer's memory or I/O address space the computer will find the I/O device and represents the 'lowest' address that will access this device. This setting must be unique for all I/O devices in the computer and cannot lie within an addressable range of any other I/O device. Memory or I/O locations on I/O devices are usually addressed in a linear range, from the base address up.

The I/O device must correctly decode the address signals, allowing access for only the proper addresses. This is usually performed by a dip switch located on the I/O device.

The unused locations in the PC's I/O map are dependent on the type of computer used (PC/XT/AT).

It is up to the user to find an unused area in the PC's I/O address space, taking into account the addresses of all other installed I/O devices.

To allow greater flexibility, it is common practice for the base address of an I/O device to be settable within a large range of I/O addresses.

Address decoding on the 24-line programmable I/O board is performed by the 74HCT688 comparator, which compares addresses on the address lines SA[9..2] to the value set by the address DIP switch. Where a switch is open, the comparators 'P' input is in the high state, pulled high by the 'pull-up' resistors connected to the 5 V rail, while a closed switch will drive the 'P' input low.

The output of the comparator will be asserted low (active) when:

- The address lines SA[9..2] at the 'Q' inputs correspond exactly to the state of the 'P' inputs, as determined by the switch settings.
- The AEN bus signal is low.

The AEN (address enable) bus signal is used so that addresses appearing on the bus during DMA cycles are not misinterpreted as I/O addresses. It is an output signal that is asserted high by the DMA controller logic during DMA cycles and when low, indicates to slave I/O devices that they may respond to addresses and I/O commands on the bus.

Thus the output of the comparator is low when there is a valid address in the board's address range being presented on the bus and a DMA cycle is not in progress. It is not necessarily true however that the I/O device should respond to this address, since there is a large number of addresses in the memory address space that would have the same SA[9..2] address settings. The CPU may be performing a 'memory read' or 'memory write' cycle to one of these locations.

The I/O control signals /IORC and /IOWC are in fact used to differentiate this.

- The /IORC signal is asserted low during an 'I/O read' cycle, to indicate to an addressed I/O device that it should drive its data onto the system data bus. The data must be held valid by the I/O device after the rising edge of /IORC to ensure that valid data is placed on the bus.
- The /IOWC signal is asserted low during an 'I/O write' cycle to indicate to an addressed I/O device that data is available on the system data bus. To ensure that valid data is received, the I/O device should 'latch' the data on the rising edge of /IOWC.

For this I/O device the comparator output is used as the chip select for all devices on the board and the /IORC and /IOWC control lines to instruct these devices what action to perform.

The base address comparison is performed for address lines SA[9..2] and is independent of the two address lines SA1 and SA0, which are used to address the bottom four I/O locations from the base address.

Therefore base addresses may be assigned any location from 0h to 7F8h, on eight byte boundaries.

4.10.3   Timing requirements

One of the most important aspects of interfacing I/O devices to the I/O expansion bus is timing considerations.

Slow I/O devices

Sometimes I/O devices, or more particularly the memory and I/O ICs on them, have slower access times than would normally be required in an I/O or memory read or write cycle.

This limitation is overcome by the insertion of 'wait states' into the read or write cycles, a 'wait state' being the condition where all bus lines remain in their current state for another full bus clock (BCLK) cycle.

Upon decoding a valid address and detecting the assertion of one of the control signals (/IORC etc), the I/O device asserts the CHRDY signal low, thus signaling the CPU that a wait state is required. The CPU will keep inserting wait states for as long as CHRDY is detected as being low at the falling edge of the BCLK cycle.

Where an I/O device does require wait states to be inserted, there should be enough flexibility to allow easy selection of a variable number of wait states. Most commonly, the number of wait states is provided by adjustable hardware settings on the I/O device (switches or links) with accompanying logic to provide the necessary CHRDY signal. This flexibility is needed, since the faster the PC to which the I/O device is interfaced, the greater the number of wait-states that will be required.

Note: CHRDY should be driven low by an open collector or tri-state driver and never be driven high. It should not be held low for more than 2.5 μs or about 10 BCLK periods, whichever is less.

Fast I/O devices

Alternatively, an I/O device may be very fast and may not need all the time that is taken in a normal bus cycle.

When this is the case, the I/O device must assert the /NOWS (no wait state) signal. This must be detected on the falling edge of BCLK for it to be recognized in that BCLK cycle.

All the memory and I/O read or write cycles can be shortened, except the 16-bit I/O read/write cycle.

Note: /NOWS should be driven low by an open collector or tri-state driver and never be driven high.

Practical timing considerations

To determine if our I/O device requires wait states (or in fact if the default number of wait states may be reduced) the access times of the ICs being addressed on the device must be investigated.

Referring to the timing diagram of a standard 6 BCLK I/O cycle, we see that the /IORC line goes low roughly halfway through T_2. This tells the addressed device to drive the bus data lines D[7..0] with its data.

The CPU latches the data present on the bus just before driving the /IORC line high again, which occurs at the end of T_6.

The period over which /IORC is low is about 4.5 machine states which, for a 10 MHz BCLK with a period of 100 ns, corresponds to a period of 450 ns.

Taking into account bus delays of 25 ns and allowing a 25 ns setup time for a safe margin of error, then the data must be presented to the data bus 400 ns after /IORC goes low.

There are two further considerations:

- The access time through the 8255 (access time is the delay between the read line going low to valid data on the 8255's data lines).
- The propagation delay through the 74LS245 transceiver of about 25 ns (see manufacturers' specification sheets).

Looking at the 8255A data sheet (available in the manufacturer's data book) we see that the read line must be low for 300 ns and the access time is 250 ns (or 200 ns for the faster 8255A-5 device).

The write cycle is considered in the same way. The 8255 data sheet indicates that the write pulse width must be 400 ns (or 300 ns for the 8255A), the data must be valid 100 ns before the write line goes high and remain valid 30 ns thereafter.

Therefore, the 8255A-5 device may be comfortably interfaced to the PC expansion bus without wait states, but the use of the slower 8255A would be marginal in general even though it just meets the timing requirements of this PC.

This is true for PCs whose BLCK speed does not exceed 10 MHz. In a PC with a 13.7 MHz BCLK, and having an I/O cycle with three wait states instead of four, the /IORC strobe is low for 3.5 machine states X 73 ns (255 ns). This would not meet the timing requirements of the 8255A. An additional wait state would be needed by the I/O device in order to use the board in this computer.

5

Plug-in data acquisition boards

5.1   Introduction

In recent years, the distinction between separate data acquisition systems and control systems has narrowed because an increasing number of real-life systems are designed not only to acquire data, but also to act on it. This is true of a wide range of plug-in data acquisition and control boards now available. Commonly used multi-purpose plug-in data acquisition boards, currently on the market, typically combine all aspects of data acquisition and control. This includes analog input circuitry for measuring and converting analog input voltage signals to digital format, analog output circuitry for generating analog output voltages from digital control signals, counter/timing circuitry and digital I/O interfaces. Depending on the number of analog inputs/outputs and digital inputs/outputs required for a particular application, multi-purpose boards represent the most cost effective and flexible solution for DAQ systems. Also available and widely used are plug-in boards that specialize in each of the data acquisition and control functions just mentioned.

Examples of these plug-in boards are:

- Analog input (A/D) boards
- Analog output (D/A) boards
- Digital I/O boards
- Counter/timer I/O boards

Computer plug-in data acquisition and control boards often represent the lowest cost alternative for a complete data acquisition and control system. As they interface directly to the host computer's I/O expansion bus, they are generally compact, and represent the fastest method of gathering data and/or changing outputs. These boards are most commonly used in applications where the computer is close to the sensors being measured or the actuators being controlled. Alternatively, they can be interfaced to remotely located transducers and actuators via signal conditioning modules known as two-wire transmitters.

For simplicity this chapter looks at the different aspects of data acquisition and control separately, briefly describing the main components of typical plug-in boards their uses in a data acquisition and control system, as well as the important specifications associated with each classification of plug-in board. Also discussed are the different techniques and important considerations when interfacing analog and digital input and output signals to plug-in boards, as well different sampling techniques.

5.2 A/D boards

Analog input (A/D) boards convert analog voltages from external signal sources into a digital format, which can be interpreted by the host computer. The functional diagram of a typical A/D board is shown in Figure 5.1 and comprises the following main components:
Input channel sample and hold circuits (for simultaneous sampling)

- Input multiplexer
- Input signal amplifier
- Sample and hold circuit
- A/D converter (ADC)
- FIFO buffer
- Timing system
- Expansion bus interface

Each of these components plays an important role in determining how fast and how accurately the A/D board can acquire data.

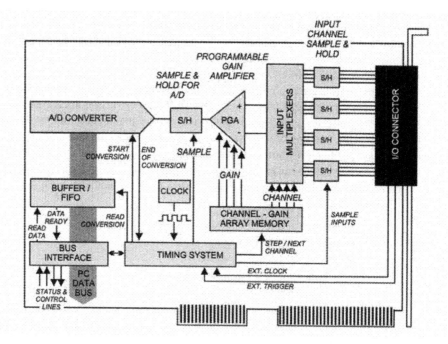

Figure 5.1
Functional diagram of a typical A/D board

5.2.1 Multiplexers

A multiplexer is a device that switches one of its analog inputs (typically up to 16 single-ended inputs) to its output; the input channel selected being determined by the binary code at the input address lines of the multiplexer. The number of address lines required is determined by the number of input channels to be multiplexed. A 16-channel multiplexer would therefore require four input address lines.

On A/D boards, multiplexers facilitate the sampling of multiple inputs on a time-multiplexed basis. The A/D converter samples one channel, switches to the next channel, samples it, switches to the next channel, samples it, and so on. This eliminates the need for a signal amplifier and an A/D converter for each input channel, thereby reducing the costs of A/D boards.

Important parameters

Two parameters that particularly affect the rate, at which the multiplexer can switch between channels, and therefore its throughput rate, are the settling time and the switching time.

Settling time

The settling time is the time the multiplexer output takes to settle within a predetermined error margin of the input when the input signal on the channel swings from –FS (negative full scale input voltage) to +FS (positive full scale input voltage) or from +FS to –FS.

Switching time

The switching time specifies how long the multiplexer output takes to settle to the input voltage, when it is switched from one channel to another.

Throughput rate

The throughput rate is considered the higher of the settling time or the switching time, since it is possible the voltage on one channel can be at –FS while the voltage on the next switched channel is +FS. The throughput rate of the multiplexer is one factor that determines the total throughput of the A/D board.

5.2.2   Input signal amplifier

To achieve the greatest resolution in the measurement of an analog input signal, its amplified range should match the input range of the A/D converter. Consider a low level signal of the order of a fraction of an mV, fed directly to a 12-bit A/D converter with full-scale voltage of 10 V. A loss of precision will result since the A/D converter has a resolution of only 2.44 mV. Some form of amplification is needed.

This is usually provided by a high performance instrumentation amplifier, typified by:

- Balanced differential inputs
- High input impedance
- Low input bias currents
- Low offset drift
- High common mode rejection ratio

Two types of amplifier are commonly included on A/D boards, depending on the cost and quality of the A/D board selected. Some A/D boards provide on-board amplification, where the amplifier gain is adjustable using hardware, while boards that provide programmable gain amplifiers (PGAs) make it possible to select, using software, different gains for different channels.

Adjustable on-board fixed gain amplifier

The gain of these amplifiers is commonly adjusted using a potentiometer or are selectable on board links. A/D boards with a fixed gain amplifier should only be used where the signal levels on each of the input channels to be sampled have comparable ranges and lie within the full scale input range of the A/D converter.

Signals with greatly different signal levels will require external signal conditioning and amplification to enable them to be used on boards utilizing fixed gain amplification. A more flexible alternative is the programmable gain amplifier discussed below.

Programmable gain amplifier (PGA)

Programmable gain amplifiers (PGAs) make it possible to program the gain of the input amplifier using software, requiring a once-only write to an on-board register to select the gain for an individual channel. This is especially advantageous where input signals on different channels have very different signal levels and input ranges. The amplifier gain for each channel can be set accordingly, so that the input range of the incoming signal is matched with the full scale range of the A/D converter, thus resulting in higher resolution and accuracy.

It is usual practice that the amplifier gain, though programmable, is selected from a specified range of gain settings, thereby maintaining the amplifier within its operating range without saturation. In some high performance boards, the gain is automatically adjusted depending on the level of the input signal.

Important signal amplifier parameters

Two parameters that particularly affect the accuracy and the rate at which the signal amplifier can amplify the input signals are amplifier drift, and the settling time.

- Calibration and drift
  Calibration of an amplifier to eliminate offset and gain errors is only valid for the temperature at which the calibration was made. Over time, and with variations in temperature, the characteristics of the amplifier change or *drift* causing offset and gain errors known as *offset drift* and *gain drift* respectively. Offset drift and gain drift in parts per million per degree Centigrade (ppm/°C) specifies the sensitivity of the amplifier to changes in temperature.

  Compounding the natural tendency of the amplifier characteristics to drift is the fact that the potentiometer settings of fixed gain amplifiers also tend to drift with time and temperature.

- Settling time
  Amplifier settling time is defined to be the time elapsed from the application of a perfect step input to the time the amplifier output settles within a pre-determined error margin of the required output value.

  A characteristic of amplifiers is that the throughput decreases with increasing gain. This is because at higher gains the signal output changes by a greater amount, therefore increasing the settling time. This applies to fixed gain and programmable gain amplifiers. If the A/D converter samples the amplified input signal before the amplifier output has settled correctly, (i.e. the time between samples is less than the settling time of the signal amplifier) then an incorrect data value may be sampled. Poor settling time is a major problem, because the level of inaccuracy varies with the gain and sampling rate and cannot be reported to the host computer.

  To allow for the lowest possible input ranges and therefore the highest required gains, A/D boards adjust their internal timing to allow for a greater settling time for the output of the amplifier. This means that the highest allowed settling time and the reduction in throughput caused by it, is imposed for all amplifier gain settings. More advanced A/D boards take into account the input range and amplifier gain required, thereby increasing throughput at higher signal level input ranges where lower gain settings are required.

5.2.3   Channel-gain arrays

On the original A/D boards the address of the channel to be sampled was written to the multiplexer, the gain setting sent to the programmable gain amplifier (PGA) and once the signal was settled an A/D conversion was initiated. The data was subsequently read and transferred to the PC's memory. This incurred a large software overhead. Background operation using interrupts is difficult and slower than polled I/O and accurately timed samples and higher speed data transfer methods such as DMA and repeat instructions are impossible in either case.

The use of channel-gain arrays (CGA) on many A/D boards overcomes these limitations. The channel/gain array is a programmable memory buffer on the A/D board, which contains the channel address and gain setting for each input channel to be sampled. The gain of the amplifier for a particular channel is set by the internal hardware preceding the sampling of the channel, based on the gain value read from the channel/gain array. Where a single PGA is provided for all channels, the gain required for each channel is stored in a channel/gain array. If there are individual PGAs for each input channel, the gains for each input amplifier are stored in a gain array. The gain of each remains the same until overwritten by the software. Channel-gain arrays vary in size from a few channel/gain pairs (one for each channel), to many thousands of pairs.

5.2.4   Sample and hold circuits

As shown in Figure 5.2, a sample-and-hold (S/H) device consists of an analog signal input and input buffer, an analog signal output and output buffer, a charge-storing device, usually a capacitor, and a control input that controls the switching circuitry, which in turn, connects the input to the output.

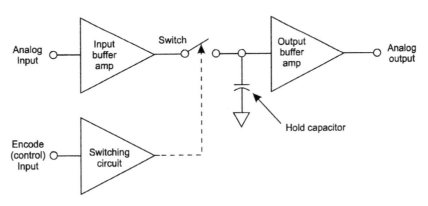

Figure 5.2
Functional diagram of a sample-and-hold device

As its name implies, a S/H has two operating states. When in sample mode, a sample command applied to the control input closes the internal switch, thereby causing the output to track the input as closely as possible. In this mode, the hold capacitor charges to the voltage level applied at the input. When a hold command is applied to the control input, the switch opens, disconnecting the output from the input. With the switch open and the high impedance of the output amplifier preventing the premature discharging of the capacitor, the hold capacitor retains the value of the input signal at the stage the hold command was applied to the control input.

With the exception of some flash A/D converters, which are very fast, most A/D converters require a fixed time period during which the input signal to be converted remains constant. When used at the input to an A/D converter, a sample and hold circuit performs this function, acquiring an analog signal at the precise time its control input is made active. The A/D converter can then convert the voltage held at the output of the sample and hold – minimizing inaccuracies in the conversion due to changes in the signal during the conversion process.

Important signal parameters

- Hold settling time
  The time that elapses from the occurrence of the sample command, to the point where the output has settled within a given error band of the input, is known as the acquisition time or hold settling time.
- Aperture time
  The time required to switch from the sample state, measured from the 50% point of the mode control signal, to the hold state (the time the output stops tracking the input), is known as the aperture time.
- Aperture uncertainty
  This value represents the difference between the maximum and minimum aperture times.
- Drop rate
  A practical sample and hold cannot maintain its output voltage indefinitely while in the hold mode. The rate at which it decays is known as the decay or drop rate.
- Aperture matching
  Data acquisition boards capable of performing simultaneous sampling (see Simultaneous sampling, p.153) require sample and hold devices on each channel. The smaller the aperture time and aperture uncertainty for each of these devices, the narrower the time range over which all the simultaneous samples will be taken. For a data acquisition board this is known as aperture matching. The lower the value, the more closely matched in time the simultaneous samples will be.

  As a point of note, A/D boards that perform simultaneous sampling still have the sample and hold circuit that precedes the A/D converter, as each channel sample still has to be multiplexed to the A/D converter. Some A/D converters have built-in sample and hold circuitry, and where this is the case, the preceding sample and hold circuit is not required.

5.2.5   A/D converters

Real-world signals are analog signals, representing some measured physical parameter for every instant in time. They must be converted to a discrete time signal to be interpreted and processed by computers. As their name would imply, analog-to-digital converters (A/D converters, ADCs) measure an analog input voltage and convert this into a digital output format. A/D converters therefore represent the heart of an A/D board or a data acquisition system.

The main types of A/D converters used and the specific and important parameters relating to their operation are detailed in the following sections.

Successive approximation A/D converters

A successive approximation A/D conversion is the most common and popular direct A/D conversion method used in data acquisition systems because it allows high sampling rates and high resolution, while still being reasonable in terms of cost. Throughput of a few hundred kHz for 12-bit ADCs is common, while 16-bit ADCs employing a hybrid conversion method (i.e. successive approximation plus a much faster method such as flash) are capable of throughput up to 1 MHz, while still being reasonable in cost. One clear advantage of this device is that it has a fixed conversion time proportional to the number of bits, n, in the digital output. If the approximation period is T, then an n-bit converter will have a conversion time of around nT. Each successive bit, which doubles the ADCs accuracy, increases the conversion time by the period T only. The functional diagram of an n-bit successive approximation A/D converter is shown in Figure 5.3.

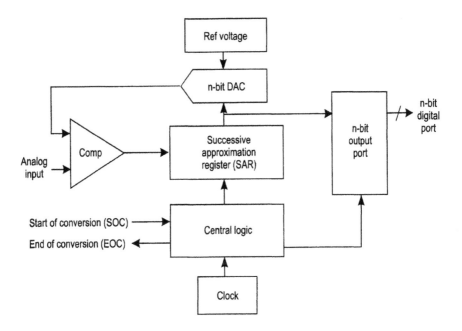

Figure 5.3
Functional diagram of an n-bit successive approximation A/D converter

The successive approximation technique generates each bit of the output code sequentially, starting with the most significant bit (MSB). The operation is similar to a binary search and is based on successively closer comparisons between the analog input signal and the analog output from an internal D/A converter.

The A/D converter starts the procedure by setting the digital input to the D/A converter, so that its analog output voltage is half the full-scale voltage of the A/D device. A comparator is used to compare the D/A analog output to the analog input signal being measured.

If the analog input signal is greater, the most significant bit (MSB) of the D/A converter input is set to logic 1 and the next most significant bit of the D/A converter input is set to logic 1, setting the analog output of the D/A at 3/4 of full scale voltage. If the analog input signal was less, the MSB of the D/A input is cleared to logic 0 and the next most significant bit of the D/A input is set to logic 1, setting the analog output of the D/A at 1/4 of full-scale voltage.

Each step effectively divides the remaining fraction of the input range in half, then again compares it to the analog input signal. This is repeated until all the n bits of the A/D conversion have been determined. It is obviously important that the analog input signal to the A/D does not change during the conversion process, hence the use of sample and hold circuits.

Flash A/D converters

Flash A/D converters are the fastest available A/D converters, operating at speeds up to hundreds of MHz. This type of device is used where extremely high speeds of conversion are required with lower resolution, for example, 8-bits.

Figure 5.4 shows the functional diagram of an n-bit flash A/D converter. Each of the 2^n-1 comparators simultaneously compares the input signal voltage to a reference voltage determined by its position in the resistor series, and corresponding to the output code of the device. Flash A/D conversion is quicker than other methods of A/D conversion because each bit of the output code is found simultaneously, irrespective of the number of bits-resolution. However, the greater the resolution of the device, the greater the number of comparators required to perform the conversion. In fact, each additional bit doubles the number of comparators, and therefore increases the size and cost of the chip.

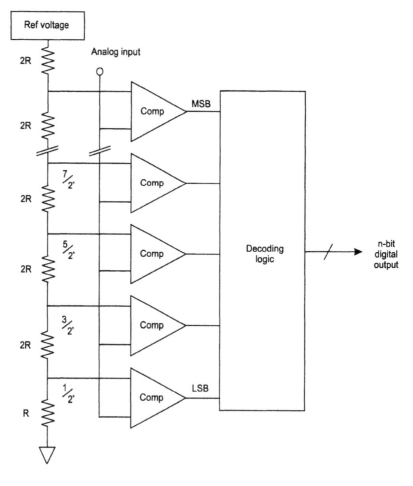

Figure 5.4
Functional diagram of an n-bit flash A/D converter

Flash A/D converters tend to be found in specialist boards, such as digital oscilloscopes, real-time digital signal processing applications and general high-frequency applications.

Integrating A/D converters

Integrating A/D converters use an indirect method of A/D conversion, whereby the analog input voltage is converted to a time period that is measured by a counter. The functional diagram of a dual-slope integrating A/D converter is shown in Figure 5.5(a).

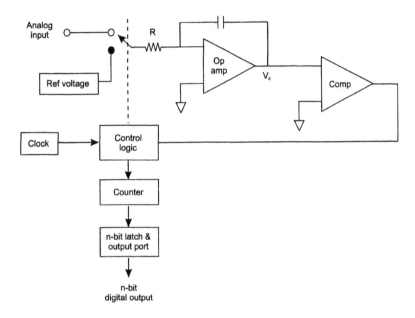

Figure 5.5(a)
Functional diagram of an n-bit dual slope integrating A/D converter

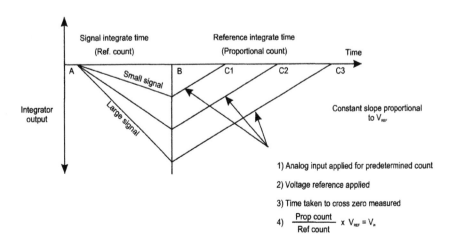

Figure 5.5(b)
Voltage appearing at V_0

The operation of a dual slope integrating A/D converter is based on the principle that the output of an integrating amplifier to a constant voltage input is a ramp whose slope is negative and proportional to the magnitude of the input voltage.

At the start of the A/D conversion, a fixed counter is cleared to zero and the unknown analog input voltage is applied to the input of the integrating amplifier. As soon as the output of the integrating amplifier reaches zero, a fixed interval count begins. After a predetermined count period, T, the count is stopped. For a positive analog input voltage, the output of the integrating amplifier has reached a negative value proportional to the magnitude of input analog signal. This is shown in Figure 5.5(b). If the analog input varies during the fixed count time interval, then the output of the integrating amplifier is proportional to the average value of the input over the fixed time interval. This is especially useful for elimination of cyclical noise and/or mains hum appearing at the input.

At this point, the count register is again cleared. A negative fixed voltage reference is now applied to the input of the integrating amplifier and the count begins. When the output of the integrating amplifier again returns to zero the count is stopped. The average value of the input analog signal is equal to the ratio of the counts multiplied by the reference voltage. This is very effective in averaging and therefore eliminating cyclical noise appearing at the analog input.

Integrating A/D converters, generally include an additional and preceding phase, during which the device carries out a self-calibrating, auto-zero operation. The stability, accuracy, and speed of the clocking mechanism, the duration of the count period, and the accuracy and stability of the voltage reference, determine the accuracy of the device.

These devices are low speed, typically a few hundred hertz maximum. However, they are capable of high accuracy and resolution at low cost. For this reason they are principally used in low frequency applications, such as temperature measurement, in digital multimeters and instrumentation.

Important A/D parameters

Analog to digital conversion is essentially a ratio operation, whereby the analog input signal is compared to a reference (full-scale voltage), converted to a fraction of this value, and then represented by a digital number. In approximating an analog value, two operations are performed. Firstly the quantization or mapping of the analog input into one of several discrete ranges, and secondly the assignment of a binary code to each discrete range. Figure 5.6(a) shows the ideal transfer function of a 3-bit A/D converter with a unipolar (0 V to FSV) input. The horizontal axis represents the analog input signal as a fraction of full-scale voltage (FSV) and the vertical axis represents the digital output. An n-bit A/D converter has 2^n distinct output codes. While not used in practical DAQ systems, a 3-bit A/D converter represents a convenient example since it divides the analog input range into $2^3 = 8$ divisions, each division representing a binary code between 000 and 111. Figure 5.6(b) shows the ideal transfer function of a 3-bit A/D converter with a bipolar (−FSV to +FSV) input. This is equivalent to the unipolar transfer function except that it is offset by −FSV.

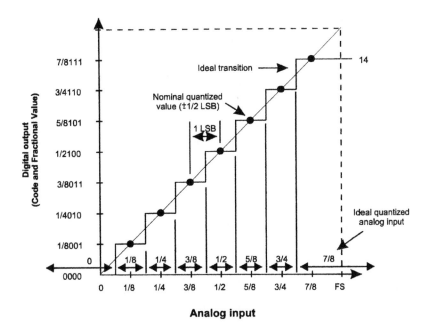

(a) Transfer function with unipolar input

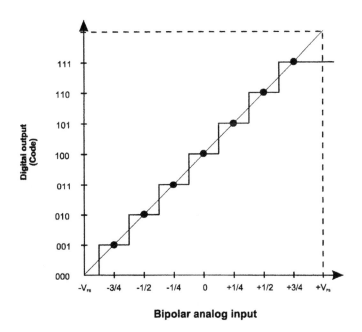

(b) Transfer function with bipolar input

Figure 5.6
Ideal transfer function of a 3-bit A/D converter

With regard to Figure 5.6, some of the important parameters of A/D converters are discussed below.

Code width

This is the fundamental quantity for A/D converter specifications, and is defined as the range of analog input values for which a single digital output code will occur. The nominal value of a code width, for all but the first and last codes in the ideal transfer characteristic, is the voltage equivalent of 1 LSB of the full-scale voltage. Therefore, for an ideal 12-bit A/D converter with full-scale voltage of 10 V the code-width is 2.44 mV. Noise and other conversion errors may cause variations in code width, however, the code width should not generally be less than 1/2 LSB or greater than 3/2 LSB for practical A/D converters.

Resolution

Resolution defines the number of discrete ranges into which the full scale voltage (FSV) input range of an A/D converter can be divided to approximate an analog input voltage. It is usually expressed by the number of bits the A/D converter uses to represent the analog input voltage (i.e. *n*-bit) or as a fraction of the maximum number of discrete levels, which can be used to represent the analog signal (i.e. $1/2^n$). The resolution only provides a guide to the smallest input change that can be reliably distinguished by an ideal A/D converter, or in effect its ideal code-width. For example, when measuring a 0–10 V input signal, the smallest voltage change an A/D converter with 12-bit resolution can reliably detect is equal to:
$1/4096 * FSV = 10/4096 = 2.44$ mV

Therefore, each 2.44 mV change at the input would change the output by ± 1 LSR or $\pm 0 \times 001$h. 0V would be represented by 0×000h, while the maximum voltage, represented by $0 \times FFF$h would be 9.9976 V. Due to the staircase nature of the ideal transfer characteristic, a much smaller change in the input voltage can still cause the A/D converter to make a transition to the next digital output level, but this will not reliably be the case. Changes smaller than 2.44 mV will not therefore be reliably detected. If the same 12-bit A/D converter is used to measure an input signal ranging from –10 V to +10 V, then the smallest detectable voltage change is increased to 4.88 mV.

Input range

Range refers to the maximum and minimum input voltages that the A/D converter can quantize to a digital code. Typical A/D converters provide convenient selection of a number of analog input ranges, including unipolar input ranges (e.g. 0 to +5 V or 0 to +10 V) and bipolar input ranges (e.g. –5 V to +5 V or –10 V to +10 V). On A/D boards, the input range is usually selectable by on-board jumpers.

Note that the transfer functions of Figure 5.6 show that the maximum input voltage is 1 LSB less than the nominal full-scale voltage (FSV). If it is essential that the A/D's input ange go from 0 to FSV, then for some A/D converters it may be possible to adjust the voltage reference to slightly above nominal FSV so that this can be achieved. This increases the real full-scale range and the LSB value by a small amount. For an input range of 0–10 V a code of 0×000h now represents 0 V while $0 \times FFF$h represents 10 V.

Data coding

While most A/D converters express unipolar ranges (i.e. 0–10 V) in straight binary, some return complementary binary, which is just the binary code with each bit inverted. Where A/D converters are used to measure voltages in bipolar ranges (i.e. –10 V to +10 V) there is an increased number of ways of representing the coded output (offset binary, sign and magnitude, one's complement and two's complement).

Most commonly, and for simplicity, A/D converters usually return offset binary values. This means that the most negative voltage in a bipolar range (–5 V for a range –5 V to +5 V) is returned as $0 \times 000h$, while the highest digitally coded value of $0 \times FFFh$ (for a 12-bit ADC), represents 4.9976 V. $0 \times 800h$ represents the mid-scale voltage of 0 V.

Conversion time

The conversion time of an A/D converter is defined as the time taken from the initiation of the conversion process to valid digital data appearing at the output. For most A/D converters, conversion time is identical to the conversion rate. Therefore, an A/D converter with a conversion time of 25 μs is able to continuously convert analog input signals at a rate of 40,000/sec. For some high-speed A/D converters, pipelining allows new conversions to be initiated before the results of prior conversions have been determined. An example of this would be an A/D converter that could perform conversions at a rate of 5 MHz (200 ns conversion time), but actually took 675 ns (1.48 MHz conversion rate) to perform each individual conversion.

Errors in A/D converters

Errors that may occur in A/D converters are defined and measured in terms of the location of the actual transition points in relation to their locations in the ideal transfer characteristic. These are discussed below.

Quantization uncertainty

Unlike a D/A converter, where there exists a unique analog value for each digital code, each digital output code is valid over a range of analog input values. Analog inputs within a given discrete range are represented by the same digital output code, usually assigned to the nominal mid-range analog value. There is, therefore, an inherent quantization uncertainty of ± 1/2 LSB (least-significant bit), in addition to any other actual conversion errors. This is shown in Figure 5 .6(b).

Unipolar offset

Note that in the ideal transfer function, the first transition should ideally occur 1/2 LSB above analog common. The unipolar offset is the deviation of the actual transition point from the ideal first transition point. This is shown in Figure 5.7(a).

Bipolar offset

As seen in Figure 5.7(b) the transfer function for an ideal bipolar ADC resembles the unipolar transfer function, except that it is offset by the negative full-scale voltage (–FSV). Offset adjustment of a bipolar A/D converter is set so that the first transition occurs at 1/2 LSB above –FSV, while the last transition occurs at –3/2 LSB below +FSV. Because of non-linearity, a device with perfectly calibrated end points may have an offset error at analog common. This is known as the bipolar offset error and is shown in Figure 5.7(b).

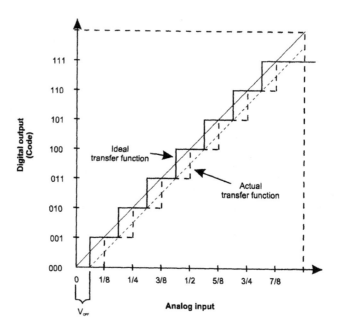

(a) Unipolar offset error

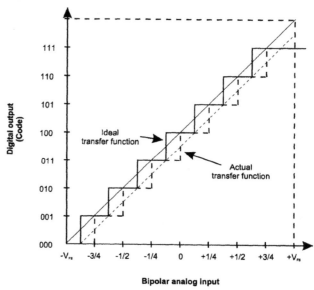

(b) Bipolar offset error

Figure 5.7
3-bit A/D converter transfer functions with offset errors

Unipolar and bipolar gain errors

The gain, or scale, factor is the number which establishes the basic conversion relationship between the analog input values and the digital output codes, e.g. 10 V full-scale. It represents the straight-line slope of the ideal transfer characteristic. The gain error is defined as the difference in full-scale values between the ideal and the actual transfer function when any offset errors are adjusted to zero. It is expressed as a percentage of the nominal full-scale value or in LSBs. Gain error affects each code in an equal ratio. Unipolar and bipolar gain errors are shown in Figure 5.8.

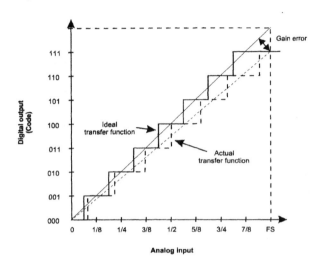

(a) Unipolar gain error

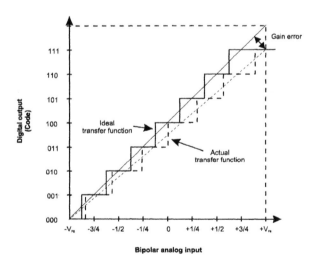

(b) Bipolar gain error

Figure 5.8
3-bit A/D converter transfer function with gain errors

Offset and gain drift

Offset and gain errors are usually adjustable to zero with calibration, however this calibration is only valid at the temperature at which it was made.

Changes in temperature result in a non-zero offset and gain error, known as offset drift and gain drift. These values, specified in ppm/deg C, represent the ADC's sensitivity to temperature changes.

Linearity errors

With most ADCs the gain and offset specifications are not the most critical parameters that determine an A/D converter's usefulness for a particular application, since in most cases they can be calibrated out in software and/or hardware. The most important error specifications are those that are inherent in the device and cannot be eliminated. Ideally, as the analog input voltage of an A/D converter is increased, the digital codes at the output should also increase linearly. The ideal transfer function of the analog input voltage verses the digital output code would show a straight line. Deviations from the straight line are specified as non-linearities. The most important of these, (because they are errors which cannot be removed), are integral non-linearity and differential non-linearity errors. The transfer characteristics of a 4-bit A/D converter showing differential and integral linearity errors are shown in Figure 5.9(a) and Figure 5.9(b).

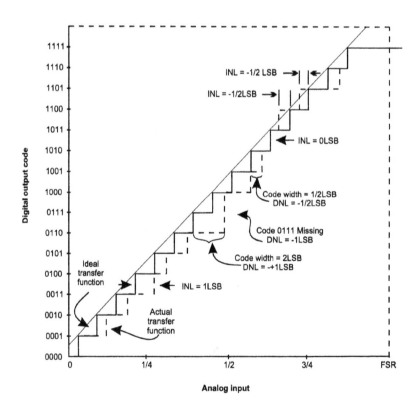

(a) Integral non-linearity errors specified as low-side transition

Figure 5.9(a)
Transfer function of a 4-bit A/D converter with integral non-linearity and differential non-linearity errors

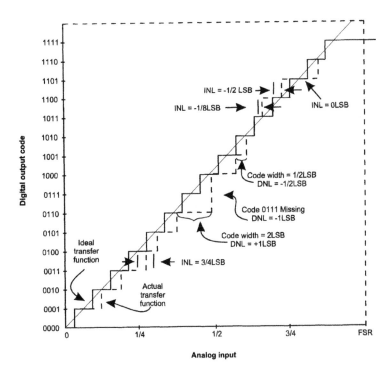

(b) Integral non-linearity errors specified as center-of-code transition

Figure 5.9(b)
Transfer function of a 4-bit A/D converter with integral non-linearity and differential non-linearity errors

Integral non-linearity (INL)

This is the deviation of the actual transfer function from the ideal straight line. This ideal line may be drawn through the points where the codes begin to change (low-side transition or LST), as shown in Figure 5.9(a), or through the center of the ideal code widths (center-of-code or CC), as shown in Figure 5.9(b). Most A/D converters are specified by low-side-transition INL. Thus, the line is drawn from the point 1/2 LSB on the vertical axis at zero input to the point 3/2 LSB beyond the last transition at full-scale input. The deviation of any transition from its corresponding point on that straight line is the INL of the transition. In Figure 5.9(a), the transition to code 0100 is shifted to the right by 1 LSB, meaning that the LST of code 0100 has an INL of +1 LSB. In the same figure the transition to code 1101 is shifted left by 1/2 LSB, meaning that the LST of code 1101 has an INL of –1/2 LSB.

When the ideal transfer function is drawn for center-of-code (CC) integral non-linearity specification, as shown in Figure 5.9(b), the INL of each transition may be different. Where the digital code 1101 previously had –1/2 LSB of LST INL, it now has 0 LSB of CC INL. Similarly, the code 1011 has –1/8 LSB of CC INL, where it previously had 0 LSB of LST INL.

The INL is an important figure because the accurate translation from the binary code to its equivalent voltage is then only a matter of scaling.

Differential non-linearity (DNL)

In an ideal A/D converter, the midpoints between code transitions should be 1 LSB apart. Differential non-linearity is defined as the deviation in code width from the ideal value of

1 LSB. Therefore, an ideal A/D converter has a DNL of 0 LSB, while practically this would be ± 1/2 LSB. If DNL errors are large, the output code widths may represent excessively large or small ranges of input voltages. Since codes do not have a code width less than 0 LSB, the DNL can never be less than –1 LSB. In the worst case, where the code width is equal to or very near zero, then a missing code may result. This means that there is no voltage in the entire full-scale voltage range that can cause the code to appear. In Figures 5.9(a) and 5.9(b), the code-width of code 0110 is 2 LSBs, resulting in a differential non-linearity of +1 LSB. As the code-width of the code 1001 is 1/2 LSB, this code has a DNL of –1/2 LSB. In addition, the code 0111 does not exist for any input voltage. This means that code 0111 has –1 DNL and the A/D converter has at least one missing code.

Often, instead of a maximum DNL specification, there will be a simple specification of monotonicity or no missing codes. For a device to be monodic, the output must either increase or remain constant as the analog input increases. Monodic behavior requires that the differential non-linearity be more positive than –1 LSB. However, the differential non-linearity error may still be more positive than +1 LSB. Where this is the case, the resolution for that particular code is reduced.

5.2.6 Memory (FIFO) buffer

A characteristic of high-speed A/D boards is the inclusion of on-board memory or I/O in the form of a FIFO (first in first out) buffer or a pair of buffers. These range in size from 16 bytes to 64 Kbytes.

The FIFO buffer(s) form a fast temporary memory area. For small FIFOs, the buffer is addressed as I/O. Larger FIFOs are actually mapped into the memory address space of the host PC. Samples can therefore be collected up to the maximum size of the buffer without actually having to perform any data transfers. Where more samples are required, existing data in the buffer must be transferred to other parts of the main memory, or written to hard disk, before it is overwritten.

FIFO buffering is particularly useful in situations where the host computer is using polled I/O or interrupts to transfer data, and might not be able to respond quickly enough to transfer the current sample, before it is overwritten by a subsequent sample. This would typically occur when the host computer is running a multitasking operating system such as Windows or OS/2, where there are inherent interrupt latencies or a large number of tasks being performed. The FIFO buffer also has the effect of evening-out variations in DMA response times, helping to guarantee full-speed operations even with substandard PC/AT clones. On-board FIFOs, when used in conjunction with specific data techniques such as polled I/O, interrupts, DMA or repeat string instructions, can greatly improve the throughput of A/D boards.

5.2.7 Timing circuitry

To perform multiple analog-to-digital conversions automatically, at precisely defined time intervals, A/D boards are equipped with timing circuitry whose principal responsibility is to generate the strobe signals that allow the components of the analog input circuitry to perform their respective functions efficiently and correctly.

Clocking circuits are made up of a frequency source, which is either an on-board oscillator between 400 kHz and 10 MHz or an external user-supplied signal, and a prescaler/divider network, typically a counter/timer chip that slows the clock signal down to more usable values. The clock frequency can be as low as 1 Hz or up to the maximum throughput of the board.

A/D conversions are started by triggers; either by a software trigger (writing to an on-board register), or an external hardware trigger. Data conversions can be synchronized with external events by using external clock frequency sources and external triggers. The external trigger event is usually in the form of a digital or analog signal, and will begin the acquisition depending on the active edge if the trigger is a digital signal, or the level and slope, if the trigger is an analog signal.

In performing an analog-to-digital conversion cycle on a single input channel, the timing circuitry must ensure that the following steps are performed:

- Once the channel/gain array has been initialized, the timing circuitry increments to the next channel/gain pair. The next channel to be sampled is output to the address lines of the input multiplexer and the required gain setting is output to the programmable gain amplifier (PGA). The sample-and-hold (S/H) is put into sample mode.
- The timing circuitry must wait for the input multiplexer to settle, then for the PGA output delay time and lastly for any S/IA delay.
- The S/H is put into hold mode. The timing circuitry must wait for the duration of the aperture time of the S/H for the signal to become stable at the output of the S/H.
- A start conversion trigger is issued to the A/D converter.
- The timing circuitry waits for the end of conversion signal from the A/D converter to become active.
- The available data is then strobed from the A/D converter into a data buffer or a FIFO, from where it is usually accessible by the host computer.
- If simultaneous sampling is available on the A/D board, the timing circuitry generates the necessary sequence of strobes to the input S/H devices, so that all channels are sampled at the beginning of the sampling cycle, before the data is passed to the rest of the analog input circuitry.

Total throughput, for multiple conversions on different channels, is often increased by overlapping parts of this cycle. For example, while the A/D converter is busy converting the S/H output, the next channel/gain pair can be output to the multiplexer and PGA, so that their settling and delay times are overlapped with the A/D conversion time.

The timing circuitry may also include a block-sampling mode, which allows blocks of samples to be collected at regular intervals at the A/D board's maximum sampling rate. This is discussed in the section on *Sampling techniques*, p. 151.

5.2.8 Expansion bus interface

The bus interface provides the control circuitry and signals used to transfer data from the board to the PC's memory or for sending configuration information (e.g. channel/gain pairs) or other commands (e.g. software triggers) to the board.

It includes:

- The plug-in connector, which provides the hardware interface for connecting all control and data signals to the expansion bus, (e.g. ISA, EISA etc), of the host computer.
- The circuitry, which determines the base address of the board. This is usually a selectable DIP switch and defines the addresses of each memory and I/O location on the A/D board.

- The source and level of interrupt signals generated. Interrupt signals can be programmed to occur at the end of a single conversion or a DMA block. The configuration of the interrupt levels used is commonly selected by on-board links.
- DMA control signals and the configuration of the DMA level(s) used. The configuration of the DMA levels used is typically selected by on-board links.
- Normal I/O to and from I/O address-locations on the board.
- Wait state configuration for use in machines with high bus speeds or with non-standard timing. The number of wait states is usually configurable by on-board links.

5.3   Single ended vs differential signals

As previously demonstrated, great care must be taken in the connection, earthing and shielding of signals, received from external transducers (or similar signal sources), to signal conditioning equipment. This is especially true where the signal levels are very small and/or the signal sources are a long way from the measuring equipment. In these cases, the effects of earth loops, induced noise, and common mode voltages can introduce errors that lead to large inaccuracies in the signal measurement.

The three basic configurations for connecting input signals to signal conditioning equipment are available on plug-in A/D boards:

- Single ended
- Pseudo differential
- Differential

5.3.1   Single ended inputs

Single ended inputs are those where the signal is transmitted over a single conductor and referenced to analog ground AGND. The single conductor is connected to the HI terminal of the amplifier while the LO terminal of the amplifier is connected to AGND. This is shown in Figure 5.10.

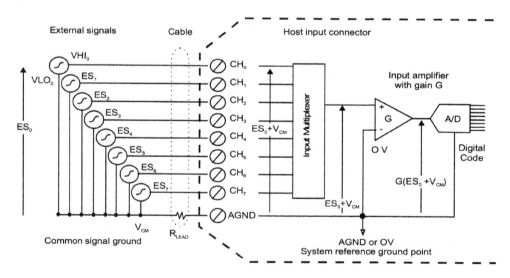

Figure 5.10
Single ended inputs

Single ended inputs usually carry high-level signals (in the order of volts), which do not require high gains ($> \times 5$), transmitted over short distances (0.5 m). Where they are required to be transmitted over longer distances, they should be shielded, and the shield connected to AGND at the instrument end only.

While this configuration allows more inputs to be multiplexed to a single A/D converter, it should only be used where there is no practical way of bringing a remote ground or an analog ground back to the measurement point.

Because the amplifier LO terminal is connected to AGND, what is amplified is the difference between $E_{sn} + V_{cm}$ and AGND. This introduces the common mode offset voltage as an error.

Plug-in boards that do not have an amplifier (i.e. where the multiplexed input is fed straight to the A/D converter) must use the single ended input configuration.

5.3.2 Pseudo-differential configuration

The pseudo-differential input configuration is a variation of the single ended input configuration, providing some degree of common mode rejection while still allowing the maximum number of multiplexed input channels.

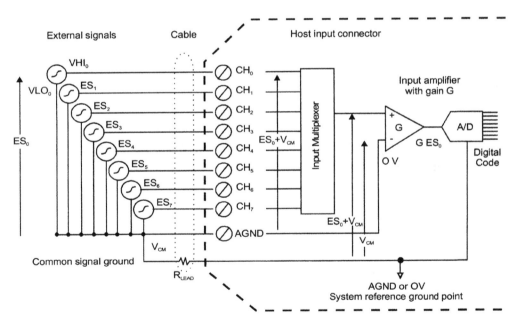

Figure 5.11
Pseudo differential inputs

In the configuration shown in Figure 5.11 the signal source LO outputs are all connected to the LO input terminal of the amplifier, while the signal source HI outputs are connected to the respective HI input for each of the channels. The LO input to the amplifier is then referenced to analog ground AGND at the signal end of the cable. This method is only possible if the LO terminal of the amplifier is brought out to the connector and the signal sources can be grounded at their signal ends. In using this configuration only the difference between the channel input $E_{sn} + V_{cm}$ and the signal ground, which has the common mode voltage on it, is amplified.

5.3.3 Differential inputs

True differential inputs, where the HI and LO outputs of the signal source are connected directly to the HI and LO terminals of the amplifier, as shown in Figure 5.12, offer the greatest noise immunity and common mode rejection.

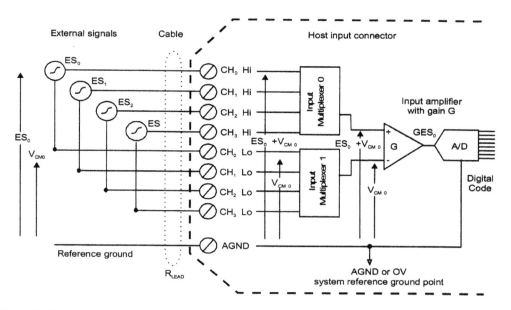

Figure 5.12
Differential input configuration

In this configuration, only the difference in the signal inputs is measured. Noise induced equally in each signal line will cancel out at the inputs of a true balanced differential amplifier, while common mode voltages (appearing at both inputs) will be rejected if the amplifier has a large CMRR. This should not preclude the added precautions of twisting differential pairs and providing earthed shields to reduce noise induced in long cables.

Differential inputs should be used:

- When measuring signals with large common mode voltages (e.g. strain gauges).
- Where several transducers with different ground points (and possibly different ground potentials) are to be measured. Connecting the Lo channel of each sensor together at a common point, as in the pseudo differential connection, can create unwanted ground currents that induce offset and noise errors at the amplifier inputs.
- When measuring signal voltages that are very small, and the signal/noise ratio is low.
- When the input transducer is physically located a large distance from the measuring device and may be susceptible to the effects of noise.

Note that for differential input configuration, two input multiplexers are needed, and for the same number of input terminals as single ended and pseudo differential inputs, half the number of input channels is available in differential mode.

Where high impedance sources are used, bias resistors may be required to return bias currents to the operational amplifier, thus preventing the floating of inputs beyond the limits of the amplifier inputs. Such bias resistors normally consist of high impedance resistors, typically between 100 KΩ and 1 MΩ, connected between the HI and LO signal lines and AGND.

5.4 Resolutions, dynamic range, and accuracy of A/D boards

5.4.1 Dynamic range

One of several considerations in determining the analog input requirements of an A/D board is the range of voltages, which each channel is required to measure. The physical parameters to be measured, the type of sensor(s) used and how they are connected, determine the input voltage ranges required.

The input range specifications quoted by board manufacturers of A/D boards refer to the minimum and maximum voltage levels that the A/D converter on the board can quantize. Typically, a selection of input ranges is provided, either unipolar (e.g. 0 to 10 V), for measuring positive voltages only, or bipolar (e.g. −10 V to +10 V), for measuring both positive and negative voltages. This allows the user to match the input signal range to that of the A/D converter, taking into account the resolution of the A/D converter and the gain required of the input amplifier.

When considering the input range, it is only the dynamic range of the input signal that needs to be taken into account. For example, consider a strain gauge setup in a Wheatstone bridge configuration. The input voltage to be read has a common mode component due to the excitation of the bridge, while the small differential voltage changes, (of interest) are due to the change in strain gauge resistance. The common mode voltages do not provide any useful information and are greatly attenuated, (almost eliminated), by using differential inputs and instrument amplifiers with high CMRR. Only the small differential voltage changes are amplified and converted by the A/D converter. The amplifier gain should therefore be selected so that the maximum differential voltage change expected at the input will be amplified to cover as much of the input range of the A/D converter as possible.

As only one of the allowable range settings can be selected at any time, typically by jumpers on the board itself, care should be taken in matching the input signal requirements where more than one channel is sampled. The A/D converter input range selected must accurately measure the signal inputs from a number of channels, possibly different sensors, and therefore potentially different input voltage levels and signal ranges. The input range should therefore cover each channel's input range with as little overlap as possible, thus giving the greatest number of data points and therefore the highest resolution and accuracy.

It should be noted that the input ranges specified do not necessarily refer to the maximum or minimum voltage levels that can be applied at any single input, or to the maximum allowable common mode voltage, which can be applied, to a differential input. These are specifications, more related to the input amplifier. If there are any doubts with regard to this, users should consult the board manufacturer.

5.4.2 Resolution

The resolution specification quoted by manufacturers of A/D boards refers to the resolution of the A/D converter used on the board. It is usually expressed by the number of bits the A/D converter uses to represent the analog input voltage (i.e. n-bit) or as a fraction of the maximum number of discrete levels, which can be used to represent the analog signal (i.e. $1/2^n$). The resolution implicitly defines the number of discrete ranges into which the full-scale voltage (FSV) input range can be divided to approximate an analog input voltage. A 12-bit

A/D board can divide the input range into (2^{12} = 4096) discrete levels, each 1/4096 the size of the input voltage range.

Together, the resolution, input range, and input amplifier gain available on the A/D board, determine the smallest detectable voltage change in the signal input. For an ideal A/D board with a resolution of *n*-bits, this is calculated using the formula:

$$\text{smallest detectable change} = \frac{\text{input range}}{\text{amplifier gain} \times 2^n}$$

For example, on a 12-bit A/D board, with a 0 V to +10 V input range, and the amplifier gain set to 1, the smallest detectable voltage change would be 10/(1 * 4096) = 2.44 mV.

Therefore, each 2.44 mV change at the input would change the output of the A/D converter by ± 1 LSB or ± 0 × 001h. 0 V would be represented by 0 × 000h, while the maximum voltage, represented by 0 × FFFh would be 9.9976 V. Due to the staircase nature of the ideal transfer characteristic of an A/D converter, a much smaller change in the input voltage can still result in a transition to the next digital output level, but this will not reliably be the case. Changes smaller than 2.44 mV will not therefore be reliably detected. If the same 12-bit A/D converter is used to measure an input signal ranging from –10 V to +10 V, then the smallest detectable voltage change is increased to 4.88 mV. This value represents the voltage equivalent of 1 LSB, of the full-scale voltage, and for A/D boards, is termed code width.

The resolution figure quoted only provides a guide to the smallest detectable change that can be reliably distinguished by the A/D board, since the value calculated is based on the ideal performance of all components of the analog input circuitry. The effects of noise, non-linearities in the A/D converter, and errors in the other components of the analog input circuitry, can mean that the true resolution of an A/D board can be as much as 2 bits lower than the manufacturer's specification. This means that a 16-bit A/D board may be accurate to only 14-bits.

5.4.3 System accuracy

The system accuracy, or how closely the equivalent digital outputs match the incoming analog signal(s), is another very important criteria, especially where the analog signal contains a lot of information, or where a small part of the signal range is to be examined in detail. As has been demonstrated, the functional components of the analog input circuitry (i.e. multiplexer, amplifier, sample-and-hold and A/D converter) of A/D boards are not ideal. The practical performance limits and errors in each of these components influence the overall performance and accuracy of the system as a whole.

The specification known as system accuracy usually refers to the relative accuracy of the A/D board and indicates the worst-case deviation from the ideal straight-line transfer function. Relative accuracy is determined on an A/D board by applying to the input a voltage at minus full-scale, converting this analog voltage to a digital code, increasing the voltage, and repeating the steps until the full input range of the board has been covered. By subtracting the theoretical analog voltage, which should cause each code transition from the analog input voltage that actually resulted in the code transition. The maximum deviation from zero is the relative accuracy of the A/D board. Board manufacturers usually quote the system accuracy in terms of LSB, since an absolute voltage value would only have meaning relative to the selected input voltage range. For example, where '*n*' = 2, the system accuracy of a 12-bit A/D board is 2/4096 (±0.048%), while for a 16-bit A/D board the accuracy is 2/65536 (±0.003%).

The tendency of analog circuits to change characteristics or *drift*, with time and temperature, requires that A/D boards be periodically calibrated to maintain accuracy within the specified range. Manufacturers specify the offset voltage and gain accuracy to be adjustable in a range ± *n* LSB. This means that where the input range to a 12-bit A/D converter is 0–10 V, and the input is set to +1/2 LSB (i.e. +1.22 mV), the digital output would read no greater than 0 × 005h. This would represent a maximum offset voltage adjustment of 5/4096 × 10 V = 12.2 mV. Where the input range to a 12-bit A/D converter is 0–10 V, and the input is set to –3/2 LSB (i.e. +9.996 V), the digital output would read no less than 0 × 99 Ah. For gain accuracy, this figure represents the maximum gain error.

Autocalibration, where the entire analog section of the board (multiplexer, amplifier, sample and hold), as well as the A/D converter, is automatically calibrated without user intervention, is provided on some A/D boards.

Several auto-calibration methods are used:

- Calibration is carried out automatically when a voltage reference is connected to the board.
- Calibration takes place as part of the conversion process.
- The accuracy on each input channel is checked for all available gain settings. A correction code for each channel/gain combination is stored, then recalled to dynamically compensate for drift in hardware.

5.5 Sampling rate and the Nyquist theorem

One of the most critical factors confronting users of data acquisition systems and A/D boards is the question of how frequently should an analog signal be sampled to be able to represent and reconstruct the input signal accurately. How fast should the A/D board be able to sample the data?

5.5.1 Nyquist's theorem

Nyquist's sampling theorem states that:

An analog band-limited signal that has no spectral components at or above a frequency of F Hz can be uniquely represented by samples of its values spaced at uniform intervals that are no more than 1/2 F seconds apart or sampled at a frequency of no less than 2 F Hz.

The maximum sampling period, $T=1/2$ F, is known as the Nyquist interval, while the minimum sampling frequency, corresponding to this period, 2 F, is known as the *Nyquist sampling frequency,* or *rate.*

Sampling at a rate higher than the Nyquist rate is called *oversampling.* This is routinely performed where it is essential to recover a true replica of the signal being sampled. When a signal is sampled at less than the Nyquist rate, this is known as *undersampling* and can lead to erroneous results.

5.5.2 Aliasing

To intuitively understand what happens when a signal is oversampled compared to a signal that is sampled less than the Nyquist sampling rate, consider Figure 5.13.

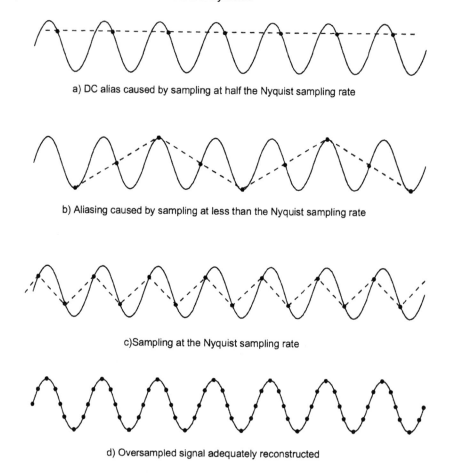

a) DC alias caused by sampling at half the Nyquist sampling rate

b) Aliasing caused by sampling at less than the Nyquist sampling rate

c)Sampling at the Nyquist sampling rate

d) Oversampled signal adequately reconstructed

Figure 5.13
Effect of sampling rate on the reconstructed input signal

Figure 5.13 (d) shows a signal that is sampled at a frequency well above the Nyquist sampling rate. In this case, the information contained in the signal, including its shape and frequency, can be correctly reproduced. If the sampling rate is reduced to below the Nyquist sampling rate, that is, the sample points are too far apart, then the input signal is misrepresented by what appears to be a much lower frequency signal. This phenomenon is known as aliasing and is demonstrated in Figure 5.13 (b).

In Figure 5.13(a), the input signal is sampled at half the Nyquist sampling rate, which is the same frequency as the frequency of the signal itself. The reconstructed waveform appears as a DC signal. When the input signal is sampled at the Nyquist sampling rate, as shown in Figure 5.13(c), the reconstructed signal has the correct frequency but incorrectly appears as a triangular waveform. Where undersampling occurs, the frequency of the reconstructed signal appears to be much lower, lying between DC and the Nyquist frequency.

Theoretically, the effects of aliasing are more easily understood by looking at the frequency spectrum of an analog signal. Without detailing the complex mathematical descriptions and frequency analysis required, it can be shown that a time varying band-limited signal can be equally represented by its spectrum in the frequency domain. Figure 5.14(b) shows the frequency spectrum of the band-limited signal shown in Figure 5.14(a). If the time varying signal is sampled using a very narrow series of square wave pulses, as shown in Figure 5.14(c), then the frequency spectrum of the sampled waveform is the original signal with exact replicas of itself spaced about multiples of the sampling frequency. Figure 5.14(d) illustrates the frequency spectrum of a signal that is sampled at exactly twice the maximum frequency of the original signal, showing that the replicas of the original signal just touch. Oversampling the original signal, as shown in Figure 5.14(e) separates the input signal bands by a wider frequency. This is shown in Figure 5.14(f). Undersampling narrows the separation between the bands so that they fold over each other and result in aliasing, as demonstrated in Figure 5.14(g) and Figure 5.14(h). Where this occurs, the resultant signal appears as an aliased signal between DC and the Nyquist frequency, and cannot be distinguished from valid data.

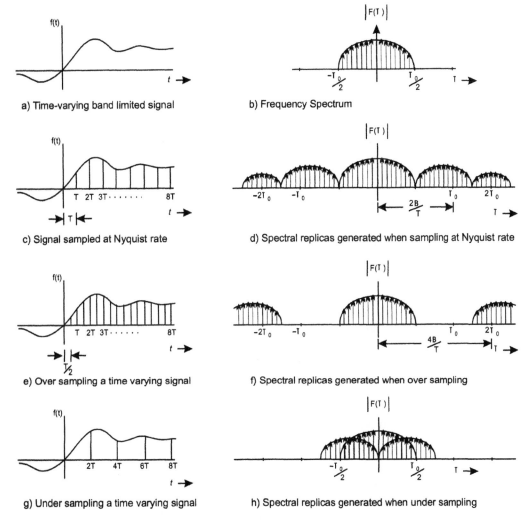

a) Time-varying band limited signal

b) Frequency Spectrum

c) Signal sampled at Nyquist rate

d) Spectral replicas generated when sampling at Nyquist rate

e) Over sampling a time varying signal

f) Spectral replicas generated when over sampling

g) Under sampling a time varying signal

h) Spectral replicas generated when under sampling

Figure 5.14
Demonstrating the effect of aliasing in the frequency domain

Consider a band-limited signal, which contains three sinusoidal waveforms, a 25 Hz waveform representing the wanted signal, a 50 Hz signal, which is unwanted mains hum, and an unwanted high frequency noise signal at 260 Hz. Figure 5.15(a) shows the frequency spectrum of this band-limited signal.

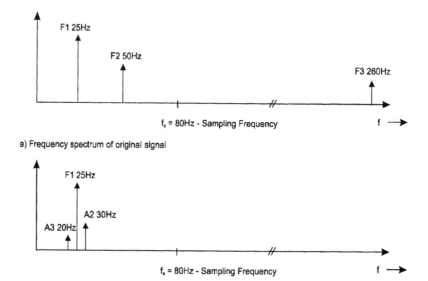

a) Frequency spectrum of original signal

b) Frequency spectrum of original and sampled signals

Figure 5.15
Frequency spectrum of original and sampled signals

The frequency spectrum of the reconstructed signal, sampled by an A/D board at $f_s = 80$ Hz is shown in Figure 5.15(b). Frequencies below the Nyquist frequency, $f_s/2 = 40$ Hz, in the original signal spectrum, appear correctly. However, replicas of the signal frequencies above the Nyquist frequency are reproduced about multiples of the sampling frequency and therefore appear as aliases. A2 and A3 are aliases of the original signals F2 and F3 respectively. The alias frequency of any signal frequency can be simply calculated by the formula:

Alias Freq = ABS (closest integer multiple of sampling frequency – signal frequency)
Alias A2 = [80 – 50] = 30 Hz
Alias A3 = [(3)80 – 260] = 20 Hz

In this example, the resulting aliases are very close to the frequency of the signal of interest and would be very difficult to remove. Once an aliased signal has been introduced, it is almost impossible to remove it by digital filtering methods.

5.5.3 Preventing aliasing

One method of preventing aliasing is by filtering the input signal with a low pass filter with a cutoff point set to the Nyquist frequency or half the sampling rate. This type of filter is known as an *antialiasing filter*. A perfect antialiasing filter would simulate the *brick-wall* response of an ideal low pass filter, as shown in Figure 5.16, rejecting all unwanted frequency components above the Nyquist frequency. Thus, by using this filter the input signal could be sampled at twice the Nyquist rate without aliasing.

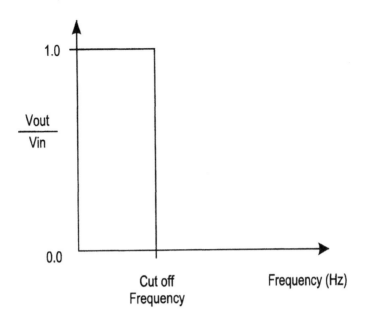

Figure 5.16
Ideal low pass filter response

Unfortunately, real filters do not simulate ideal filters, and in fact exhibit some attenuation (dB/octave) near the cutoff frequency. As shown in Figure 5.17, this roll-off may not be steep enough to totally eliminate all the higher frequency components. Although attenuated, these higher frequency components can, and will, fold down to the signal band of interest.

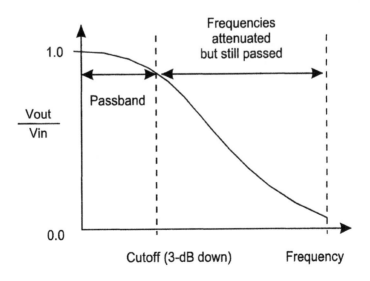

Figure 5.17
Practical low pass filter response

Therefore, to accommodate the filter cutoff frequency and roll-off, the sampling rate should be increased. Using simple passive antialiasing filters, it is recommended that the sampling rate be a minimum of about five times the cutoff frequency. Non-periodic wave-forms can be oversampled by about ten times.

High performance antialiasing filters with very steep roll off near the cutoff frequency, as shown in Figure 5.18, allow the signal to be sampled at two to three times the filter cutoff frequency.

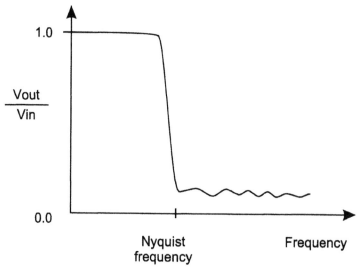

Figure 5.18
Steep roll-off antialiasing filter

5.5.4 Practical examples

A common data acquisition application is machine vibration analysis. All machines resonate at certain frequencies, both under normal operation and when driven by an external source. In this example, strain gauges were placed on the machine and the output signal sampled, digitized (yielding a time domain plot, see Figure 5.19(a) and converted into the frequency domain (for example, using FFT).

The spectrum resulting from sampling at 50 kHz is shown in Figure 5.19(b). It has two resonant frequency peaks, one around 4 kHz, and another slightly above 5 kHz. The machine vibration analyst knows that the 4 kHz component corresponds to the machine's rotational speed, but the 5 kHz component is a mystery. Passing the input signal through a 10 kHz cutoff antialiasing filter with subsequent resampling, yields the spectrum in Figure 5.19(c), clearly revealing the 5 kHz component to be an alias. Indeed, sampling the original signal (without the antialiasing filter) at 100 kHz yields the spectrum in Figure 5.19(d) and shows that an actual frequency component, present in the vibration signal, of 45 kHz has been aliased down to 5 kHz when sampled at 50 kHz.

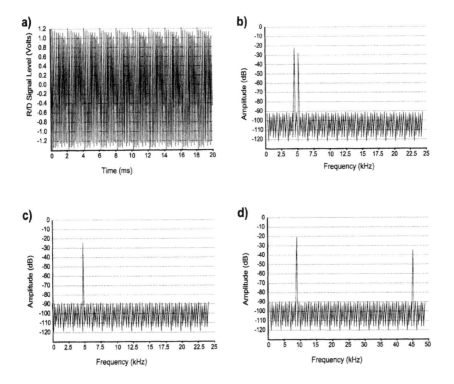

Figure 5.19
Aliasing due to undersampling

In the example of machine vibration analysis, the frequency components were clearly visible and constant. However, in the case of speech digitization or speech analysis, the desired signal consists of many frequency components that vary quickly and unpredictably. An application may require spoken messages to be digitized and stored for later playback.

As most speech is composed of frequency components below 5 kHz, digitizing the incoming signal at 10 kHz appears to be adequate and places only low demand on memory usage. Unfortunately, an attempt to digitize a message signal from a microphone in this way resulted in the message so buried in extraneous hums, pops, and whines that it could hardly be used. The frequency spectrum of the sampled signal is shown in Figure 5.20(a).

In the assumption that high frequencies present on the input were aliasing down, a 5 kHz, antialiasing filter was put in place, leading to the spectrum in Figure 5.20(b). The spectrum shows little difference from the unfiltered signal's spectrum. Increasing the sample rate (to 100 kHz, Figure 5.20(c)) shows why: although attenuated, components above the filter's cutoff point are still present and do alias down. The filter had a roll-off of 24 dB/octave and the real-world properties of the filter allowed the attenuated high-frequency components to fold down into the band of interest. Practically, solutions are using filters with greater roll-off (reducing the magnitude of high-frequency components that might alias) or sampling at a higher rate (frequencies in the new sampling band do not fold down).

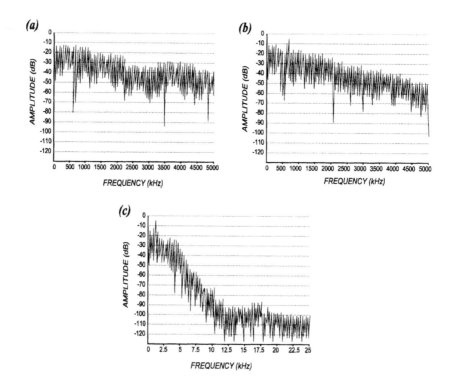

Figure 5.20
Many aliases combined with a speech signal

It could be assumed that aliasing is a phenomenon associated with high frequencies, and that low frequencies (such as thermocouple temperature signals) are immune to this effect. Temperature changes so slowly that the input signal is almost DC; it seems therefore reasonable to sample it extremely slowly and not be concerned with frequency analyses.

However, if the input signal contains a noise spike as shown in Figure 5.21(a), the resulting spectrum results in a noise floor around –60 dB, shown in Figure 5.21(b). This is because an impulse spike in the time domain spreads itself out evenly in the frequency domain. Thus, when sampled at a low frequency, the high-frequency components of the noise spike alias down and add to the low frequency components. The extra energy of these added frequencies causes the temperature application to oscillate. If a low pass filter is used, the spike – with its equivalent high-frequency components – is removed (as shown in the time domain in Figure 5.21(c). The spectrum corresponding to this (Figure 5.21(d)) now has a noise floor of –80 to –90 dB, which does not affect the readings obtained by the A/D board.

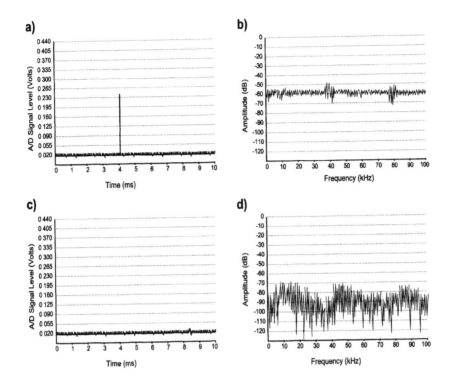

Figure 5.21
A spike causes wideband aliasing

5.6 Sampling techniques

These techniques are discussed in the following sections:

- Continuous channel scanning
- Simultaneous sampling
- Block mode operations

5.6.1 Continuous channel scanning

The method of sampling that facilitates the connecting of the required input channel to the A/D converter at a constant rate is known as continuous channel scanning. Continuous channel scanning allows channels to be sampled in a pre-determined and arbitrary order (e.g. channel 5, channel 1, channel 11), as well as at different sampling rates. An example of this would be the sampling of three channels in the following order (channel 5, channel 1, channel 11, channel 1). Channel 1 is being sampled at twice the rate as channels 5 & 11, which for an A/D board with throughput of 100 kHz represents a sampling rate of 50 kHz. Channels 5 & 11 are sampled at 25 kHz. There are two methods of continuous channel scanning, either under software control or by on-board hardware control.

Software channel scanning

Where continuous channel scanning is performed by software the address of the channel to be sampled is written to the multiplexer and the gain setting sent to the programmable gain amplifier (PGA), where one is fitted. Once the signal is settled, an A/D conversion is initiated. The data is subsequently read and transferred to the PC's memory. This incurs a large software overhead. Background operation using interrupts is difficult and slower than polled I/O and accurately timed samples and higher speed data transfer methods such as DMA and repeat instructions are impossible in either case.

Hardware channel scanning

Continuous channel scanning is implemented in hardware using channel-gain arrays (CGA). These programmable memory buffers contain a list of the channels and the gain settings required for each input channel to be sampled. When the A/D board begins sampling, input channels are sampled in the sequence loaded into the channel-gain array.

The use of on-board channel-gain arrays (CGA) overcomes many of the limitations associated with channel scanning using software and has the following advantages:

- The channel sequence information may be setup once and then sampling initiated (and repeated) with a single command. Once initiated, the sampling process is controlled by the A/D board's hardware.
- Arbitrary sample sequences may be defined.
- Within the limitations on the size of the CGA, different sampling frequencies may be specified for different channels.
- The speed of software-transfer methods such as interrupt and polled I/O is greatly increased, in many cases doubled. This is due to the fact that delays caused by the host computer transferring channel and gain information before each sample is taken, are avoided.
- Very accurate timing is achievable since the board hardware is optimized to control the individual sub-systems on the board.
- Advanced transfer methods such as DMA and repeat instructions are possible. DMA transfer is controlled directly by the hardware on the A/D board and the host computer. This is not a very flexible arrangement, since it does not allow intervention by software to change the channels being scanned once a DMA transfer has been initiated. A/D boards, which are capable of DMA but do not have channel-gain arrays may only perform DMA transfers from a single input channel, whose address and required gain is setup by software before the DMA transfer is initiated. Where channel-gain arrays are implemented, the on-board hardware will automatically change the address and gain settings during the DMA transfer. Where repeat instructions are used to transfer information, usually from an on-board FIFO, the sampling of multiple channels must continue to be performed in the background. On A/D boards, which do not have channel-gain arrays, repeat instruction transfers may only be performed on a single input channel whose address and required gain is setup by software before the repeat instruction transfer is initiated.

Practical applications

Some of the practical applications, which utilize the flexibility in the selection and throughput of individual channels, using hardware channel scanning, are detailed below.

Sampling different channels at different frequencies

When signals with different frequencies are sampled (for example, a heart rate electrocardiogram (ECG) with 300 beats/min and an electroencephalogram (EEG) with a frequency of 5 kHz), it is much more memory-efficient to sample each channel at around its

Nyquist rate instead of scanning all channels at the rate of the highest frequency channel. In the example above, if the ECG rate is 2 kHz and the EEG rate is 20 kHz, then the ratio of the two channels is 1:10. The CGA could therefore be setup with ten EEG readings followed by one ECG reading with a scan clock of, say, 22 kHz. (Note that this introduces a phase shift in the EEG readings since every eleventh reading is an ECG reading. This can be alleviated by ensuring that all channels are sampled evenly.)

As a common ECG waveform is determined from three electrodes, a sequence of EEG, ECG_1, EEG, ECG_2, EEG, ECG_3 with scan clock of 40 kHz yields an EEG rate of 20 kHz, and the sampling rate for each of the three ECG electrodes is 6.667 kHz.

Variable timing and scan rate

If the board has an external trigger and/or sample clock input, variable timing and scan rates are possible. The CGA is programmed and the board is simply left to wait for the external trigger. When the trigger occurs, the board automatically takes the required number of samples.

In machine analysis, the sample clock may be connected to an output on the machine proportional to its rotational speed. As the machine speeds up, the sample rate increases. Both these schemes allow for more efficient data collection by reducing the amount of redundant data that is acquired.

Post acquisition auto-ranging

While the CGA allows different gains to be set, sometimes the optimum gain is not known in advance. Using a gain factor that is certain not to saturate the A/D will result in poor accuracy when the input signal is at a low level, while a high gain may saturate the A/D when the opposite occurs, yielding meaningless readings. What can be done is programming the CGA to take two or more readings for the channel; for example, one might be taken at unity gain and the next at a gain of 10. If the first reading is less than 1/10 of the full-scale, the second is used for greater precision. An example of an application would be the measuring of an object's response to destructive testing – for example, to shockwaves. As the amplitude of the response is unpredictable and the test often renders the object unusable for further testing, it is imperative that the first, and perhaps only, set of readings are useable.

While the use of channel-gain arrays on A/D boards has greatly increased the flexibility in the selection and throughput of individual channels, continuous hardware channel scanning does not provide adequate results for applications that require the simultaneous sampling of multiple channels. This requirement is discussed in the following section.

5.6.2   Simultaneous sampling

When the input multiplexer switches between channels, a time skew is generated between each channel sampled. On an A/D board being sampled at its maximum total throughput of 200 kHz, the minimum channel-to-channel time skew between samples on different channels is 5 µs. Since the skew is additive from channel to channel, the total time skew between the first and last samples, when 16 channels are being sampled, is 80 µs. Time skew between signal measurements taken on different channels can lead to an inaccurate portrayal of the events that generated the signals as demonstrated in Figure 5.22.

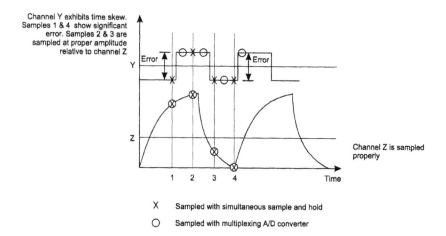

Figure 5.22
Time skew between channels

In Figure 5.22, channel 1 is sampled properly since it is deemed the reference channel. Channel 2 exhibits time skew as samples 1 and 4 show significant errors relative to their actual values at the time channel 1 was sampled.

Where the time relationship between each channel sampled is unimportant, or the skew is negligible compared to the speed of the channel scan rate, such delays are not significant. In many applications, however, such as those dealing with accurate phase measurements or high-speed transient analysis, time skew between channels is unacceptable, since it is crucial to determine the output of several signals on different channels, at precisely the same time.

To avoid the timing errors introduced when continuously sampling from one input channel to the next, special applications require A/D boards capable of simultaneous sampling. These A/D boards are fitted with so-called simultaneous sample and hold devices on all input channels. The sample and hold device on each input channel holds the sampled data until the A/D converter can scan each channel.

The maximum possible difference in sampling time between the channels, usually introduced as a result of variations in the aperture time of the individual sample and hold devices, is the time variable known as the aperture matching, or sometimes known as aperture uncertainty. This measurement reflects the maximum possible difference in sampling time between channels. The aperture uncertainty can be calculated from the maximum input frequency of the signal to be sampled. For an error of less than 1-bit on a 12-bit A/D board, this is about 800 ns at 100 kHz, 1.6 ns at 50 kHz, and so on. Boards with aperture matching of the order of 0.4 ns (±0.2 ns) are available.

Dedicated plug-in boards that perform the function of simultaneous sampling, when it is not available on the main A/D board, can be interfaced easily to the A/D board with the necessary conversion strobe signals.

5.6.3 Block mode operations

Where channel-gain arrays are available on an A/D board, an additional method by which a group of channels can be sampled almost simultaneously becomes available. Block mode triggering, sometimes known as burst mode triggering or interval scanning, creates the effect of simultaneous sampling, while maintaining the lower cost benefits of continuous channel scanning.

When operating in continuous scanning mode, conventional A/D conversion triggering works as follows: The sample trigger source, either from software, an on-board pacer clock, or an external clock, is programmed for a specified sample rate. Each sample trigger initiates a single A/D conversion on the next channel in the channel/gain array and every sample is evenly spaced in time.

Block mode triggering initiates an A/D conversion on all the required input channels at the maximum sampling rate of the A/D board, every time a sample trigger pulse occurs. A second counter is used to trigger the sampling of each of the channels at the maximum sampling rate. The number of samples to be taken in each block is typically stored by software in an on-board buffer, while the channel and gain for each sample in the block is read from the channel/gain array. The scan sequence is repeated at the next sample trigger pulse.

Consider an example where four channels are being sampled at a total throughput rate of 20 kHz, corresponding to a channel scan rate of 5 kHz. Figure 5.23 shows that in continuous scanning mode, the total scan time is 200 µs, with the samples evenly spaced every 50 µs. In block trigger mode, the four samples are taken in a single scan sequence at the maximum throughput of the board. Assuming the board is capable of taking samples at 200 kHz, the time between each of these four samples is 5 µs, while the total time taken for all the samples is 20 µs instead of 200 µs.

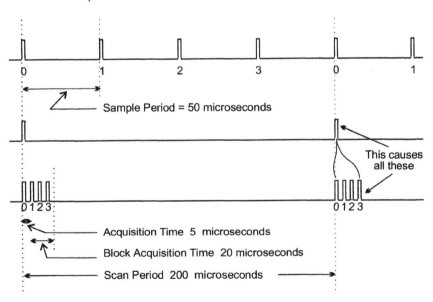

Figure 5.23
Conventional and burst trigger scanning

Where the sampling rate remains the same, that is, a sample trigger occurs every 50 µs, the throughput of the board is increased by the number of samples taken in each sample block. In this case, the throughput would be increased to 80 kHz. To maintain the total throughput rate at 20 kHz the sampling rate must be reduced by the number of channels sampled in each block. This is called the burst trigger rate and can be calculated by dividing the throughput required by the number of channels to be sampled.

$$\text{Burst trigger rate} = \frac{\text{Required total throughput}}{\text{No channels}}$$

For each burst trigger, the A/D board generates the required number of A/D conversion triggers at the maximum speed of the board. Even though the samples in a block (except the first sample) are taken at different times to the conventional triggered samples, the throughput of each channel and the time between samples on the same channel remains the same.

When using block mode triggering, data appears at the output of the A/D converter at the maximum throughput of the board. Therefore, for high-speed boards, the data transfer rate and therefore the method of transferring the data may need to be considered. For a large block count, DMA transfer, where available, will need to be used, while for small block counts, polled I/O or interrupt I/O, may be used where an on-board FIFO is utilized.

A little care must be taken when using block mode triggering with variable channel rate sampling, where some channels are sampled more often than others. It is possible that very large phase shifts are introduced because of the different times at which the two methods sample the data. Where variable channel rate sampling is used, conventional continuous channel scanning should be used.

5.7　Speed vs throughput

Throughput, or the speed at which an A/D board can acquire data, is always a consideration in data acquisition systems. There is, however, some confusion in the throughput figures quoted by board manufacturers when relating the performance of a particular board. Often these figures relate to the maximum data acquisition rate and can depend on the particular method of storing values into memory. Herein lies the real key to the throughput of data acquisition boards.

Strictly speaking, the throughput specification of an A/D board indicates the total number of analog signal input samples that can be converted to their digital equivalents per second. As it is usual for several analog input circuits to share a common A/D converter, the number of input channels in use also affects throughput. Therefore, the sample rate of each channel is the total throughput divided by the number of channels sampled.

$$\text{Maximum throughput / channel} = \frac{\text{Total throughput}}{\text{\# Channels used}}$$

For example, if you wish to sample four channels at 50 kHz each, you need an A/D board with a throughput of at least 200 kHz (four inputs X 50 KHz / input).

A/D board conversion throughput is determined by:

- Acquisition time: the time needed by the signal conditioning and acquisition circuitry (multiplexer, amplifier, filter and sample / hold), to obtain and present an accurate analog input signal to the A/D converter.
- Conversion time: the time needed to perform the actual A/D conversion and have the digital output available in a register or buffer to be transferred to memory. Here the speed (determined by the type of A/D conversion process or combination of processes) is paramount. High speed, high quality A/Ds with low drift and requiring less calibration will obviously increase the cost of a data acquisition board.

Total throughput is also determined by:

- Transfer time: the time required to transfer data to and/or from the data acquisition board to memory, where software can determine if it is to be displayed and/or transferred to a permanent storage location. The transfer rate of the data acquisition system, is the slowest of either the board throughput, the data transfer rate or, if relevant, the storage rate.

In most data acquisition systems the storage rate is not significant, since the amount of data involved is small enough to allow it all to be stored in memory, left there, displayed if necessary, and stored later. The transfer rate of the link from memory to permanent data storage is therefore not important.

Where the volume of data acquired is greater than the memory available to it, and particularly when the required throughput is very high, then alternative methods of storing the data must be found.

5.8   D/A boards

D/A boards convert digital signals from a host computer into an analog format for use by external devices such as actuators in controlling or stimulating a system or process. The principle component of all D/A boards is the digital-to-analog converter (D/A Converter).

D/A boards fall into two main categories:

- Waveform generation boards
- Analog output boards

Waveform generation D/A boards

As their name would imply waveform generation D/A boards are used in the high speed generation of analog waveforms, typically in a laboratory environment for the reproduction or simulation of noise, audio signals, power line signals and also for many other control applications.

The functional diagram of a waveform generation D/A board is shown in Figure 5.24 and comprises the following main components:

- D/A converter (DAC)
- Output amplifier/buffer
- FIFO buffer
- Timing system
- Expansion bus interface

Each of these components plays an important role in determining the speed, accuracy, and flexibility with which the D/A board can generate analog waveforms.

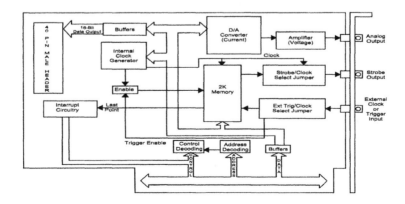

Figure 5.24
Functional diagram of a waveform generation D/A board

Analog output D/A boards

Unlike high speed, high resolution, waveform generation boards, more typical analog output D/A boards, as shown in Figure 5.25 and used for example in industrial control, are not designed for outputting precise waveforms. Instead, they maintain constant output levels unless instructed otherwise. While multi-function data acquisition boards often include two or more analog output channels, applications which require many dedicated analog outputs are most efficiently provided for by dedicated analog output boards.

Whether part of a multi-function DAC board or a dedicated analog output D/A board, the D/A conversion sub-system is straightforward in design and can be divided into two main functional components:

- D/A converter
- Output amplifier and buffer

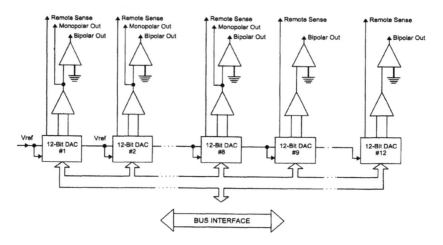

Figure 5.25
Functional diagram of an analog output D/A board

Analog output D/A boards typically have between two and sixteen dedicated output channels, each with its own D/A converter and where required output buffer/amplifier.

5.8.1   Digital to analog converters

Digital to analog converters (D/A converters or DACs) accept an n-bit parallel digital code as input and provide an analog current or voltage as output. The primary output value is a current, however, this is easily converted to a voltage using an operational amplifier. A D/A converter consists principally of a network of analog switches, controlled by the input code, and a network of precision weighted resistors. The switches control currents or voltages derived from a precise reference voltage and provide an analog output current or voltage. The output current/voltage represents the ratio of the input code to the full-scale voltage of the reference source. The main types of current output DACs and their specific important parameters are discussed in the following sections.

Weighted-current source D/A converters

The weighted-current source method of implementing a D/A converter is shown in Figure 5.26.

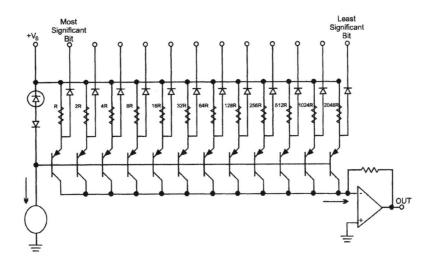

Figure 5.26
N-bit weighted-current source D/A converter

This method creates an output current, I_T, which is the summation of the weighted currents from each of the parallel transistor sources; the current contributed by each transistor set by the resistances R, $2R$, $4R$, $8R$, etc. The selection of the currents to be summed is determined by the digital code appearing at the input. For example, if the digital voltage at the MSB is logic low, current will flow through the forward biased diode rather than through the collector of the transistor, and the transistor will remain off.

When the digital voltage at the MSB is logic high, the current flowing through the collector and emitter of the transistor is equal to V_{REF}/R. A stable reference voltage with suitable temperature compensation (base-to-emitter for each transistor) ensures that each transistor produces a constant emitter current inversely proportional to the collector resistance.

Since the output from the inverting summing amplifier is $V_0 = -I_T R/2$ the output voltage is directly proportional to the voltage reference according to the equation

$$V_0 = V_{REF} (B_0 2^{-1} + B_1 2^{-2} \ldots + B_{n-1} 2^{n-1})$$

Weighted codes other than straight binary can be converted by proper choice of the weighting resistors.

R-2R ladder D/A converters

A D/A converter which uses resistors of only two values, R and $2R$, is shown in Figure 5.27.

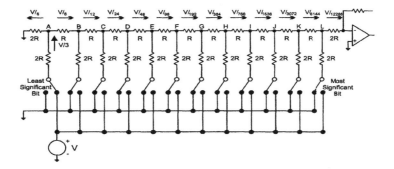

Figure 5.27
N-bit R-2R ladder D/A converter

Like the weighted-current source network, this DAC produces an output current I_T, proportional to the input code and the voltage reference source. The principle of operation of the ladder network relies on the binary divisions of the current as it flows through the ladder resistance network. A simple resistance calculation at point A shows that the resistance to the right adds up to $2R$, and the resistance to the left is $2R$. Hence the current flowing in the resistive leg of the MSB is $I_0 = V_{REF}/3R$. At node A, this current splits, half flowing to the left of node A and half flowing into node B. At node B the current splits in half again, half flowing into node C and half flowing to ground through the resistance $2R$ in the leg of this, the next most significant bit. This continues, with the current from the LSB being divided by 2^n when it reaches the summing junction of the operational amplifier. The same analysis can be applied for each switch that connects the voltage reference source to the ladder network, with the current contributed by each finally being added at the operational amplifier's summing junction.

The main advantages, which make this type of DAC popular, are the easy matching of resistances (R or $2R$), the constant input resistance for the output amplifier, and the fact that low resistor values can be used, thus ensuring high-speed operations.

5.8.2   Parameters of D/A converters

Most of the performance parameters and errors associated with A/D converters are applicable to D/A converters. In addition, several specifications for D/A converters determine the quality of the output signal produced. These are settling time, slew rate, and resolution.

Resolution

This is a measure of the size of the output step associated with a change of 1 LSB at the input. A greater number of bits, in the digital input code generating the analog output, reduce the magnitude of each output voltage increment. This allows the D/A converter to generate a more smoothly changing output signal for applications, where there is a wide dynamic output range.

Output range

Output from a D/A converter can be in two forms, current, and voltage. If a DAC produces a current output where the application requires an output voltage, an external operational amplifier is required.

The feedback resistors that would be used to set the offset, gain, and therefore range of the output, are usually provided within the D/A converter. Internal resistors are provided which track the temperature characteristics of the internal resistors of the ladder network. This eliminates the need to use an external resistance, which may introduce tracking errors. If more than one feedback resistor is provided, a choice of analog output ranges is available. Bipolar output voltage ranges are usually obtained by simply on-setting the unipolar offset voltage, with an internal bipolar offset resistor.

The selection and range of a unipolar or bipolar output range is commonly made with jumper connections.

Input data codes

There are a number of ways in which the digital data can be presented to D/A converters. The type of coding (i.e. binary, binary offset, two's complement, BCD, arbitrary etc), and its sense (positive true and negative true) must be applicable to the D/A converter used.

Settling time

In a practical D/A converter, there is a limit to the rate at which the converter can acquire new analog output values, because the analog output signal produced takes a finite time to settle to a new value, in response to a change in digital input. The settling time is defined as the time required for the output to reach, and remain, within a given error margin of the final value, usually a percentage of full-scale or ±1/2 LSB, following a prescribed change at the input (usually a full-scale change). This figure takes into account all internal factors affecting the settling time, i.e. turning the switches on and off; current changes within the resistor network, and the time required by the op-amp or buffer outputs to settle within their error bands.

The settling time of the D/A converter, especially of high speed DACs, is prolonged by the occurrence of sometimes-large transient errors in the output. Glitches are spikes in the output of a D/A converter that may result when, due to the occurrence of an intermediate state, the output is driven toward a value opposite to its final value. An intermediate state is the result of one or more switches in the DAC being faster than the others are. As an example, consider the most major transition of a DAC, when the input changes from 100...000 to 011...111. If the MSB switch changes faster, an intermediate state of 000...000 could occur, momentarily driving the output of the DAC to 0 V before returning the correct value. This is shown in Figure 5.28.

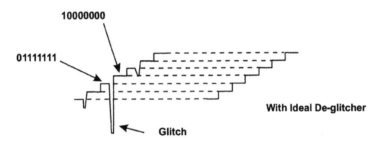

Figure 5.28
Glitch occurring at the output of a DAC during the major transition

The better matched the switching times and the faster the switches, the smaller will be the energy contained in the glitch. As the size of the glitch is not proportional to the signal change, linear filtering may be unsuccessful and can in fact make matters worse. De-glitchers, in the form of a sample and hold circuitry, are often included as part of the D/A converter, holding the outputs constant at the previous value until the switches reach equilibrium, then sampling and holding the new value. The de-glitcher circuitry, though cleaning up the output, will result in a reduction of the update rate.

Slew rate

The slew rate is the maximum rate of change that the DAC can produce on the output signal, usually limited by the slew rate of the amplifier used at its output.

Update rate

The speed or update rate of a DAC is a function of both the settling time and the slew rate and is critical in determining the maximum frequency of an output waveform that can be produced. Therefore, a DAC with a small settling time and high slew rate can generate high frequency signals, because little time is needed to accurately change the output to a new voltage level.

The generation of high frequency signals in the audio range is one application where a high slew rate and small settling time are required to reduce over-tones and interference generated as the output stabilizes. In motion control applications, where the system responds more slowly to the output voltage, the settling time and slew rate are less critical. The motor acts like a damper and reduces the effect of the oscillating output. Another application that does not require fast D/A conversion is a voltage source, which controls a heater, since the heater responds relatively slowly to a voltage change.

5.8.3 Functional characteristics of D/A boards

The important functional characteristics of D/A Boards are given below.

Double buffered input

Double buffering of input latches internal to the D/A converter allows a complete data word to be stored in the first register buffer, and then transferred to the second buffer for conversion. This prevents invalid partial data from reaching the DAC and generating spurious output, especially when updating a 12-bit DAC via an 8-bit data bus.

Simultaneous update

Double buffering of the inputs of DACs used commonly on D/A boards allows the simultaneous update of the outputs of the DAC of each channel. When programmed for simultaneous update, the data written to the registers of the D/A converters has no effect on the output value until the board is commanded to update the output of all channels simultaneously.

Remote sensing

A remote sensing facility allows the D/A converter of each channel to compensate for voltage drops over long output leads.

Offset and gain adjustment

Where the output operational amplifier is not provided on the DAC itself, an external amplifier is used. As for instrumentation amplifiers, the settling time and slew rate are the most important parameters to consider, since it is these that affect the performance (update rate) of the analog outputs.

Offset and gain errors from the D/A converter are most commonly adjusted using the offset and gain trims of the output amplifier.

5.8.4 Memory (FIFO) buffer

One of the features that differentiate a waveform generator board from an analog output board is the inclusion of on-board memory, or I/O in the form of a FIFO (first in first out) buffer. This ranges in size from 1024 bytes to 64 Kbytes.

The FIFO buffer(s) form a fast temporary memory area, addressed as I/O that holds a predefined array of data points. This gives great flexibility in creating arbitrary waveforms. Once stored in the FIFO, a single cycle of the waveform can be output, or the waveform can be continuously repeated without intervention from the PC. This allows full processor power to be dedicated to other tasks, including calculation of waveform data. It takes only a few milliseconds to load a modified or alternate waveform from memory. In continuous cycling mode, a delay can be programmed between cycles.

Waveform generation boards with more than one channel to be output, either simultaneously or sequentially, require that the digital information stored in the FIFO must also include the address of the channel to which the data is to be output.

5.8.5 Timing circuitry

Analog output boards, which do not have FIFOs do not require timing control circuitry. Output of any generated wave-shapes is performed by polled I/O. Using this method, to output waveforms, does not guarantee strict or accurate timing between the update of the outputs. The maximum update rate of the waveform output is determined by the maximum transfer rate of data to the D/A converters on the board.

Where accurate frequency and amplitude control are required, the update rate of the board must be very accurate and well controlled. High-speed waveform generation boards are provided with either on-board programmable high-speed counter/timer circuitry, to generate precise high-speed conversion probe signals, or facilities at least, to utilize an external signal as the pacer clock. Clocking circuits are made up of a frequency source, which is either an on-board oscillator between 400 kHz and 10 MHz or an external user supplied signal, and a pre-scaler/divider network, typically a counter/timer chip, that slows the clock signal down to more usable values. The clock frequency can be as DC Hz or up to the maximum update rate of the board.

D/A conversions can be initiated by triggers – either by a software trigger (writing to the D/A converter directly) or an external hardware trigger. Data conversions can be synchronized with external events with the use of external clock frequency sources and external triggers. The external trigger event is usually in the form of a digital or analog signal and will begin the acquisition depending on the active edge if the trigger is a digital signal, or the level and slope, if the trigger is an analog signal.

5.8.6 Output amplifier buffer

Operational amplifiers connected to the output of D/A converters are most commonly used where the on-board D/A converter produces a current output and the application requires an output voltage. An operational amplifier connected in the configuration shown in Figure 5.26 and Figure 5.27 can be used to convert the current to a voltage. Operational amplifiers are also used at the output of D/A converters to provide alternative voltage output ranges or higher current output. The feedback resistors that would be used to set the offset, gain, and therefore range of the output, are usually provided within the D/A converter. This allows accurate tracking of the temperature characteristics of the internal resistors of the ladder network of the DAC and eliminates the need to use an external resistance that may introduce tracking errors. If more than one feedback resistor is provided, a choice of analog output ranges is available. Bipolar output voltage ranges are usually obtained by simply offsetting the unipolar offset voltage, with a bipolar offset resistor internal to the D/A converter.

5.8.7 Expansion bus interface

The expansion bus interface provides the control circuitry and signals used to transfer data from the PC's memory, either directly to the D/A converter or the on-board FIFO, for sending configuration information (e.g. number of times to repeat a waveform or setting the clock source and frequency) or other commands (e.g. software triggers) to the board.

It includes:

- The plug-in connector, which provides the hardware interface for connecting all control and data signals, to the expansion bus (e.g. ISA, EISA or PCI) of the host computer
- The circuitry, which determines the base address of the board – this is usually a selectable DIP switch and defines the addresses of each memory and I/O location on the D/A board
- The source and level of interrupt signals generated. Interrupt signals can be programmed to occur at the end of a single conversion or at the end of one cycle of a waveform – the configuration of the interrupt levels used is commonly selected by on-board links
- The DMA control signals and the configuration of the DMA level(s) used – the configuration of the DMA levels used is typically selected by on-board links
- Normal I/O to and from I/O, address locations on the board

5.9 Digital I/O boards

Digital I/O interfaces are commonly used in a PC based DAQ systems to provide monitoring and control for industrial processes, generate patterns for testing in the laboratory and communicate with peripheral equipment such as data loggers and printers which have parallel digital I/O capabilities.

The digital I/O interface of any DAQ board is that component of the board which consists of ICs capable of input or output of TTL-compatible signals. A signal is defined to be TTL-compatible if its logic low level is between 0 V and 0.8 V and its logic high level is between 2.2 V and 5.5 V. Typically, the digital interface is a number of digital I/O lines grouped into ports, each port usually consisting of four or eight lines, although this is specific to the particular board used. It is most common for all the digital I/O lines of a particular port to be configured for either input or output, although there are circumstances where the direction of individual lines of a port can be configured independently. By reading from or writing to a port, the logical states of multiple digital lines can be simultaneously retrieved or set. The important parameters of digital I/O interfaces include the number of digital lines available and how many are configurable for input and output (or both), the rate at which data can be transferred on the digital lines, and the device drive capability of the digital output lines.

Many multi-purpose plug-in DAQ boards currently on the market, including A/D boards, D/A boards, and counter/timer boards, provide digital I/O interfaces with a varying number of digital I/O lines. Where the digital I/O capabilities of these DAQ boards does not meet the requirements of a specific application, or, a single digital I/O board is all that is required, then specialized and dedicated plug-in digital I/O boards are used. A typical digital I/O board is shown in Figure 5.29.

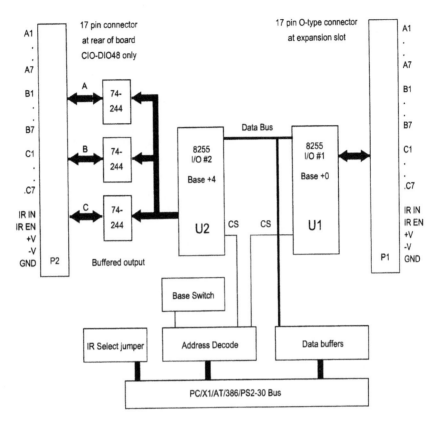

Figure 5.29
Typical digital I/O board block diagram

Figure 5.29 shows a typical digital input/output circuit. A data acquisition card or device will often combine the inputs and outputs on the same channel. The I/O channels will do either inputs or outputs but not both at the same time. The software configures the I/O channel on the card or device as either an input or output. The voltage applied to the channel, defined as an input, is usually in the form of a ground or common. This ground biases the LED of the opto-coupler on. The LED in the opto-coupler shines on the base of the transistor and turns it on. This in turn tells the input circuitry in the card that the input is 'on'.

When the channel is configured as a digital output the outgoing ground turns on the transistor. This supplies a ground to the output channel. This ground is used by the device in the field, a relay or sold state relay, to turn something on. The chips that are used to do digital output often have either diode or capacitor snubber networks on their outputs. These built in snubber networks are not intended to replace external networks that are required on long lines.

Non-latched digital I/O

Non-latched digital I/O is the mode of operation in which the state of a digital output line is updated immediately a digital value is written to the digital I/O port. In addition, for digital I/O lines configured as input lines, the current digital value present on the line when the port is read is the value that is returned. Non-latched digital I/O is the most common and simplest implementation used in digital I/O interfaces and is supported by all boards with digital I/O lines. The direction of the digital lines of a digital I/O port is conveniently set by software and can be changed as many times as required.

Latched digital I/O

For applications that require handshaking of digital data, latched digital I/O is used. In this mode of operation, an external signal determines when the data is either input to or output from the digital I/O port. The signals that are used to control the transfer of data are sometimes known as handshake lines. They are used to ensure that the digital interface is ready to input digital information appearing at the input lines, sent from a remote device or instrument, or a remote device or instrument is ready to receive data available to be sent on the output lines of a board's digital I/O interface. They could also provide digital control, in particular, to switch AC or DC power relays or alarm relays, or provide the PC tremendous power for a variety of industrial control applications.

Where digital I/O lines are used to drive panel LED displays or switch AC and DC power through relays, a high data transfer rate is not required. What is more important, however, is that the number of output lines should match the number of processes that are controlled, and that the amount of drive current required to turn the devices on/off are less than the available drive current from the output lines of the digital I/O interface.

5.10   Interfacing digital inputs/outputs

5.10.1   Switch sensing

In many applications, and particularly in industrial monitoring and control, switches form a primary interface for control actions that must be initiated by an operator. Operator controlled panel switches can be used to indicate that an action should be performed by the system. Alternatively, where switches have multiple contacts, one contact can actually perform the action required (i.e. turning on a pump), while another contact can be used to indicate that the action was actually initiated. The monitoring of abnormal system conditions can also be made easier by using limit switches to indicate that an alarm condition has been reached. In each of these cases, and in many other applications, the condition of the switch contact must be determined, requiring that the switches be interfaced and sensed by DAQ hardware.

Since switches are passive devices with no power source, they must be made to emit TTL signal levels for direct connection to a TTL compatible digital I/O interface. The open/closed position of the switch is then deduced by the TTL logic level read at the digital input. This can be carried out quite easily, as demonstrated in the two switch-sensing connections shown in Figure 5.30.

In the first switch-sensing connection, a pull-up resistor connected to one side of the switch is pulled up to the supply voltage level, which is normally available from the DAC board. The open position of the switch contact is deduced by the high logic level read at the digital input. When the switch contact is in the closed position the digital input is connected to digital ground. This configuration has higher noise immunity and has the added advantage that one terminal is connected straight to ground and can be grounded at a convenient point near the location of the switch.

In the second switch-sensing configuration, a pull-down resistor is used to present a low logic level (digital ground) at the digital input, when the switch contact is open. When the switch contact is in the closed position, the digital input is connected directly to the 5 V supply voltage. The value of the pull-up or pull-down resistor is determined by the supply voltage and the digital input current sink capability.

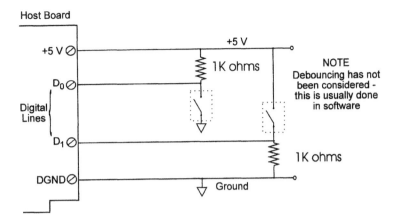

Figure 5.30
Switch position sensing circuits

Where it is likely that the signal source will be a button, switch or contact that bounces or glitches, or the signal may be a voltage higher than TTL levels, additional de-bounce and/or voltage divider circuitry is required.

5.10.2   AC/DC voltage sensing

In industrial monitoring and control, the throwing of a switch is used to begin or end an action, such as switching power to a motor or other machinery. In critical processes, the action of turning the switch is not necessarily enough to confirm that the motor has received power. This would require the sensing of the AC/DC voltages present at the motor inputs. As the AC/DC voltages involved could be quite high, any sensing circuitry directly connected to the digital I/O interface would need to provide high isolation, in addition to compatibility with the TTL digital I/O interface. A very simple AC/DC voltage sensing circuit that performs these tasks is shown in Figure 5.31.

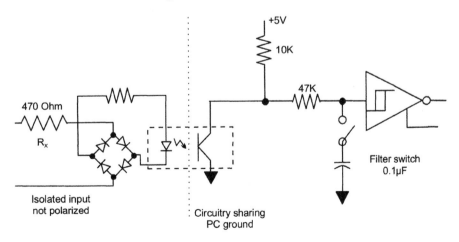

Figure 5.31
AC/DC voltage sensing circuit

There are several advantages of this low cost circuit. It is polarity insensitive and can be driven from 12 or 24 volt AC control transformers, the input voltage range can be extended by increasing the resistance of R_x and the use of the opto-isolator, guarantees high isolation, typically 500 V.

Detection of AC voltages requires the use of a filter to smooth out the AC pulses from the opto-isolator and provide a continuous signal level to the digital inputs. This slows down the response to AC voltages (typically 1 ms), however, if the capacitor is switchable, it can be de-selected, thus allowing faster response times (typically 20 μs), for DC input voltages.

5.10.3 Driving an LED indicator

Where it is necessary to provide a visual indication of an action being performed or to inform an operator of the status of a system or process, and a low level indicator is acceptable, light emitting diodes (LEDs) provide an easily implemented solution. As the standard TTL outputs from devices such as the 8255 PPI chip may not have sufficient drive to operate an LED, special driver circuitry is required, as shown in Figure 5.32.

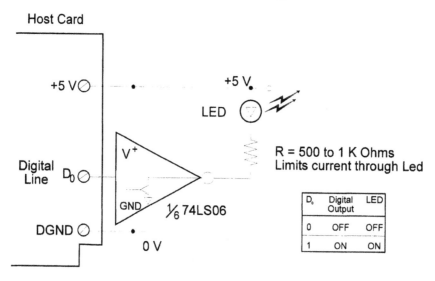

Figure 5.32
Driving an LED

5.10.4 Driving relays

Using a plug-in board for digital control, in particular to operate relays that control AC or DC power or alarms, gives the PC tremendous power for a variety of control applications. Where digital I/O interfaces are required to operate relays, special circuitry is usually required because the relays typically cannot be driven directly from TTL signal levels. In addition, the drive current required to operate either electromechanical or solid-state relays is much greater than that provided by normal TTL circuits, such as the 8255 PPI chip.

To accommodate any special signal level and current drive requirements of relays several options are available:

- Specialized rack mounted relay boards that interface directly, via ribbon cable, to the output connectors of common digital I/O boards. The circuitry necessary for the higher current drive, required to operate the relays, is provided on the relay boards. Individual relay modules may be fitted to the boards to meet the contact configuration and rating required for a specific application.

- Specialized plug-in digital I/O boards with higher current drive capability, designed especially for interfacing directly to relay modules.
- Specialized plug-in digital I/O boards, that contain both the drive circuitry and relays on a single board.
- External drive circuitry is provided by the user for each relay requiring higher drive current than that provided by the digital outputs of the digital I/O board being used.

Whether included on specialized digital I/O boards, or provided by the user, circuitry is required to interface to electromechanical and solid-state relays. This is discussed in the following sections.

Electromechanical relays

Electromechanical relays, by nature of their construction, provide a degree of isolation from the voltages switched through their contacts. Where a failure occurs, the contacts can usually withstand an AC voltage of ten thousand volts for a short time.

Coils of electromechanical relays provide an inductive load to drive circuitry. When the coil current is removed, the back EMF generated by the energy in the coil can cause damage to driving circuitry. Extra protection is required, usually in the form of a freewheel diode, across the relay control terminals, whereby the back EMF generated is dissipated through the diode's current path.

Electromechanical relays are also prone to contact arcing while switching inductive loads. Continued contact arcing causes contact degradation, high contact impedance, and eventual failure. In addition, contact arcing causes electromagnetic radiation, which may cause interference in digital circuitry. (To prevent degradation of contacts from the back EMF induced when switching inductive loads, a freewheel diode (and possibly a resistor in series to dissipate energy) should be placed across the contacts.)

The drive current required to operate this type of relay depends on the rated coil voltage and coil resistance. As an example, a relay with a rated coil voltage of 5 V and coil resistance of 100 Ω would require 50 mA drive current.

Clearly, the TTL compatible outputs of the 8255 chip could not provide this. Buffering with a high drive current chip is required. The ULN2003 Darlington Transistor Array chip serves two purposes, providing up to 500 mA of open collector drive current, as well as internally providing the freewheel diode required to dissipate the back EMP created when de-energizing the coil. The circuit configuration for driving this type of relay is shown below in Figure 5.33.

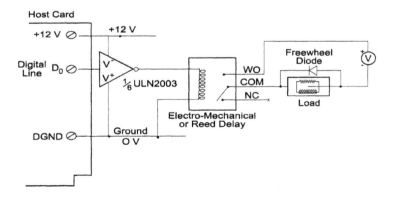

Figure 5.33
Driving electromechanical relays and preventing back-EMF damage

One advantage of this type of relay is that multiple pole relays are available which allow the switching of several contacts at the same time.

Solid state relays

Solid state relays (SSRs) are fabricated from power semiconductor devices. Their opto-isolated inputs provide a degree of isolation (up to 4000 V AC) from high voltages appearing at their output.

Solid state relays able to switch loads up to 3.5 A at rated voltages of 0–200 DC and 0–220 V AC are capable of starting small motors, switching larger capacity motor starter relays, electric appliances, sprinkler valves, alarms and enunciator beams.

These relays require greater than TTL levels of current to switch on. As a result, a simple 8255 circuit does not have the power to turn on a solid state relay. An output buffer, a chip capable of sinking at least 16 mA of drive current, is required between the 8255 digital output lines and the solid state relay. The ULN2003 Darlington Transistor Array chip is more than capable of driving these types of relay. Figure 5.34 below shows the configuration for driving an SSR.

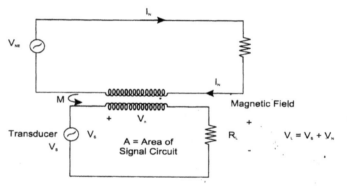

Figure 3.34
Driving solid state relays

There are several advantages of solid state relays over reed and electromechanical relays:

- Solid state relays do not have the problems of contact arcing and there is no back EMF to drive circuits when switched off.
- Two solid state relays can be connected in parallel with each other and handle twice the current of a single relay's rating. Built-in circuitry allows the units to share the current with no additional wiring.
- Switching of high current SSRs is performed at the AC voltage zero crossover point, thus preventing surge currents and electromagnetic interference.

5.11   Counter/timer I/O boards

Counter/timer circuitry is useful for many applications, including digital event counting, digital pulse timing, one shot and continuous clocked outputs and generating waveforms with complex duty cycles.

All of these applications can be implemented using a simple counter, which comprises a *source* input and *gate* input, a single *output* and an internal *n*-bit count register, as shown in Figure 5.35.

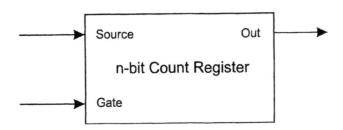

Figure 5.35
Simplified model of a counter

A counter is a digital device that responds to and outputs TTL compatible signals, counting input signal transitions at its source input by incrementing its internal count register every time a transition occurs. The source input therefore provides the time-base for the operation of the counter. The counter can be configured to count either negative going (high-to-low) transitions, or positive going (low-to-high) transitions, of signals occurring at the source input. The internal count register can be read by software at any time.

The gate input can be used to enable/disable the function of the counter, by enabling or disabling counting dependent on the current signal level at the gate input. In this mode, the active level of the gate input can be configured to enable counting when the gate input is at a high level and disable counting when it is at a low level, or vice-versa. The counter may also be configured to begin counting input transitions only after a transition of the signal at the gate input, that is, the gate acts as a count trigger. In this mode, the active edge of the gate input signal can be configured to allow counting only after a rising-edge (low-to-high) transition of the gate input, or alternatively, a falling-edge (high-to-low) transition of the gate input. In all modes, counting begins at the next active clock edge after the active gate signal. When the counter is used with no gating, software initiates the counting sequence.

The output signal can be configured, depending on the mode of operation of the counter, to either toggle states or pulse, when the count register reaches its terminal count (TC). If the signal output from another counter is used as the gate input, thereby enabling and disabling the count as required, complex waveforms with specified duty cycles and frequency could be generated. This is dependent on the mode of operation of the counter, since in some modes the gate input has no effect.

Figure 5.36 shows the general operation of a counter configured for high-level gating and positive-edge triggered gating. The output signal is configured for positive polarity and is shown for both pulse-output and toggle mode on reaching terminal count (TC). For simplicity, the counter is operating in count-up mode and the count is set to five (5).

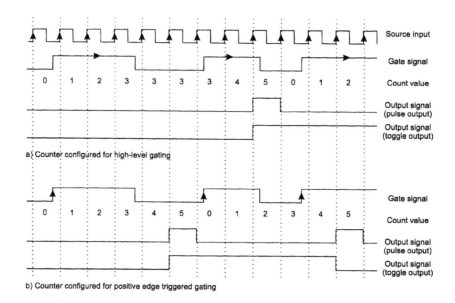

Figure 5.36
Waveforms showing general operation of a counter

Two commonly used counter/timer chips are available which provide counters with the signal functions described above. The 8254 counter/timer chip contains three independently programmable 16-bit counters. It is regularly used on plug-in A/D boards as a pacer clock and pulse trigger for accurate timing of A/D sampling rates and D/A conversion rates. It is often included on multi-purpose DAQ boards as an uncommitted counter/timer to perform any necessary counting or timing functions required of an application. A more frequently used counter/timer chip for dedicated timer/counter I/O boards is the AM9513 chip, a powerful and flexible device, consisting of five independent 16-bit counters. A typical counter/timer I/O board employing the AM9513 chip is shown in Figure 5.37.

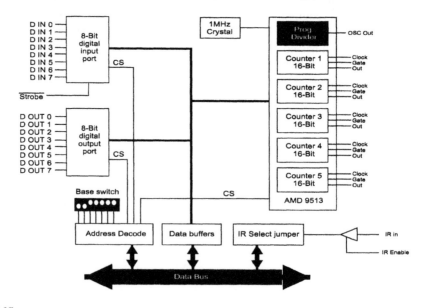

Figure 5.37
Typical counter/timer I/O board

As details of the AM9513 operation are quite complex they will not be detailed here, however, the most important specifications of a counter/timer chip are its resolution, which is simply the bit width of the internal count registers, and clock frequency. A counter/timer chip with 16-bit resolution means that each of its counters can count up to 65,535. A higher resolution simply means that each of the counters can count higher. Greater count resolution can be achieved by connecting the output of one counter to the source input of another counter. When two or more counters are cascaded together, extremely large counts can be performed. For example, two 16-bit counters, when cascaded to make a 32-bit counter, can count to over four billion.

The clock input to a counter is a physical connection to a high frequency stable clock source, which is usually internally-divided to more suitable clocking frequencies. The frequency of the clock source becomes important since new counts are loaded and the internal counter is decremented on a transition of the clock pulse, usually on the falling edge of the clock. Consider a signal applied to the source input that has a higher frequency than the clock source. The input signal may perform several transitions before the next clocking edge allows the internal counter to be decremented, leading to inaccurate measurement of the source input signal. Therefore, the higher the clock frequency, the faster the internal count register can be decremented and the higher the frequency signals on the source input, which can be detected and accurately measured. In addition, the higher the frequency of the clock source, the higher the frequency of pulses, square waves and complex waveforms, which can be generated on the output.

Counter/timers can be configured to operate in many different modes of operation, the number of modes, and the functions performed in each mode dependent on the manufacturer of the counter/timer chip. Several of the most common functions that can be performed by counter/timers are demonstrated below.

Generating waveforms

Generating waveforms of variable duty cycles is straightforward. Consider a counter on the 8254 counter/timer chip, configured for mode 1 operation, in which the count is triggered by the active edge of the signal on the source input. The next clock pulse after this count-trigger, the internal counter is loaded with the initial count (N), the output goes low, and the count begins. The output will remain low for N clock pulses before returning high and remaining high until the next clock pulse after the next active edge of the source input signal.

By using this method, output waveforms of specified duty cycle and frequency can be generated. The frequency of the waveform is exactly the same as the frequency of the signal applied at the source input. The duty cycle is determined by dividing the period T_1 for which the output signal remains high, divided by the period of the output signal $T_3 = T_1 + T_2$.

When a waveform is generated in this way there will be an uncertainty in the period of the output waveform compared to the period of the signal on the source input, of up to one clock period. This error depends on when the count trigger occurs in relation to the active edge of the clock signal.

Measuring pulse width

Counters can be used to measure the width of a pulse by applying the unknown pulse signal to the gate input of the counter and counting the number of cycles of a known frequency clock signal applied to the source input. The known clock signal can be derived from the clock input of the counter/timer chip, an external clock source, or from the output of another counter configured to produce a periodic waveform of the required frequency.

When a counter is configured to enable counting on the active high level of the gate input, the internal counter starts counting the source input transitions at the next active transition, after the gate input pulse goes high and stops counting at the end of the pulse. The duration of the gate input pulse (T_{pw}) is found by reading the count register contents, determining the number of known clock transitions that occurred (N), and multiplying this by the time between each active transition of the clock (T_S). In this case, it does not matter whether the count occurs on the positive or negative going edge of the source input. What is more important is the frequency of the known clock signal applied to the source input. As shown in Figure 5.38, an error can occur in the counting of the clock transitions, depending on when the pulse begins and ends in relation to the active edge of the clock input. This error can be almost two full clock cycles. Clearly, the higher the frequency of the clock signal, the smaller the counting error will be, and the higher the resolution of the pulse measurement. Care must also be taken not to choose too high a frequency clocking source input as the counter may reach its terminal count before the end of the pulse.

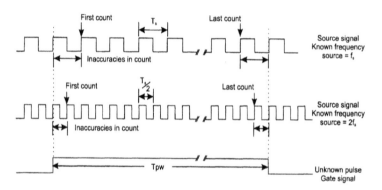

Figure 5.38
Measuring an unknown pulse width

Consider a 500 kHz clock signal, with a clock period of 2 µs, applied to the source input. As a 16-bit counter can count up to $2^{16}-1 = 65,535$ transitions of the clock input, the maximum measurable pulse width will be $65,535 * 2$ µs $= 131$ ms. Decreasing the frequency of the clock input source increases the pulse width that can be measured.

Selection of the frequency of the clocking source input is therefore a compromise between the resolution and accuracy required of the measurement and the pulse width that must be measured.

Measuring an unknown frequency

Counters can also be used to measure the frequency of a periodic square wave, irrespective of its duty cycle. This is accomplished by applying the unknown signal to the source input of the counter and counting the number of cycles of the signal during a fixed duration pulse applied to the gate input. The fixed duration gate input signal can come from an external source or from the output of another counter, configured to produce a pulse of the required duration.

As was the case for pulse width measurement, the counter is configured to enable counting on the active high level of the gate input. The internal counter starts counting the source input transitions at the next active transition after the gate input pulse goes high and stops counting at the end of the pulse. The frequency of the signal at the source input, (f_S), is found by determining the number of signal transitions (N), which occurred and dividing this by the period of the fixed duration gate input pulse, $T_{pw} \cdot f_S = N / T_{pw}$. This is shown in Figure 5.39.

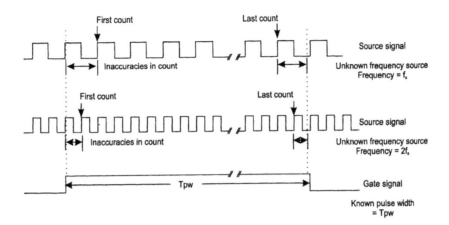

Figure 5.39
Measuring an unknown frequency

The lower the frequency of the signal that must be measured, the greater should be the fixed duration pulse width to achieve the same resolution and accuracy. If however, the duration of the fixed pulse, (T_{pw}), is too long compared to the clock period, (T_S), of the unknown frequency signal being measured, the counter may reach its terminal count before the end of the pulse. Therefore, the selection of the duration of the fixed pulse at the gate input is a compromise between the resolution and accuracy required and the frequency of the input signal being measured.

6

Serial data communications

The standardization of the RS-232 serial port as part of the IBM PC and its compatibles has led to this communications interface being used for many stand-alone loggers and other instruments that have interfaced to the PC. With the advent of smart instrumentation such as digital transmitters and their use in a distributed data acquisition and control system, the requirements of interfacing multiple devices on a multi-drop network has led to the extensive use of the RS-485 communications interface.

This chapter reviews the fundamental definitions and basic principles of digital serial data communications. It details two of the most popular and common interface standards used in data acquisition and control systems (RS-232 and RS-485). In addition, the most common industrial protocols are examined, including methods for detecting errors in communication, an important consideration in noisy industrial environments. A section on troubleshooting and testing serial data communications circuits has been included for completeness.

In the past, Ethernet has been usually thought of as an office networking system. Now many manufacturers are using Ethernet and industrial fieldbuses as communication systems to interconnect data acquisition devices. This can take the form, among others, of connecting computers that are using plug-in data acquisition cards or data loggers that are networked together. Fieldbuses such as Profibus and Foundation Fieldbus are being used to interconnect devices that are doing data acquisition. There are currently several hybrid analog and digital standards available for communication between field devices and between field devices and a master system. Only a fully-compatible digital communication standard will provide the maximum benefits to end users and one such standard, currently being proposed, is the Foundation FieldBus

6.1   Definitions and basic principles

All data communications systems have the following components:

- The *source* of the data (e.g. a computer). Also required is circuitry that converts the signal into one that is compatible with the communications link, called a transmitter or line driver

- The *communications link* (twisted-pair cable, coaxial cable, radio, telephone network etc), which transfers the message to the receiver at the other end
- The *receiver* of the data where the signal is converted back into a form that can be used by the local electronics circuitry

Both the receiver and the transmitter must agree on a number of different factors to allow successful communications between them, the most important being:

- The type of *electrical signals* used to transmit the data
- The type of *codes* used for each symbol being transmitted
- The *meaning* of the characters
- How *the flow* of data is controlled
- How errors are detected and corrected

The hardware rules that apply to the physical interface and its connections are known as *interface standards,* while the software rules which apply to the format and control of data flow and the detection and correction of errors are generally referred to as the *protocol.*

All data communication is based on the same binary system used in computers. Each basic unit of information is called a BIT (BInary digIT), and it can be one of two values: '0' or '1'. These are termed logic 0 and logic 1. Inside a computer, a bit is usually represented by a voltage; typically, 5 V is logic 1, and 0 V is logic 0. In data communications, a logic 1 is also referred to as a *mark* and a logic 0 as a *space.*

In a data communications link, the 1s and 0s may be indicated by a +/– voltage. Other methods can involve audio tones, with one frequency for logic 1 and a different frequency for logic 0. Some advanced techniques allow several bits to be encoded in a single voltage change: this is how you can use a 14.4 kbps modem on a telephone line with a nominal band-width of only around 3 kHz.

A string of bits is a binary number. It can be interpreted simply as a numerical value (in binary, hexadecimal or some other numbering system) or it may be translated into a character according to an agreed code. For example, in the ASCII system the binary value 1000100 represents 'D'.

6.1.1 Transmission modes – simplex and duplex

In any communication system connecting two devices, data can be sent in one direction only, or in both directions. A simplex system is one that is designed for sending messages in one direction only. This is shown in Figure 6.1.

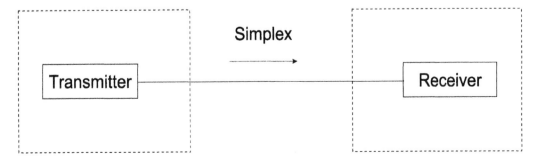

Figure 6.1
Simplex communications

A duplex system is designed for sending messages in both directions. Two types of duplex systems exist:

- Half duplex occurs when data can flow in both directions but only in one direction at a time as shown in Figure 6.2. First one end transmits and the other end receives. After what is called the line turnaround time, the roles are reversed.

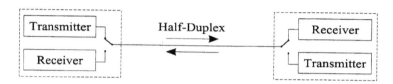

Figure 6.2
Half duplex communications

- In a full duplex system, shown in Figure 6.3, the data can flow in both directions at the same time.

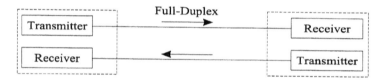

Figure 6.3
Full duplex communications

6.1.2 Coding of messages

To transfer a message across a communications interface, both the sender and receiver must, among other things, agree on the meaning of the binary digital patterns transferred, or the code. Encoding is the process of converting the message data into a standard binary code for transfer over the data communications link. The number of bits in a code determines the total number of unique characters that are possible.

The most common character set used for digital data communications in the Western world is the American Standard Code for Information Interchange, or ASCII code (see Appendix E).

This code assigns a binary field of 7-bits to represent each character, giving 128 (2^7) unique characters made up of:

- Upper and lower case letters, and numerals 1 to 9
- Various punctuation marks and symbols
- A set of control codes (the first 32 characters) that are used by the communications link itself and are not printable

On a computer keyboard, ASCII control codes are generated by pressing the Control key [Ctrl] and another key. For example, [Ctrl]-[A] generates the ASCII code SOH.

A communications link set up for 7-bit data strings can only handle hexadecimal values from 00 to 7F. For full hexadecimal data transfer, an 8-bit link is needed, with each packet of data consisting of a byte (two hexadecimal digits) in the range 00 to FF. For this reason, an 8-bit link is often referred to as *transparent* because it can transmit any value. In this case, a character can still be interpreted as an ASCII value if required (in which case the eighth bit – the most significant bit is ignored).

It is worth mentioning that the full hexadecimal range can in fact be transmitted over a 7-bit link by representing each hexadecimal digit as its ASCII equivalent. Thus the hexadecimal number 8E would be represented as the two ASCII (hexadecimal) values 38 45 ('8' 'E'). The disadvantage of this technique is that the amount of data to be transferred is almost doubled, and extra processing is needed at each end.

ASCII (American Standard Code for Information Interchange) is the most commonly used code for encoding characters for data communications. It is a 7-bit code with only $2^7 = 128$ possible combinations of the seven binary digits (bits).

Each of these 128 codes is assigned to a specific control code or character, as specified by the these standards:

- ANSI – X3.4
- ISO – 646
- CCITT alphabet #5

An ASCII table records the bit value of every character defined by the code. There are many different forms of the table, but they all contain the same basic information. The values in the table may be expressed in decimal (DEC), ranging from 0–127, or in binary (BIN), ranging from 0000000 to 1111111, or in hexadecimal (HEX) numbers, ranging from 00 to 7F. (See Table 6.2 for examples of binary to hexadecimal conversions.)

A condensed form of the table showing characters and control codes is presented in Table 6.1. The control codes and their meanings are listed in Table 6.3.

Table 6.1 shows the code for each character in hexadecimal and binary values. It takes the form of a matrix in which the MSB (most significant bits) are along the top and the LSB (least significant bits) are down the left-hand side.

			MSB							
		HEX	0	1	2	3	4	5	6	7
	HEX	BIN	000	001	010	011	100	101	110	111
LSB	0	0000	(NUL)	(DLE)	Space	0	@	P	'	p
	1	0001	(SOH)	(DC1)	!	1	A	Q	a	q
	2	0010	(STX)	(DC2)	"	2	B	R	b	r
	3	0011	(ETX)	(DC3)	#	3	C	S	c	s
	4	0100	(EOT)	(DC4)	$	4	D	T	d	t
	5	0101	(ENQ)	(NAK)	%	5	E	U	e	u
	6	0110	(ACK)	(SYN)	&	6	F	V	f	v
	7	0111	(BEL)	(ETB)	'	7	G	W	g	w
	8	1000	(BS)	(CAN)	(	8	H	X	h	x
	9	1001	(HT)	(EM)	)	9	I	Y	I	y
	A	1010	(LF)	(SUB)	*	:	J	Z	j	z
	B	1010	(VT)	(ESC)	+	;	K	[	k	{
	C	1100	(FF)	(FS)	,	<	L	\	l	|
	D	1101	(CR)	(GS)	-	=	M	]	m	}
	E	1110	(SO)	(RS)	.	>	N	^	n	~
	F	1111	(SI)	(US)	/	?	O	–	o	DEL

Table 6.1
ASCII table

Some examples of the HEX and BIN values are given below:

Character	HEX	BIN
A	41	100 0001
M	4D	100 1101
m	6D	110 1101
@	40	100 0000
?	3F	011 1111

Table 6.2
Some examples of binary and hexadecimal values

The meanings of standard abbreviations used in the table are as follows:

ACK	Acknowledge
BEL	Bell
BS	Backspace
CAN	Cancel
CR	Carriage return
DC1	Direct control 1
DC2	Direct control 2
DC3	Direct control 3
DC4	Direct control 4
DLE	Data link escape
EM	End of medium
ENQ	Enquiry
EOT	End of transmission
ESC	Escape
ETB	End of transmission block
ETX	End of text
FF	Form feed
FS	Form separator
GS	Group separator
HT	Horizontal tab
LF	Line feed
NAK	Negative acknowledge
NUL	Null
RS	Record separator
SI	Shift in
SO	Shift out
SOH	Start of heading
STX	Start of text
SUB	Substitute
SYN	Synchronous idle
US	Unit separator
VT	Vertical tab

Table 6.3
ASCII control codes and their meanings

6.1.3   Format of data communications messages

It is not reasonable to expect that a device can send a string of characters across a communications link and that the receiver at the other end will know what to do with the data. Any message must be presented according to some pre-arranged rules. Consequently, data is usually arranged in a particular format, with additional information added so that the message can be effectively transmitted and understood at the receiving end. Consider a simple asynchronous system such as RS-232, for which it is common practice to send one character at a time. The format of a typical character frame is indicated in Figure 6.4.

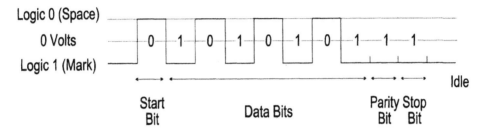

Figure 6.4
Format of a typical serial asynchronous data message

Initially the data communications link is in the idle state: the line is in the mark state, held to a constant negative voltage.

The transmitter then sends a start bit to indicate that it is transmitting a character. The start bit is in the opposite voltage state to the idle voltage and allows the receiver to synchronize to the character that follows the start bit.

The receiver reads in the individual bits of the character as they arrive. At the end of the data bits, a parity bit may be included to allow the receiver to detect any possible errors in the character frame. After the parity bit, optionally 1, 1½ or 2 stop bits follow (1½ stop bits is a logic 1 held for 50% longer than 1 stop bit). The stop bit effectively puts the com-munication line back into an idle state. After the last stop bit is transmitted down the line, a start bit can be transmitted for the next character.

The parity bit included at the end of the character is effectively a fingerprint of the character to enable the receiver to identify whether any errors have occurred in the transmission.

Even parity means that the total number of logic 1 bits in the data together with the associated parity bit must be an even number. The UART (see p. 210) works out if there is an even or odd number of 1 bits and sets the parity bit to 0 or 1 to make the total count, including the parity bit, even. Odd parity is done in a similar way, except that the UART sets the parity bit to be 1 or 0 to ensure that the total number of bits in the message, again, including the parity bit, is an odd number.

Two more options are available for the parity bit. Mark parity always sets the parity bit to 1, whilst space parity sets the parity bit to a 0. While mark and space parity obviously do not allow the receiver to detect any errors in the character frame, they are sometimes used when there are timing problems.

Statistically, in selecting even or odd parity, there is a 60% chance of the receiver detecting a parity error. The parity scheme works based on an odd number of bits being in error in the character frame. An even number of bits in error will not be detected by the parity detection scheme. Thus due to the availability of more sophisticated error-checking schemes today, (such as block check calculation and cyclic redundancy check), the parity bit is sometimes set to no parity (that is, there is no parity bit at all in the character frame).

In summary, the optional settings for asynchronous transmission of characters are:

- Start bits   1
- Data bits   5, 6, 7, 8
- Parity bits   even, odd, mark, space or none
- Stop bits   1, 1½ or 2

As there cannot be half a bit, 1½-stop bits means that the mark length is 50% longer than for one stop bit.

6.1.4   Data transmission speed

The maximum rate at which data can be transferred from the source to the receiver on a communications interface depends on a number of factors:

- Type and complexity of the circuitry at each end (interface)
- Communication link (twisted-pair, coaxial cable, radio etc)
- Distance between the sender and receiver
- Amount of data being transferred
- The overhead associated with the data transfer
- The acceptable rate of error

The lower the data rate, the less complex are the requirements of the communication link, the source and receiver circuitry and the lower the errors due to timing and noise problems.

Data transfer rates are usually measured in bits per second or **bps**. This is an indication of the useful data that has been transmitted to the receiver. For example, in Figure 6.4 the useful data is only seven bits, whilst the total number of bits transmitted was ten. The additional three bits are viewed as overhead bits for the data communication.

The baud rate (named in recognition of Maurice Emile Baudot) can be considered the physical rate, or signaling speed at which data bits can be transmitted and correctly received on the communications interface. Referring to Figure 6.4, if each bit occupied a time of 1 millisecond (ms), the total baud rate would be 1 / 1 ms = 1000 baud. This is the signaling speed. The data transfer rate, on the other hand, can be calculated as:

7 data bits / 10 ms = 700 bits per second (bps)

The actual data transfer rate is therefore 30% less than the baud rate for this example of a total of 10 bits in the frame. Baud rates are usually quoted in standard values such as 50, 110, 300, 600, 1200, 2400, 4800, 9600, 19200, 38400, 57600, and 115200 baud. It is common practice in industry to use the terms baud rate and data transfer rate interchangeably, unless it is specifically noted that they are not equal.

When using modems and sophisticated encoding techniques, a single signal change on the line can indicate several encoded bits. This means that the data rate (in bps) is far greater than the baud rate (the reverse of the situation quoted in the previous example).

6.2   RS-232-C interface standard

The Electronic Industries Association (EIA) RS-232 interface standard is probably the most widely known of all serial data interface standards. It was developed for a single purpose clearly stated in its title, and defines 'Interface between Data Terminal Equipment (DTE) and Data Communications Equipment (DCE) employing serial binary data interchange'. It was issued in the USA in 1969 by the engineering department of the EIA. Bell Laboratories and leading manufacturers of communications equipment also cooperated to clearly define the interface requirements when connecting data terminals to the Bell telephone system. Almost immediately, shortcomings resulted in minor revisions to the standard, becoming EIA-232-C still widely used today. The current revision is EIA/TIA-232-E (1991), which brings it into line with the international standards CCITT V.24, CCITT V.28, and ISO-2110. Some users contest the statement that EIA-232 is a 'standard', because its interpretation has been responsible for many problems in interfacing equipment from different manufacturers. It should be emphasized that EIA-232 standard defines the electrical and mechanical details of the interface and does not define a protocol.

The EIA-232 standard consists of three major components, which define:

- **Electrical signal characteristics**
  Electrical signals such as the voltage levels and grounding characteristics of the interchange signals and associated circuitry.

- **Interface mechanical characteristics**
  The mechanical characteristics of the interface between the DTE and DCE. This section dictates that the interface must consist of a plug and socket, and that the socket will normally be on the DCE. The familiar DB-25 connector is specified together with a smaller 26-pin alternative connector.

- **Functional description of the interchange circuits**
  This section defines the function of the data, timing and control signals used at the interface between DTE and DCE. Very few of the definitions in this section are relevant to applications for data communications for data acquisition and control.

6.2.1   Electrical signal characteristics

The RS-232 standard is designed for the connection of two devices: data terminal equipment (DTE), such as a computer or printer, and data communications equipment (DCE), such as a modem. DCEs are now also called data circuit-terminating equipment in the RS-232-D/E standard. This definition was required since the RS-232 standard is often used to interface items of equipment that are not communication devices.

A typical connection between a computer (DTE), which transmits data on pin 2 and receives data on pin 3 of a 25-way DB connector, to a stand-alone controller (DCE) that receives data on pin 2 and transmits data on pin 3 is shown in Figure 6.5.

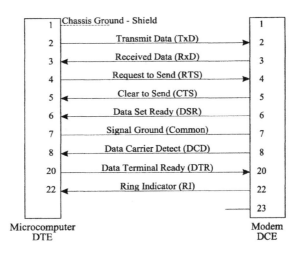

Figure 6.5
Connections between a computer (DTE) and a stand-alone controller (DCE)

At the RS-232 receiver the following signal voltage levels are defined:

- +3 V to +25 V for transmission of logic 0
- −3 V to −25 V for transmission of logic 1
- +3 V to −3 V for an undefined logic level

To meet these voltage requirements at the receiver and to overcome any voltage drops that occur along the communications lines, the RS-232 transmitter must produce slightly higher voltages.

These are in the range:

- +5 V to +25 V for transmission of logic 0
- −5 V to −25 V for transmission of logic 1
- +5 V to −5 V for an undefined logic level

In practice, many EIA-232 transmitters operate very close to their margin of safety, e.g. at +7 and −7 volts. This can be acceptable for short cable runs, where it is hoped that there will be no voltage problems. Unfortunately, increased error rates can be expected at the receiver because of induced external interference voltages.

The voltage levels associated with the internal electronics of DTE and DCE devices are commonly −5 V to +5 V and are therefore not directly compatible with the signal levels associated with the communications interface. Consequently, at the transmitting end, a **Line Driver** is necessary in each data and control line to amplify this voltage to the higher level required on the EIA-232 interface. Modem power supplies, such as those used in PCs, usually have a standard +12 V to −12 V voltage output that can then be used for the line driver output. This falls within the voltage range specified by the EIA-232 standard and is the voltage level most commonly used these days.

At the receiving end, a **Line Receiver** is necessary for each data and control line to reduce the voltage level to the −5 V to +5 V level required by the internal electronics.

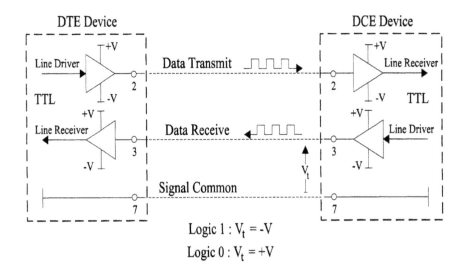

Figure 6.6
EIA-232 transmitters and receivers

The EIA-232 standard defines twenty-five (25) electrical connections, which are each described in more detail in Section 6.2.3.

The electrical connections are divided into the four groups shown below:

- Data lines
- Control lines
- Timing lines
- Special secondary functions

The data lines are used for the transfer of data. Pins 2 and 3 are used for this purpose. Data flow is designated from the perspective of the DTE interface. Hence, the 'transmit line', on which the DTE transmits (and DCE receives), is associated with pin 2 at the DTE end and pin 2 at the DCE end. The 'receive line', on which the DTE receives (and DCE transmits), is associated with pin 3 at the DTE end and pin 3 at the DCE end. Pin 7 is the common return line for the transmit and receive data lines.

Control lines are used for interactive device control, commonly known as 'hardware handshaking', and regulate the way in which data flows across the interface.

The four most commonly used control lines are as follows:

- RTS – request to send
- CTS – clear to send
- DSR – data set ready (or DCE ready in EIA-232-D/E)
- DTR – data terminal ready (or DTE Ready in EIA-232-D/E)

It is important to note that the handshake lines operate in the opposite voltage sense to the data lines. When a control line is active (logic 1), the voltage is in the range +3 to +25 volts and when deactivated (logic 0), the voltage is zero or negative.

Hardware handshaking is usually the cause of most of the interfacing problems. Manufacturers sometimes omit certain of these control lines from their EIA-232 equipment or assign unconventional purposes to them. Consequently, many applications do not use hardware handshaking but instead use only the three data lines (transmit, receive and signal common ground) with some form of software handshaking. The control of data flow is then part of the application program. Most of the systems encountered in data communications for data acquisition, instrumentation and control, use some sort of software based protocol in preference to hardware handshaking. Simple examples of software handshaking protocols are the ETX/ACK, where the transmitter software is in control of the handshake, and XON/XOFF, where the receiver software controls the handshake.

There is a relationship between the allowable speed of data transmission and the length of the cable connecting the two devices, on the EIA-232 interface. Briefly, as speed increases, the quality of the transition of the data signal from one voltage level to another (e.g. from –25 V to +25 V) becomes increasingly dependent on the capacitance and the inductance of the cable. The rate, at which voltage can 'slew' from the one logic level to the other, depends mainly on the cable capacitance, which increases with cable length. The length of the cable is thus limited by the number of data errors that are acceptable during transmission. The EIA-232-D&E standard specifies the limit of total cable capacitance to be 2500 pF. With typical cable capacitances of 50 pF/ft, the maximum cable length would appear to be 50 ft, but in practice, longer cable lengths appear to be possible with lower data transmission rates.

The common data transmission rates used with EIA-232 are 110, 300, 600, 1200, 2400, 4800, 9600 and 19200 baud. Based on field tests, Table 6.4 shows a practical relationship between selected baud rates and the cable length supporting reliable transmission. Table 6.4 indicates that much longer cable lengths are possible at lower baud rates. These values do not take into account the effects of noise, which can adversely affect the maximum cable length supported at a given baud rate.

Baud rate	Cable length (m)
110	850
300	800
600	700
1200	500
2400	200
4800	100
9600	70
19200	50

Table 6.4
Demonstrated maximum cable lengths with EIA-232 interface

6.2.2 Interface mechanical characteristics

Although not specified by RS-232C, the DB-25 connector (25 pin D-type) has become so closely associated with RS-232 that it is accepted as the de facto standard. On some RS-232 compatible equipment, where not all the control lines are required for handshaking, the smaller DB-9 connector (9 pin D-type) is commonly used.

The pins of the DB-9 connector are usually allocated as follows:

- *Pin 2* – receive data
- *Pin 3* – transmit data
- *Pin 7* – signal ground

While this pin configuration is likely to be adhered to by manufacturers at the computers communications interface, it is possible (and often likely) that the data receive and transmit lines on remote stand-alone systems are on different pins of the DB-9 connector. It is therefore wise to consult the manufacturers' data sheets.

The common RS-232 pin assignments for both the DB-9 and DB-25 connectors are shown in Table 6.5, below (continued on the following page).

Pin no.	DB-9 connector IBM 232 pin assignment	DB-25 connector EIA-232 pin assignment
1	Received line signal	Shield
2	Received data	Transmitted data
3	Transmitted data	Received data
4	DTE ready	Request to send
5	Signal common/ground	Clear to send
6	DCE ready	DCE ready
7	Request to send	Signal ground/common
8	Clear to send	Received line signal
9	Ring indicator	+ Voltage (testing)
10		– Voltage (testing)
11		Unassigned
12		Sec rec'd line signal detector/data

		signal
13		Sec clear to send
14		Sec transmitted data
15		Transmitter signal DCE element timing
16		Sec received data
17		Receiver signal DCE element timing
18		Local loopback
19		Sec request to send
20		DTE ready
21		Remote loopback/signal quality detector
22		Ring indicator
23		Data signal rate
24		Transmit signal DTE element timing
25		Test mode

Table 6.5
Table of the common DB-9 and DB-25 pin assignments for EIA-232

6.2.3   Functional description of interchange circuits

The EIA circuit functions are defined, with reference to the DTE, as follows:

- **Pin 1: Protective ground (shield)**
  A connection is seldom made between the protective ground pins at each end. Their purpose is to prevent hazardous voltages, by ensuring that the DTE and DCE chassis are at the same potential at both ends. However, there is a danger that a path could be established for circulating earth currents, so, usually the cable shield is connected at one end only.

- **Pin 2: Transmitted data (TXD)**
  This line carries serial data from pin 2 on the DTE to pin 2 on the DCE. The line is held at MARK (or a negative voltage) during periods of line idle

- **Pin 3: Received data (RXD)**
  This line carries serial data from pin 3 on the DCE to pin 3 on the DTE.

- **Pin 4: Request to send (RTS)**
  The RTS line is a request to send from the DTE to the DCE. This line is used in conjunction with the CTS line to do hardware control. The DCE will not enable the CTS line until the DTE enables its RTS line.

- **Pin 5: Clear to send (CTS)**
  When a half-duplex modem is receiving, the DTE keeps RTS inhibited. When it becomes the DTE's turn to transmit, it advises the modem by asserting the RTS pin. When the modem asserts the CTS, it informs the DTE that it is now safe to send data.

- **Pin 6: Data set ready (DSR)**
  This is also called **DCE ready**. In the answer mode, the answer tone and the data set ready are asserted two seconds after the telephone goes off hook.

- **Pin 7: Signal ground (common)**
  This is the common return line for the data transmit and receive signals. The connection, pin 7 to pin 7 between the two ends, is always made.
- **Pin 8: Data carrier detect (DCD)**
  This is also called the **received line signal detector**. Pin 8 is asserted by the modem when it receives a remote carrier and remains asserted for the duration of the link.
- **Pin 20: DTE ready (or data terminal ready)**
  DTE Ready enables, but does not cause, the modem to switch onto the line. In originate mode, DTE Ready must be asserted in order to auto dial. In answer mode, DTE Ready must be asserted to auto answer.
- **Pin 22: Ring indicator**
  This pin is asserted during a ring on the line.
- **Pin 23: Data signal rate selector (DSRS)**
  When two data rates are possible, the higher is selected by asserting pin 23.

6.2.4 The sequence of operation of the EIA-232 interface

The following description of one particular operation of the EIA-232 interface is based upon a half duplex data interchange. It should be noted that full duplex communication is generally used today; however, a half duplex description is given as it encompasses that of full duplex operation. Figure 6.7 gives a graphical description of the operation with the initiating user terminal (or DTE) and its associated modem (or DCE) on the left of the diagram and the remote computer and its modem on the right.

The following sequence of steps occur:

- The initiating user manually dials the number of the remote computer.
- The receiving modem asserts the ring indicator line (RI-pin 22) in a pulsed ON/OFF fashion as per the ringing tone. The remote computer already has its data terminal ready line (or DTR-pin 20) asserted to indicate that it is ready to receive calls. (Alternatively, the remote computer may assert the DTR line after a few rings.) The remote computer then sets its request to send line (RTS-pin 4) to ON.
- The receiving modem then answers the phone and transmits a carrier signal to the initiating end. It also asserts the DCE ready (DSR-pin 6) after a few seconds.
- The initiating modem then asserts the data carrier detect line (DCD-pin 8). The initiating terminal asserts its DTR (if it is not already high). The modem then responds by asserting its data set ready line (DSR-pin 6). The receiving modem then asserts its clear to send line (CTS-pin 5), which permits the transfer of data from the remote computer to the initiating side.
- Data is then transferred from the receiving DTE on pin 2 (transmitted data) to the receiving modem. The receiving remote computer can then transmit a short message to indicate to the originating terminal that it can proceed with the data transfer. The originating modem transmits the data to the originating terminal on pin 3.
- The receiving terminal then sets its request to send line (RTS-pin 4) to OFF. The receiving modem then sets its clear to send line (CTS-pin 5) to OFF as well.
- The receiving modem then switches its carrier signal OFF.
- The originating terminal detects that the data carrier detect signal has been switched OFF on the originating modem and then switches its RTS line to the ON

state. The originating modem then indicates that transmission can proceed by setting its CTS line to ON.

- Transmission of data then proceeds from the originating terminal to the remote computer.
- When the interchange is complete, both carriers are switched OFF (and in many cases the DTR is set to OFF). This means that the CTS, RTS and DCE ready (or DSR) lines are set to OFF.

Note that full duplex operation requires that transmission and reception occur simultaneously. In this case, there is no RTS/CTS interaction at either end. The RTS line and CTS line are left ON with a carrier to the remote computer.

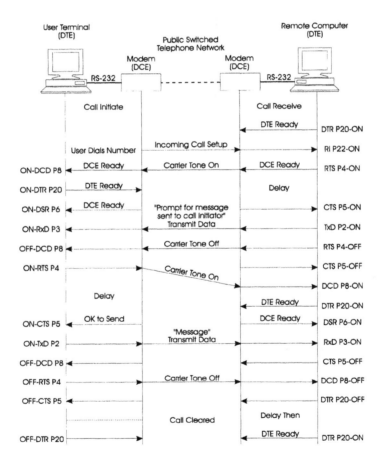

Figure 6.7
Example operation of an EIA-232 data interchange

Break detect

To gain the attention of the receiver, a transmitter may hold the data line in a space condition (+ voltage) for a period longer than that required for a complete character. This is called a 'break' and receivers can be equipped with a break detect to detect this condition. It is useful for interrupting the receiver even when it is in the middle of a stream of characters being sent. Obviously, the break detect time is a function of the baud rate.

6.2.5   Examples of RS-232 interfaces

Although, the RS-232-C interface standard defines only a point-to-point connection on the communications link, there are circumstances when more than one device is required to be connected to the PC. An example of this would be the connection of multiple digital transmitters to a PC that has only one RS-232-C standard communications port. Whilst the RS-232 interface is not designed to be used in a multi-drop system, it is possible to connect the modules to the PC using a daisy-chain network.

In a daisy-chain configuration, each transmit-signal line output is wired to the 'receive input' signal line of the next digital transmitter in the daisy chain. This wiring sequence must be followed until the output of the last digital transmitter in the chain is wired to the receive input of the host. For the daisy-chain network to work correctly all digital transmitters must be set to the same baud rate, they must be uniquely addressable and be able to echo any data received at the receive input to its transmit output. All characters transmitted by the host computer are received by each digital transmitter in the chain, and then passed on until the information is echoed back to the host. Commands sent from the host are examined by all digital transmitters. The device which is addressed, responds by transmitting its response on the daisy-chain network, rippling through any other modules before reaching its destination at the receive input of the host computer.

The daisy-chain must be carefully implemented to avoid pitfalls inherent in its structure. As the daisy-chain is a series-connected structure, any break in the communications link or failure of any of the devices connected will make the network inoperable.

Where the distance between the host computer and several remote devices is beyond the capacity of a normal RS-232 communications link, the network shown in Figure 6.7 can be implemented. The communications link at either end is still RS-232.

6.2.6   Main features of the RS-232 interface standard

The following are some of the main features of equipment that use the EIA-232 interface standard:

- Communication is point-to-point
- They are suitable for serial, binary, digital, data communication (data is sent bit by bit in sequence)
- Most EIA-232-C communications data is in the ASCII code, although that is not part of the standard
- Communication is asynchronous (fixed timing between data bits, but variable time between character frames)
- Communication is full-duplex (both directions simultaneously) with a single wire for each direction and a common wire
- Voltage signals are:
    1. Logic 1: –3 volts to –25 volts
    2. Logic 0: +3 volts to +25 volts
- The communications signal voltages are 'unbalanced', making them more susceptible to noise
- They provide reliable communication up to about 15 m (50 ft)
- Data rates of up to about 20 kbps are possible. In spite of its popularity and extensive use, it should be remembered that the EIA-232 interface standard was originally developed for interfacing data terminals to modems. In the context of modem requirements, EIA-232 has several weaknesses, most of which have arisen as a result of the increased requirements for interfacing other devices such

as PCs, digital instrumentation and other peripheral devices in industrial plants – such as digital variable speed drives, power system monitors, etc.

The main limitations of EIA-232 when used for data communications or data acquisition and control in an industrial environment are as follows:

- The point-to-point restriction is a severe limitation when several 'smart' instruments are used
- The distance limitation of 15 m end to end is too short for most control systems. The 115 kbps rate is too slow for many applications
- The –3 to –25 volts and +3 to +25 volts signal levels are not directly compatible with the modem standard power supplies in computers of ±5 volts and ±12 volts

As a result of these limitations, a number of other communications interfaces have been developed. The RS-422 and RS-485 interface standards are increasingly being used for data acquisition and control systems. The most popular of these, the RS-485 standard (of which RS-422 is effectively a limited version), is discussed in the following sections.

6.3   RS-485 interface standard

The EIA RS-485 is the most versatile of the EIA standards, and is an expansion of the RS-422 standard. The RS-485 standard was designed for two-wire, half duplex, balanced multidrop communications, and allows up to 32 line drivers and 32 line receivers on the same line. It incorporates the advantages of balanced lines with the need for only two wires (plus signal common) cabling.

RS-485 provides reliable serial communications for:

- Distances of up to 1200 m
- Data rates of up to 10 Mbps
- Up to 32 line drivers permitted on the same line
- Up to 32 line receivers permitted on the same line

The line voltages range between –1.5 V to –6 V for logic '1' and +1.5 V to +6 V for logic '0'. The line driver for the RS-485 interface produces a 5 V differential voltage on two wires. For full-duplex systems, four wires are required. For a half-duplex system, only two wires are required.

A major enhancement of RS-485 is that a line driver can operate in three states (called tri-state operation), logic '0', logic '1' and high-impedance. In the high-impedance state, the line driver draws virtually no current and appears to be disconnected from the line. This 'disabled' state can be initiated by a control pin on the line driver integrated circuit. This feature allows 'multidrop' operation where up to 32 line drivers can be connected on the same line, although **only one line driver can be active** at any one time. Each terminal in a multidrop system must therefore be allocated a unique address to avoid any conflict with other devices on the system. RS-485 includes current limiting in cases where contention occurs.

The RS-485 interface standard is very useful for data acquisition and control systems where many digital transmitters or stand-alone controllers may be connected together on the same line. Special care must be taken in software to co-ordinate which devices on the network become active. Where there is more than one slave device on the network, the host computer acts as the master, controlling which transmitter/receiver will be active at any given time.

The two-wire transmission line does not normally require special termination. On short lines however, the leading and trailing edges of data pulses will be much sharper if terminating resistors, approximately two times the characteristic impedance (Z_o) of the line, are fitted at the extreme ends. These resistors are typically between 100 and 500 Ω and reduce the effect of reflections at the ends of the line. For twisted pairs, the characteristic impedance is typically between 100 to 120 Ω.

During operation, there maybe periods of time when all RS-485 line drivers are off and the communications lines are in 'idle'. In this condition, the lines are susceptible to noise pickup, which can be interpreted as random characters on the communications line. A solution to this problem is to incorporate 1 kΩ bias resistors as indicated in the terminated multidrop network of Figure 6.8. These resistors will maintain the data lines in a mark condition when the system is 'idle'.

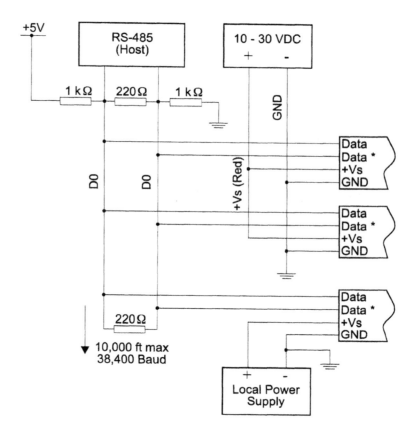

Figure 6.8
RS-485 multidrop networks

6.3.1   RS-485 repeaters

RS-485 line drivers are designed to drive up to 32 receivers on a network. This limitation can be overcome by employing an RS-485 repeater. The repeater is a two-port device that re-transmits data received on one side, at full voltage levels, to the network on the other side. Consequently, another 31 devices may be connected for each repeater used, as shown in Figure 6.9.

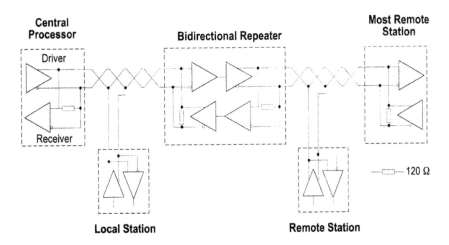

Figure 6.9
RS-485 multidrop networks

6.4 Comparison of the RS-232 and RS-485 standards

The main features of the four most common EIA interface standards are compared below:

Transmitter		EIA-232	EIA-485
Mode of operation		Unbalanced	Differential
Max no. of drivers & receivers on line		1 Driver 1 Receiver	32 Drivers 32 Receivers
Maximum cable length		15 m	1,200 m
Maximum data rate		20 kbps	10 Mbps
Maximum common mode voltage		25 V	+12 V to –7 V
Driver output signal		± 5.0 V min ± 25 V max	± 1.5 V min ± 6.0 V max
Driver load		>3 kΩ	60 Ω
Driver output resistance (high-Z state)	Power 'On'	n/a	100 microA –7V≤Vcm ≤12V
	Power 'Off'	300 Ω	100 microA –7V≤Vcm ≤12V
Receiver input resistance		3 kΩ to 7 kΩ	>12 kΩ
Receiver sensitivity		±3.0 V	± 200 mV –2V≤Vcm ≤12V

Table 6.6
Comparisons of main features of EIA-232 and EIA-485

6.5   The 20 mA current loop

Another commonly used technique, based on EIA-232 but **NOT** part of the standard, is the **current loop**. This uses a current signal rather than a voltage signal. As shown in Figure 6.10 a separate pair of wires is used for the transmitter current loop and receiver current loop.

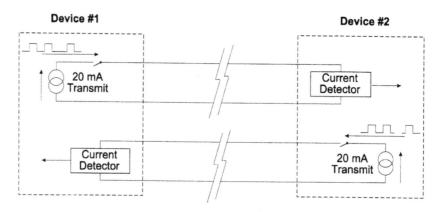

Figure 6.10
The 20 mA current loop interface

A current level of 20 mA, or sometimes 60 mA, is used to indicate a logic-1 and 0 mA, a logic-0. The use of a current signal enables a far greater separation distance to be achieved than with a standard EIA-232 voltage connection. This is due to the higher noise immunity of the 20 mA current loop that can drive long lines (up to 1 km) but at reasonably slow bit rates (up to a typical maximum of 9600 baud). This interface is mainly used between printers and terminals in an industrial environment.

6.6   Serial interface converters

Interface converters are increasingly important today with the movement away from RS-232C to the industrial interface standards such as RS-485. Since many industrial devices still use RS-232 ports, it is necessary to use these converters to interface the device to other network standards. In addition, interface converters are sometimes used to increase the effective distance between two RS-232 devices, especially in noisy environments. The block diagram of an RS-232 / RS-485 converter is shown in Figure 6.11.

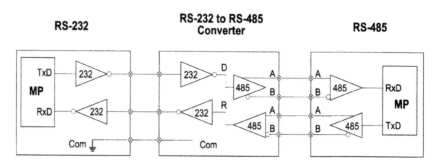

Figure 6.11
Block diagram of an RS-232 / RS485 converter

The RS-232/RS-485 interface converters provide bi-directional full-duplex conversion for synchronous or asynchronous transfer between an RS-232 and RS-485 node. These converters may be powered from an external AC source, typically a wall mounted transformer. Some smaller units can be powered from the handshaking pins 9 and 10 (+12V) of the RS-232 port, although for industrial applications externally powered units are recommended.

When the communications network is operating over long distances, a useful feature of interface converters is optical isolation. This protects both the computer and the equipment, remotely attached to the communications equipment, from power surges picked up over long communication lines.

Typical specifications for an RS-232/RS-485 converter are:

- Data transfer rate up to 1 Mbaud
- DCE/DTE switch selectable
- Converts all data and control signals
- LEDs for status of data and control signals
- Powered from AC source or self-powered from pins 9 & 10 of EIA-232 port
- Optically isolated (optional)
- DB-25 connector (male or female)
- DB-37 connector (male or female)

The signal flow diagram for an RS-232 / RS-485 converter is shown in Figure 6.12.

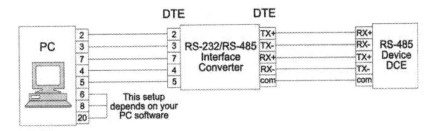

Figure 6.12
Signal flow diagram for RS-232 / RS-485 converter

6.7   Protocols

A protocol is essentially a common set of rules governing the exchange of data between the transmitter and receiver of a communications network, and is normally associated with the packaging of data transmitted on the communications interface.

A protocol is essential to the correct operation of the communication system and determines a number of important features including:

- **Initialization**
  This initiates the protocol parameters and starts the transmission of data across the link.

- **Framing and frame synchronization**
  This defines the beginning and end of a frame and ensures that the receiver can synchronize with the frame.

- **Flow control**
  This ensures that the rate at which the receiver reads the data in from the transmitter is matched and no data is lost.

- **Line control**

  This applies to half duplex links where the transmitter indicates to the receiver when it can 'turn the line around' and commence transmitting in the opposite direction.

- **Error control**

  Typical techniques used here are block redundancy checks and cyclic redundancy checks.

- **Timeout control**

  This applies to the transmitter when it doesn't receive an acknowledgement within a pre-defined period of time and assumes the receiver never received the original message.

Just as there are a number of human languages, there are a considerable number of communication protocols that have been defined at different times by computer vendors and international bodies. Protocols on serial communications hardware linking the PC to data acquisition and control hardware are typically ASCII based. This allows much easier troubleshooting of communication problems where the level of understanding of industrial communication systems may be fairly low. In addition, a high level of integrity of data transfer is required where the PC controls critical equipment over the communications link. In an industrial environment where there may be a lot of electrical noise, a high degree of error checking, such as cyclic redundancy checks, is used. Several of the most commonly used protocols are discussed below.

6 7.1   Flow control protocols

Cooperative flow control, in which the transmitter and receiver operate under a common set of rules, is called a flow control protocol. Below are described the two most popular flow control protocols.

Character flow protocols (XON/XOFF)

This is a popular flow control protocol that has two characters assigned as XON (start) or XOFF (stop). Typically, the ASCII characters DC3 (Ctrl-S) and DC1 (Ctrl-Q) are assigned to XOFF and XON respectively. For example, consider a transmitter, a PC, sending a stream of characters to a printer. When the printer buffer fills to a certain predefined level, say 66%, it transmits an XOFF character back to the PC, instructing it to stop transmitting characters. Once the printer buffer has emptied to a preset level, say 33%, the printer sends an XON character to the PC, which then resumes transmission of the characters. A variation of this protocol is that the PC will resume transmission of the character stream when any character is received from the printer.

Whole line protocols (ETX/ACK)

The ETX/ACK protocol, designed by IBM, is based on the transmitter appending an ETX character after each line of data and waiting for the receiver's ACK, requesting the next line of data.

6.7.2   ASCII-based protocols

The use of ASCII-based protocols is popular because of their simplicity and ease of troubleshooting. Their main disadvantage is that they are slow and unwieldy, especially when the system requires considerable amounts of data to be transmitted at high speeds.

Consequently, the ASCII protocol is normally only used for slow systems with one master talking to a limited number of slaves. ASCII protocols are also popular for stand-alone instruments where a serial interface has been added, with no major design changes, to the existing system. Essentially, this means that the additional serial port is treated like another keypad by the instrument.

Protocol structure

A simple command/response ASCII protocol, used for communications between a personal computer and a digital transmitter is shown in Figure 6.13. The host computer always generates the command sequence. Communications are initiated by using command messages containing the address of the device and a two character ASCII command code. All analog data is returned as a nine-character string consisting of a sign, five digits, a decimal point, and two additional digits. The $ character is used to indicate a request from the master and the * character a response from the slave device. Both the command and response messages are terminated by a [CR] character.

Command from Host

$	1	R	D	[CR]

Response from the Transmitter Module

*	+	0	0	2	7	5	.	0	0	[CR]

Figure 6.13
Short form command and response messages

In this example, the command above reads from the digital transmitter at address 1 and receives a value of 72.10 in the response message.

A variation of the short form command and response messages is their long form equivalents. To ensure greater message integrity, and increase reliability, long form messages are included with a block checksum at the end of the message. In addition, the command message is echoed back within the response message from the slave device. The long form command is initiated using the # character instead of the $ character. An example of long form messages is shown in Figure 6.14.

Command from Host

Response from the Transmitter Module

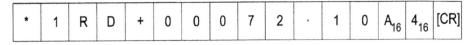

Figure 6.14
Long form command and response messages

The calculation of the block checksum is performed simply by adding the hexadecimal values of all the ASCII characters in the message and converting the resulting hexadecimal number into its ASCII equivalent digits. This is shown in Figure 6.15 below. Note that where the summation is greater than $0 \times FF$, the most significant digit is discarded.

ASCII character	HEX value
*	2A
1	31
R	52
D	44
+	2B
0	30
0	30
0	30
7	37
2	32
.	2E
1	31
0	30
SUM	2A4

Figure 6.15
Block checksum calculation

Errors

If the remote device receives a message with an error it will respond with the ? character. Alternatively, there may be no response at all if an incorrect address or a command prompt has been used. Typical error response messages are indicated in Figure 6.16.

?	1	[SP]	B	A	D	[SP]	C	H	E	C	K	S	U	M	[CR]

?	1	[SP]	S	Y	N	T	A	X	[SP]	E	R	R	O	R	[CR]

Note: [SP] is an ASCII space character

Figure 6.16
Typical error response message for an ASCII based protocol

6.8 Error detection

There are three popular forms of error checking used in many protocols. These are, in order of increasing error-detecting capability:

- Character redundancy checks
- Block redundancy checks
- Cyclic redundancy checks

6.8.1   Character redundancy checks

Character redundancy checks rely on the transmitter and receiver agreeing to use *even* or *odd* parity to calculate the parity bit to append to each character. For example, if *even* parity is defined for a link, the 7-bit data byte ASCII 0110001 becomes 01100011, that is, a 1 is appended to the preceding seven bits to ensure that there are even numbers of 1 in the byte. The receiver checks that the arriving 8-bit byte has *even* parity. If it does, it extracts the first seven bits as data. If the received byte has *odd* parity, the receiver reports an error.

6.8.2   Block redundancy checks

In this method, an additional character called the *block check character* is calculated and added to the stream of characters transmitted down the communications channel.

For example, transmission of the three characters A B Z would have a block check character (BCC) calculated, which is added to the end. The two different techniques for calculating this block check character are indicated in Table 6.7. The techniques are:

Vertical longitudinal block redundancy check

This method relies on the calculation of *even* or *odd* parity for each individual character and then for all the characters in a block. The mechanism is indicated in the following table.

Arithmetic checksum

This checksum is calculated by adding all the bits and then discarding the carry bits. A parity bit is also calculated for each individual character.

Transmitted character	ASCII equivalent	
	Vertical / Longitudinal block redundancy check (even)	Arithmetic checksum
A	10000010	10000010
B	10000111	10000111
Z	10110100	10110100
BCC	10110001	10111101 Carry 1
BCC character	X	]
Transmitted stream	X A B Z	] A B Z

Table 6.7
Two techniques for block redundancy checks

6.8.3   Cyclic redundancy checks

A more effective way of checking for errors is the cyclic redundancy check scheme that has a worst-case error checking ability of 99.9969%. There is thus minimal likelihood of errors slipping through undetected by the receiver.

The mechanism of operation of the CRC is fairly straightforward and is based on the following approach for a typical message which can be of variable length:

- Take the MESSAGE and multiply by 2^{16}.
- Divide (using modulo 2 arithmetic) by an arithmetic divisor (typically the CRC-CCITT which is 1000 100 00001 00 001) to obtain a quotient and remainder. The remainder is the CRC checksum.
- Append the CRC checksum to the message.

The receiver carries out the same calculation and compares the result with the checksum received.

6.9 Trouble shooting & testing serial data communication circuits

When trouble shooting a serial data communications interface, a logical approach needs to be followed, to avoid wasting many frustrating hours trying to find the problem.

A procedure similar to that outlined below is recommended:

- **Check the basic parameters**
  Are the baud rate, stop/start bits and parity set identical for both devices? These are usually set on DIP switches in the device. However, the modern trend is towards using software configuration from a terminal for these basic parameters.

- **Identify which is DTE or DCE**
  Examine the documentation to establish what actually happens at pins 2 and 3 of each device. At the DTE device, pin 2 is used for transmission of data and should have a negative voltage (Mark), whilst pin 3 is used for the receipt of data (passive) and should be at approximately 0 volts. Conversely, at the DCE device, pin 3 should have a negative voltage, whilst pin 2 should be at 0 volts. If no voltage can be detected on either pin 2 or 3, then the device is probably not EIA-232 compatible and could be connected according to another interface standard, such as EIA-422, EIA-485, etc.

- **Clarify the needs of the hardware handshaking**
  When used, this causes the greatest difficulty and the documentation should be carefully studied to yield some clues about the handshaking sequence.

- **Check the actual protocol used**
  This is seldom a problem but, when the above three points still do not yield results, it is possible that there are irregularities in the protocol structure between the DCE and DTE devices.

From a testing point of view, section 2.1.2 in the EIA-232-C interface standard sounds too good to be true. It states that:

'...*The generator on the interchange circuit shall be designed to withstand an open circuit, a short circuit between the conductor carrying that interchange circuit in the interconnecting cable, and any other conductor in that cable including signal ground, without sustaining damage to itself or its associated equipment...*'

In other words, any pin may be connected to any other pin, or even earth, without damage and, theoretically, one cannot blow anything up! This does not mean that the EIA-232 interface cannot be damaged. The incorrect connection of incompatible external voltages can damage the interface, as can *static charges*!

When a data communication link won't work, the following five very useful devices can be used to assist in analyzing the problem:

- A digital multimeter (and paper clips?)
- An LED
- A breakout box
- PC based protocol analyzer (including software)
- Dedicated protocol analyzer (e.g. Hewlett Packard)

6.9.1 The breakout box

The breakout box, as shown in Figure 6.17 is an inexpensive tool that provides most of the information necessary to identify and fix problems on data communications circuits.

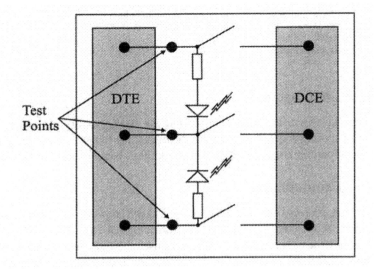

Figure 6.17
Breakout box showing test points

A breakout box is connected into the data cable, to bring out all conductors in the cable to accessible test points. Many versions of this equipment are available on the market, from 'homemade' versions using a back-to-back pair of male and female DB-25 sockets, to fairly sophisticated test units with built in diodes, switches and test points.

Breakout boxes usually have a male and a female socket, and by using 2 standard serial cables, the box can be connected in series with a communication link. The 25 test points can be monitored by LEDs, a simple digital multimeter, an oscilloscope, or a protocol analyzer. In addition, a switch in each line can be opened or closed while trying to identify where the problem is.

The major weakness of the breakout box is that, while one can interrupt any of the data lines, it does not help much with the interpretation of the flow of bits on the data communication lines. A protocol analyzer is required for this purpose.

6.9.2 Null modem

Null modems look like DB-25 'straight-through' connectors and are often used when interfacing two devices of the same gender (e.g. DTE–DTE, DCE–DCE) or devices from different manufacturers with different handshaking requirements. A null modem, as shown in

Figure 6.18, has appropriate internal connections between handshaking pins that 'trick' the terminal into believing conditions are correct for passing data. A similar result can be achieved by soldering extra 'loops' inside the DB-25 plug. Null modems generally cause more problems than they cure and should be used with extreme caution (preferably avoided!).

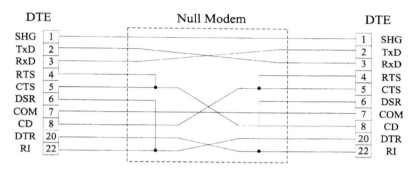

Figure 6.18
Null modem connections

6.9.3   Loop back plug

This is a hardware plug that loops back the transmit data pin to the receive data pin and similarly for the hardware handshaking lines. This is another quick way of verifying the operation of the serial interface without connecting to another system.

6.9.4   Protocol analyzer

A protocol analyzer is used to display the actual bits on the data line, as well as the special control codes, such as X-ON, X-OFF, LF, CR, etc. The protocol analyzer can be used to monitor the data bits as they are sent down the line and compared with what should be on the line. This helps to confirm that the transmitting terminal is sending the correct data and that the receiving device is receiving it. The protocol analyzer is useful in identifying incorrect setting of baud rate, parity, stop bit, noise or incorrect wiring and connection. It also makes it possible to analyze the format of the message and look for protocol errors.

When the problem has been shown not to be due to the connections, baud rate, bits, or parity, then the content of the message will have to be analyzed for errors or inconsistencies. Protocol analyzers can quickly identify these problems.

Purpose-built protocol analyzers are expensive devices and it is often difficult to justify the cost when it is unlikely that the unit will be used very often. Fortunately, software has been developed that enables a normal PC to be used as a protocol analyzer. The use of PCs, as test devices for many applications, is a growing field.

6.9.5   The PC as a protocol analyzer

The PC is proving to be a useful tool for monitoring serial line activity. The general means of configuring such a package is discussed in the following paragraphs.

Basic setup parameters

The initial step is to configure the package appropriately. This requires the following basic parameters to be appropriately set up:

- Baud rate, parity, stop and data bits

- Handshaking requirements (e.g. Request to Send and Data Terminal Ready)
- Base address and interrupt settings of board.

Generally, these items can be saved to a file for future use (where one can easily recall the required configuration).

Other items that require attention are:

Time stamping

This allows the user to put a time value next to all activity on a port. This is especially useful where the protocol analyzer may be receiving data from two different sources at the same time and the exact time of arrival of each data item needs to be determined.

Archiving

Here one can save the data coming in on the serial port, (to a hard disk for example), by specifying the destination of the archive file and giving the length of the archival file.

Trigger pattern setup

This simply means that incoming characters are sequentially scanned against a predefined pattern, called the trigger pattern, until a match is detected. When the correct sequence of characters is detected, the data is saved to a file on the disk. This avoids the user being swamped by the vast quantity of data that can be encountered in protocol work.

Display of the incoming data

This is probably the most commonly used feature of a protocol analyzer. A specific display menu is selected to show the data coming into or leaving the protocol analyzer.

Hardware features

The protocol analyzer is normally used in conjunction with a breakout box to effectively perform any hardware debugging required. Figure 6.19 indicates the typical connections used for a PC based protocol analyzer. It should be noted that a PC should have two serial ports in order to adequately perform two way signal monitoring for two devices, both transmitting and receiving.

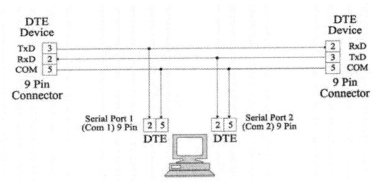

Figure 6.19
Typical PC based protocol analyzer connection

7

Distributed and stand-alone loggers/controllers

7.1   Introduction

As with other forms of data acquisition hardware, stand-alone logger/controllers are designed to measure and record real world signals, as well as to act on these signals to provide control of a system or process. In addition, stand-alone logger/controllers have many features that distinguish their operation and use from other data acquisition hardware, such as plug-in boards and distributed I/O.

This chapter looks at the hardware and software configurations of stand-alone logger/controllers, the system configurations for which they can be used, and the features that allow these devices to meet specific requirements in the field of data acquisition and control.

7.2   Methods of operation

Stand-alone logger/controllers are intelligent devices, capable of performing complex data acquisition and control functions, as well as making decisions based on current system or process conditions. To do this they must first be programmed, typically by a sequence of ASCII-based commands formatted by the host PC, which are interpreted and executed by the device so that it knows what actions to take at any point in time.

Once programmed, the stand-alone device can continue to operate, taking sensor measurements, logging the data to memory and performing control functions, even when the host computer is not connected or functional. From an operational point of view, it is this important feature that distinguishes stand-alone logger/controllers from other data acquisition hardware, such as plug-in boards and distributed I/O.

Two methods of programming – the stand-alone logger/controller, and uploading logged data to the host PC, are available, either by the RS-232 serial communications interface or by using portable and re-usable memory cards.

This flexibility allows stand-alone logger/controllers to be operated in a number of ways, depending on the required location, volume of data to be stored and availability of power:

- Stand-alone operation with periodic data recovery (and programming where required) using memory cards or a portable laptop PC
- On line to a host PC with periodic uploading of data (and programming where required)
- On line to a host PC via modem, with periodic uploading of data (and programming where required), initiated by either the host PC or the remote device

Where an application requires many more sensors than can be provided by a single stand-alone controller, and these are distributed over a large area, a distributed logger/controller network may be required. Each mode of operation, employing only a single logger/controller, is also applicable when more devices are connected as part of a distributed network.

7.2.1   Programming and logging data using PCMCIA cards

The credit card size portable memory card provides a reliable media for transporting data and programs, but requires a memory card interface connected to the RS-232 serial port of the computer. This is shown in Figure 7.1.

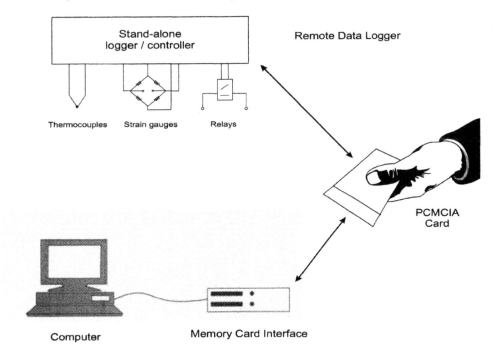

Figure 7.1
Using memory cards to program and log data from a stand-alone logger/controller

Programming the operation of a logger/controller and recovering data using the PCMCIA card is especially useful when the logger/controller is remotely located and/or not connected to a host PC. Even when connected to a host PC, the storage capacity of the logger/controller can be greatly increased by leaving the PCMCIA card permanently inserted in the device. This is its intended use since it also increases system reliability because data is logged directly to a semi-permanent storage medium.

7.2.2   Stand-alone operation

As their name implies stand-alone logger/controllers are specifically suited to be operated independently of the host PC. This makes them especially useful where the device must be located in a remote and/or particularly harsh environment, or where they are unable to be continuously connected to a host PC, either directly or via modem.

Special applications, such as temperature monitoring of a refrigerated truck, or weather reporting at a remote weather station, make use of a logger/controller in a stand-alone configuration. In these real life applications the stand-alone device can be either programmed, in the office or with a portable laptop, then left to operate, powered from a local power supply. Data stored in the device's memory can be periodically uploaded using a portable laptop PC or memory cards.

When operating as a stand-alone device there are several important considerations. Where it is necessary to power the unit from a battery supply, irrespective of whether the battery supply is rechargeable, battery power is not unlimited. This requires that the batteries be either recharged, or replaced where they are not rechargeable. Another important factor is that stand-alone units have a limited amount of memory. The greater the number of channels and the faster the sampling rate on each channel, the greater the number of samples that will be taken in a given time period. In time, the memory will become full. Care must be taken to keep the sampling rate of each channel to the minimum necessary, while still obtaining the information required. The memory capacity of a device can be greatly increased by leaving a higher capacity memory card in the device and logging data directly to the memory card.

7.2.3   Direct connection to the host PC

The most common system configuration, and one which provides the highest system reliability, is a direct connection to the host PC via the RS-232 communications interface as shown in Figure 7.2. This setup allows frequent uploading of data, constant monitoring of alarm conditions and on-line system control. It is most likely implemented in industrial plants or factories, where critical processes must be constantly monitored and controlled. The maximum distance that the logger/controller can be located away from the host PC is dependent on the baud rate of the communications interface. When a single logger/controller is directly connected to the host PC, it can be configured to return data as soon as it is available.

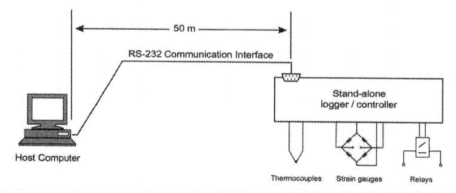

Figure 7.2
A direct connection to a stand-alone logger/controller via an RS-232 serial interface

Where an application requires more than one logger/controller, and each unit is distributed over a large physical area, for example, in an industrial plant or factory, the logger/controllers can be configured as part of a distributed RS-485 multidrop network. A single unit, deemed to be the host unit or local unit, can be connected directly to the host computer via the RS-232 serial interface, as shown in Figure 7.3, thus avoiding any requirement for an RS-232 to RS-485 serial interface card.

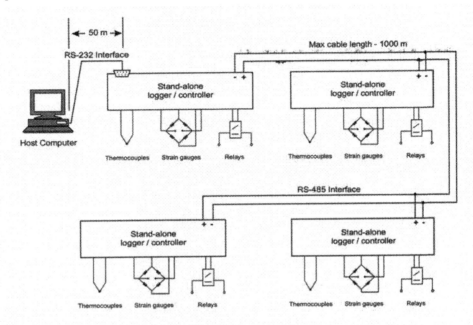

Figure 7.3
Distributed logger/controller network

An advantage of this implementation is that other host PCs, printers or terminals can be connected to the RS-232 ports of other logger/controllers further increasing system reliability. This system configuration is shown in Figure 7.4.

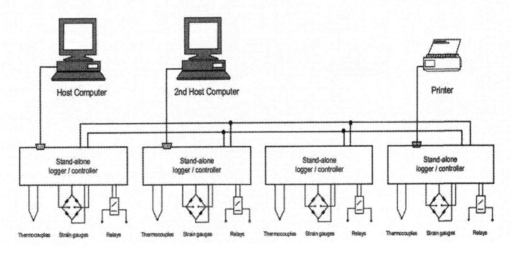

Figure 7.4
Distributed logger/controller network with additional host

How frequently logged data is uploaded depends firstly on how critical the immediate analysis of data is to the system or process being controlled, and secondly on how much memory is available and how quickly it will become full.

How quickly the memory will become full is important for two reasons. During failure of the host PC or communications interface, there must be enough memory to allow data logging to continue without loss of data. In addition, a device connected to the host PC via a multidrop network can only return data when requested by the host PC. Where a large number of units are connected to the host, the memory of each unit must be large enough to allow data logging to continue without loss of data, until the next time the host requests a data upload.

Aside from this specific limitation, it is good practice to recover data as often as possible since any sensor errors, power supply failures or problems with the unit itself will be detected early, thereby increasing system reliability. In addition, frequent data recovery will help to minimize the chance that data may be lost due to device failures such as battery-backed memory failure.

7.2.4   Remote connection to the host PC

Another useful configuration is the connection of remote logger/controllers to the host PC using modems via either a telephone network or radio communications. In large factories or industrial plants, where one or more devices are distributed over a wide physical area, the closest logger/controller to the host PC may be too far away or too greatly affected by noise to allow connection to a host PC via the RS-232 communications interface. In such applications, the use of radio communications is a practical solution. When radio communications take place between the host PC and the distributed network, all communications must go through the logger/controller to which the remote radio modem is connected. This is shown in Figure 7.5.

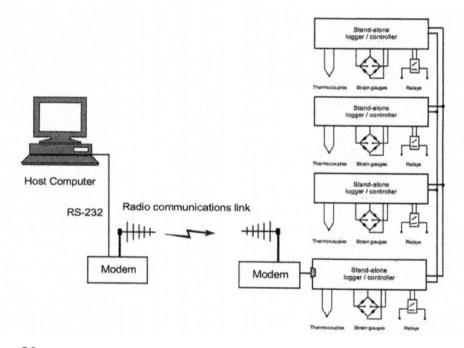

Figure 7.5
Remote connections to a logger/controller network via radio communications

In many applications, the stand-alone logger/controllers are not contained within the same factory or industrial plant, but located at a distance beyond the capabilities of radio modem communications. An example of this would be a remote electricity sub-station used to monitor alarm conditions, provide on-line voltage, current, and power readings to a central control room. Communications between the host PC and the remote units via the telephone network is shown in Figure 7.6. A dedicated phone line allows frequent uploading of data to the host PC, constant monitoring of alarms and on-line system control, where required.

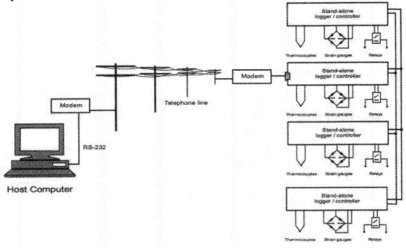

Figure 7.6
A remote connection to a logger/controller network via the telephone network

7.3 Stand-alone logger/controller hardware

The important features that give stand-alone logger/controllers the power and flexibility to operate, either as stand-alone devices, or as part of a distributed network, fundamentally lie in their relatively complex hardware structure. The simplified hardware schematic of a typical stand-alone logger/controller is shown in Figure 7.7.

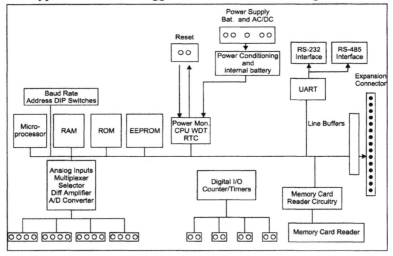

Figure 7.7
Simplified hardware schematic of a stand-alone logger/controller

The following hardware components discussed in this section are:

- Microprocessor
- Memory
- Real time clock
- Universal asynchronous receiver and transmitter (UART)
- Counter/timer circuitry
- Input multiplexer and elector
- Power supply
- Power management circuitry
- Analog and digital I/O circuitry

7.3.1 Microprocessors

At the heart of the stand-alone system is the microprocessor or microcontroller. In conjunction with the embedded software (firmware), it provides the control and functionality of the system.

It is important to clarify the distinction between microprocessors and microcontrollers. A microprocessor is just the central processing unit (CPU) part of a computer, without the memory, input/output circuitry, and peripherals needed for a complete system. The Intel 8088 and 80286 chips are microprocessors. All other chips in the PC are there to add features not found within the microprocessor chip itself. However, when a microprocessor is combined with on-chip I/O, memory and peripheral functions, the combination is called a microcontroller.

The microcontroller is probably the most popular choice for stand-alone systems, as it provides the necessary peripheral functions on chip. The advantages of microcontrollers include reduced cost, a reduction in chip count and hence reduction in printed circuit board 'real estate'.

7.3.2 Memory

Non-volatile memory, used for the storage of sensor measurements and control parameters, is an important feature of a stand-alone system. Typically, random access memory (RAM) is used for data storage and requires some form of battery backup to maintain the contents during power loss.

Manufacturers of stand-alone controllers are now incorporating memory card readers that allow measurement data to be stored directly onto memory cards. The memory card can subsequently be removed and the data transferred to a host computer.

ROM and EPROM

The embedded operating system or firmware of a stand-alone device is stored in either read only memory (ROM) or erasable programmable read only memory (EPROM). Once-only programmable ROM technology is typically used in systems where high volume manufacturing is involved.

EPROM technology is therefore more popular in low to medium volume stand-alone systems, as it allows manufacturers to modify firmware, incorporating new features or enhancements, without committing to the volume requirements of ROM technology. To allow easy installation and replacement of the ROM or EPROM chip during the lifespan of the device, these chips are usually mounted on the circuit board via a socket.

Random access memory (RAM)

Random access memory (RAM) is generally used in stand-alone systems for the storage of measurement data, control, and system parameters. The two most popular types of RAM technology are static and dynamic. Dynamic RAM requires periodic updating or refreshing, whereas static RAM does not require refreshing. However, the advantage of dynamic RAM over static RAM is a far greater memory capacity for a given area of silicon.

Dynamic RAM is suitable for a personal computer used in an office environment where memory capacity is an important requirement. However, in a stand-alone system the advantage of static RAM lies in its ability to maintain the data contents, using a backup battery, in the absence of main power. This can be achieved with relative ease, because static RAM does not require refreshing, even in standby mode.

EEPROM and FlashPROM

Electrically erasable programmable read only memory (EEPROM) is a non-volatile memory technology, generally used for the storage of limited configuration data and control parameters. The moderate memory capacities and slow write cycle of EEPROM (typically 10 milliseconds) limit its application.

Flash programmable read only memory (FlashPROM) is also a non-volatile memory technology, and is used for both mass data and program storage. FlashPROM is available in memory capacities ranging from 32 Kbytes to 2 Mbytes. The much shorter write cycle of FlashPROM is achieved at the expense of having to erase data on the chip in fixed-size blocks rather than a byte at a time.

Memory cards

Similar to RAM, plug-in memory cards are also used in stand-alone systems for the storage of measurement and control data. Although there are a number of memory card manufacturers, the Personal Computer Memory Card International Association (PCMCIA) standard cards, have become very popular for use with notebook computers, particularly PCMCIA modems.

PCMCIA memory cards are available in ROM, one-time programmable ROM, static RAM, UV EPROM, flashPROM and EEPROM technologies. The static RAM PCMCIA cards are the obvious choice for data storage in stand-alone systems, and are currently available in memory capacities ranging from 64 Kbytes to in excess of 8 Mbytes.

An important advantage of memory cards in a stand-alone system is their ability to be removed and replaced with another blank card in the field, providing a convenient data transfer mechanism. Additionally, memory cards allow the user to purchase and install only the memory capacity required for a particular application.

7.3.3 Real time clock

The real time clock (RTC) is an important part of any stand-alone system. It not only provides the necessary date and time information, but also provides periodic and alarm functions for triggering the reading of sensors and controlling outputs under program control.

The RTC will be connected to the associated power management circuitry, allowing the system to remain in standby mode conserving power, until the RTC periodic event or alarm event wakes the system up. The control software is then able to read and record sensor data and manipulate control outputs, before returning the system to the low power standby mode (sleep mode).

In a typical stand-alone data acquisition application, sensor readings are taken at periodic intervals, allowing the system to return to the standby mode conserving power, during these periods of inactivity. For example, sensor readings may only be required once every 500 milliseconds. The RTC would, therefore, be programmed for an alarm wake-up event every 500 milliseconds (see low power mode). The system activity could be reduced to approximately 10 milliseconds in every 500 milliseconds, providing a substantial power reduction, which is very important for battery-operated systems.

7.3.4   Universal asynchronous receiver/transmitter (UART)

The start, stop, and parity bits used for checking data integrity in asynchronous transmission are physically generated by the universal asynchronous receiver/transmitter (UART), located between the microprocessor bus and the line driver, which interfaces to the actual communications link.

The main purpose of the UART is to look after all the routine housekeeping matters associated with interfacing between the parallel bus to the microprocessor and the serial communications link to the host computer.

When transmitting, the UART performs a number of functions. These are to:

- Set the correct baud rate for transmission
- Interface to the microprocessor data bus and accept characters one at a time
- Generate a start bit for each character
- Add the data bits in a serial stream
- Calculate and add the parity bit to the data stream
- Terminate the serial group with the required stop bit(s)
- Advise the microprocessor it is ready for the next character

The receiving circuitry of the UART performs the following functions. These are to:

- Set the correct baud rate for receipt of data
- Synchronize the incoming data with the start bit
- Read the data bits in a serial stream
- Read the parity bits and confirm the parity
- Read the stop bits
- Transfer the character as a parallel data onto the microprocessor data bus
- Interface the handshaking lines
- Observe and report any errors associated with the character frame received.

Typical errors that the UART can detect are:

- Receiver overruns – bytes received faster than they are read
- Parity errors – mismatch between parity bits and character frame
- Framing error – all bits in the frame are zero or a break condition is reported

A break condition occurs when the transmitter that holds the data line is in a spaced (or positive voltage) state for a period of time longer, than that required for a complete character. This is a method of getting the receiving UART to react immediately and perform some other task.

7.3.5   Power supply

Due to the nature of their operation and the purposes for which they can be used, stand-alone devices have a variety of power sources:

- Low voltage AC (9–15 V AC)
- Low voltage regulated DC (11–17 V DC)
- 9 V alkaline battery (6–10 V)
- 6 V gel cell battery (5.6–8 V)

A simplified power supply schematic of a typical power supply circuit is shown in Figure 7.8.

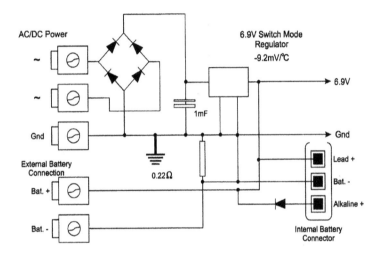

Figure 7.8
Simplified power supply schematic

When both an internal non-rechargeable alkaline battery and an external AC or DC power supply are connected, the output of the regulator is increased to a voltage greater than the alkaline battery voltage (i.e. 10 V), so that power is drawn from the external supply and not from the internal battery. In this situation, the in-line diode connected to the alkaline + terminal prevents charging of the alkaline battery.

It is not recommended that both internal and external batteries be connected. If two batteries are required, it is better if the external battery is 12 V and connected as external DC power.

Extreme care should be taken to ensure that external batteries are connected with the correct polarity otherwise, damage might occur. In addition, when an external DC supply is grounded it must be a negative (–) ground. AC supplies should never be grounded.

Battery charging

Where an internal gel cell rechargeable battery is connected, an external AC or DC power supply can provide temperature compensated charging with voltage set by the output of the switch mode regulator and charging current limited by the 0.22 ohm charging resistor. Sealed gel cell batteries may also be charged via a 12 V solar panel connected to the AC/DC power input terminals. The size of the solar panel required depends on the hours of full sunlight that can be expected. As a rule, one day in seven should be regarded as a charge day and the charge must be able to fully replenish the batteries on that day.

Battery life

The maximum battery life that can be achieved depends on:

- How often the input channels are scanned
- The number of analog channels and how many are connected to sensors
- The number of digital channels and how many are driving outputs
- Sensor excitation power draw
- Complexity of any calculations

A precise calculation of the battery life is extremely involved, however manufacturers can provide battery life charts, based on the number of channels and the time between each scan of all the channels.

7.3.6   Power management circuitry

All microprocessor systems need some supervisory functions that are analog rather than digital in nature. For a typical system, these functions include power reset, battery backup switching for RAM and real time clock, and the watch dog timer (WDT).

Reset circuitry

The reset circuit ensures that the microprocessor is in the reset state in the absence of power. Embedded systems that may need to function in hostile environments require advanced reset circuitry that provides voltage threshold detection, independent of the rate of rise of the system power.

Battery backup

The battery backup circuit ensures that the RAM and real time clock components receive a constant source of power. It also ensures that the RAM and real time clock are write-protected and in the low power mode, when the main power supply fails.

If the voltage from the main power supply falls below a preset level, then the battery backup circuit switches the RAM and real time clock power source to a supplementary battery supply. Additionally, the battery backup circuit ensures that the RAM and real time clock are write-protected, and in the low power sleep mode.

Low power mode

Two power states, wake and sleep (standby), ensure that the minimum amount of power will be used when the unit is not required to perform any data acquisition function. For example, the device will wake up when:

- RTC periodic interrupt signals a scan of the input channels is due
- A memory card is inserted
- Characters are received at the communications port
- A key is pressed (where fitted)

When an internal or external battery is being powered from an AC or DC supply then the low power sleep mode is not required to be entered.

Watch dog timer

The watch dog timer (WDT) circuit is intended to detect software-processing errors. During normal operation the software is responsible for periodically resetting the WDT. Failure to reset the WDT on a periodical basis indicates that the software is no longer executing the intended sequence, and accordingly the WDT initiates a system reset.

In essence, the WDT is a fail-safe mechanism that resets the system if for some reason it 'runs off the rails'. Although the WDT would seem a perfect fail safe mechanism for static or electrical noise induced problems, there is still the possibility of the software erroneously entering a loop, which continuously resets the WDT and hence never initiates a system reset.

7.3.7 Analog inputs and digital I/O

Analog input channels

Logger/controllers typically have multiple analog input channels. A special feature of these devices is that each channel can be configured for operation with a variety of sensors and signals. The simplified schematic of a typical input channel is shown in Figure 7.9.

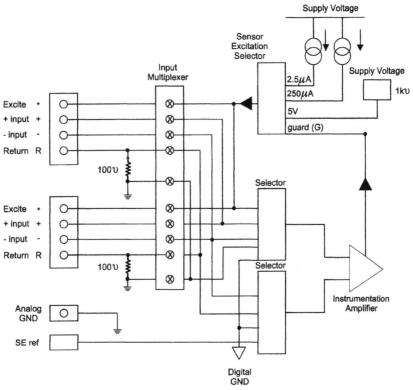

Figure 7.9
Simplified schematic of analog input channels

The versatility, that allows each channel to be configured for a wide range of sensors, different excitation requirements and either differential or single-ended input terminations, is provided by the analog signal selector. The configuration of each channel is provided by software commands that are interpreted by the logger/controller to switch the analog signal selector to the required settings.

Sensor excitation is typically provided in the form of a low level constant current source (250 µA), for measuring resistance, a higher level constant current source for RTD and Wheatstone bridge measurements or a voltage source (usually unregulated) via an internal resistance, useful for powering some sensors.

Input termination resistors, typically of 1 MΩ can be switched into the circuit to provide a return path for instrumentation amplifier bias currents. Where the termination resistors are not switched into the circuit, the input impedance seen by the sensor is of the order of 100 MΩ.

Digital I/O channels

Logger/controllers typically have multiple dual-purpose digital I/O channels that share the same terminations and act as both digital inputs and outputs. This is shown in Figure 7.10.

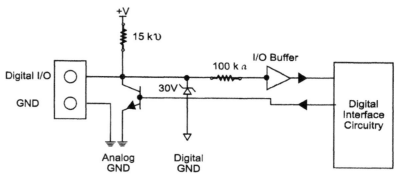

Figure 7.10
Schematic of digital I/O channel

Digital inputs have a high impedance input resistance and are buffered to protect the more sensitive CMOS digital interface circuitry from damage from current surges. A 30 V zener diode provides input over-voltage protection by limiting the incoming voltage to below the transient voltage threshold of the input buffer.

The most commonly implemented form of digital output available on stand-alone loggers/controllers is the open collector output configuration capable of sinking 200 mA at 30 V. In this configuration, the zener also acts as a voltage limiter when the channel is used as an open collector output.

The schematic of a typical digital counter channel is shown in Figure 7.11. Counter input channels are provided with a Schmitt trigger input buffer with threshold voltage set to 2 volts. This prevents spurious noise below the threshold value causing a count transition. The capacitor at the input to the Schmitt trigger input buffer provides filtering but limits the count rate to approximately 1 kHz (=1/RC). When it is removed, the count rate can be as high as 500 kHz.

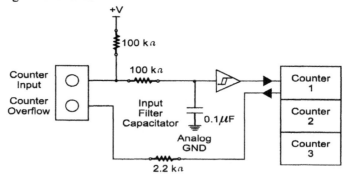

Figure 7.11
Schematic of digital counter channel

7.3.8   Expansion modules

Expansion modules provide increased localized channel capacity for a data acquisition system using stand-alone logger/controllers. Expansion connectors provide an extension of the internal data and control bus lines of the logger/controller. When connected, additional analog input channels, digital I/O channels, and counter channels, on the expansion module, are treated as if they were part of the logger/controller to which they are attached. This is shown in Figure 7.12.

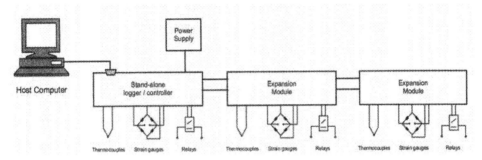

Figure 7.12
Connection of expansion modules

7.4   Communications hardware interface

Communications interface standards define the electrical and mechanical details that allow communication equipment from different manufacturers to be connected together and to function efficiently. Two standards are commonly employed for communications between PCs and stand-alone or distributed logger/controllers:

- RS-232 standard
- RS-485 standard

7.4.1   RS-232 interface

As the RS-232 communications interface is standard on most IBM PCs and compatibles (i.e. COM1 and COM2 ports), stand-alone devices first used this interface for communication to the PC. The RS-232 interface is discussed in detail in Chapter 6, however some of the most basic setup parameters for stand-alone logger/controllers are discussed below.

Comms port parameters

Usually, the comms port parameters (i.e. start bits, data bits, stop bits) are fixed. The only user setable parameter is the baud rate, which is typically set by dip switches on the device. Commonly used baud rates are 300, 1200, 2400, 4800, 9600 and 19200. The optimum baud rate setting is a compromise between the speed necessary to transmit the necessary amount of data over the communications interface, and the speed required so that data can be transferred without error over the distance that the host PC is located from the local stand-alone logger/controller.

Communications port connections

Stand-alone loggers/controllers are typically equipped with a DB-9 connector. The pin allocations commonly used with the DB-25 and DB-9 connectors for the RS-232C interface are not quite the same and often provide a trap to the beginner.

Figure 7.13 shows the standard connection between both the DB-25 and DB-9 connector of the host PC and the DB-9 connector of a stand-alone logger controller.

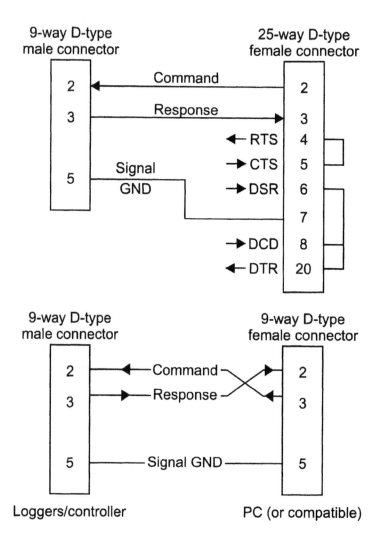

Figure 7.13
Communication cable connections to IBM host PC

Great care should be taken when connecting stand-alone devices to the host PC. While the standard pin configuration is likely to be adhered to by manufacturers at the PC's communications interface, it is possible (and often likely) that the data receive and transmit lines on remote stand-alone systems are on different pins of the DB-9 connector. It is therefore wise, to consult the manufacturers' data sheets.

Data handshaking is not usually employed on stand-alone logger/controllers as their use often leads to communication problems. Instead, the handshaking lines are connected at the host PC, as required by the communications software, and left unconnected at the logger/controller interface.

7.4.2   RS-485 standard

With a growing need for a distributed logger/controller network, RS-485 interfaces have been added to the hardware. The RS-485 interface typically operates as a balanced two wire, half duplex and un-terminated network (see Chapter 6). However, the protocols used for communications on the network are often proprietary, with different manufacturers using undisclosed protocols, and error detection/correction methods, between devices. This does not alter the fact that communications to devices on the RS-485 network still occur via a single logger/controller, known as the local device, which communicates to the PC via the RS-232 interface.

An RS-485 repeater is used where more than 32 stand-alone devices are required on one network. A further 32 devices may be connected for each repeater used.

7.4.3   Communication bottlenecks and system performance

When logger/controllers are constantly logging data to memory, data gathered can be sent to the host PC via the communications interface at any convenient time before the memory becomes full. This allows great flexibility in obtaining the data from a stand-alone device or a network of devices. However, when operating in real time, that is, data is continuously returned to the host PC from a single stand-alone logger/controller or a network of logger/controllers, an important consideration is 'can the volume of data obtained, be transmitted over the serial communications link?' This depends on a number of factors:

- Baud rate
- The number of channels being scanned
- How often the channels are scanned
- Whether the device is stand-alone or part of a distributed network

Stand-alone logger/controller

Consider first, a stand-alone logger/controller connected to the host PC via the RS-232 interface operating at 9600 baud. As we have seen previously, data sent over the communications interface is sent in a 10-bit frame consisting of 1 start bit, 8 data bits and 1 stop bit. The time to transmit each byte of data at 9600 baud is 1.042 ms (t = 10 bits/9600).

Therefore, to transmit the maximum amount of data at the required baud rate, the maximum time between each data byte, being ready to be sent, is 1.042 ms.

Consider a logger/controller that is scanning 10 channels. If, for each channel, seven bytes of data are sent (on average), plus there are another ten bytes for each scan of the input channels, then the total number of bytes to be sent for each channel scan is 80 bytes. The maximum time each channel scan could take is 83.36 ms (80 bytes × 1.042 ms). Therefore, all channels could be scanned at the maximum rate of approximately 12 Hz (1/83.36 ms).

This calculation assumes that there are no hardware factors, such as multiplexer settling time, input amplifier-settling time etc, which limit this rate even further. Irrespective of the performance limitations imposed by the stand-alone communications interface, logger/controllers are not designed for high-speed data measurement.

Distributed logger/controller network

When operating as part of a distributed network, the considerations that determine the performance of the system are different. Despite RS-485 being an extremely reliable interface, even at high speed, the potential speed at which the network can operate is limited by a number of factors:

- Each device in the system has a unique address and must be polled by the host computer for information.
- Only one device can be polled at any time.
- As the RS-485 network is half-duplex, the host PC must wait for a response to each request for data before polling the next device.
- There is an inherent delay, or turnaround time, in responding to a host request irrespective of whether one byte or one hundred bytes of data are returned in the response. This is because the device must interpret, process, then act on the command received before returning its response.
- Where the RS-485 network is operating much faster than the RS-232 interface (i.e. more than twice as fast) the potential speed at which the data can be sent to the host PC is limited by the baud rate of the RS-232 interface. Where this is not the case, the baud rate of the RS-485 network is the limiting factor.

It has been shown in the example for a stand-alone device, that a device scanning 10 input channels and returning 80 bytes of data to the host PC for each scan, would take 83.36 ms to transmit the data. If the time required to transmit a ten-byte poll command is 10.42 ms, and assuming a turnaround time of 1 sec, the total time to obtain the data from a single device is 1093.78 ms. The time taken for 10 devices operating in the same manner would be approximately 11 sec. Clearly, the system would not be able to operate in real time unless each device scanned its input channels and returned data to the host once every 11 sec.

Where the channels of one or more of the devices must be scanned at a faster rate, the data should be logged to memory and returned at a more convenient time.

7.4.4   Using Ethernet to connect data loggers

Data loggers have traditionally used RS-485 as a type of networking system. RS-485 works very well for multi-dropping up to 32 data loggers. As requirements have expanded in the plant environment, we have seen a need to connect data loggers to expanding networks of systems. This has seen the rise of data loggers being connected via existing Ethernet networks. The advantages are obvious. One of the main problems with connecting data loggers on an RS-495 network is the limited range of access to the system. By having access to the data loggers over the Ethernet, the user can view and even change data anywhere the network is connected. This brings rise to the use of data loggers using the Internet. Intra- and inter-networking of data loggers also raises security issues. How safe is the data? Will someone that does not have authorization be able to access the hardware? These problems and their solutions will be the subject of much discussion when it comes to connecting Ethernet to data loggers.

7.5   Stand-alone logger/controller firmware

The hardware represented by stand-alone, or distributed loggers/controllers, and used as a part of a data acquisition system, provides the physical interface, which allows the PC to measure data from and control real-world signals. The software that is stored and executed from the ROM or EPROM of the stand-alone device, and known as the firmware, controls the continuous operation of the stand-alone device. However, the firmware does not initiate any data acquisition and control functions by itself. Instead, the firmware can only interpret and execute the commands it receives from the host PC so that it knows what actions to take at any point in time.

The firmware performs many functions including:

- Overseeing the correct operation of all peripheral hardware devices (i.e. memory card, display, keyboard)
- Interpreting, checks for error, then acting on commands received via the communications interface or from memory cards
- Sending responses to the computer via the communications interface, including any errors that occur in the communication of the command and in the device itself
- Performing the necessary data acquisition and control functions as specified by the programming commands received from the host PC

The firmware for a stand-alone device is often upgraded by the manufacturer to provide new features and enhancements, and in some cases, bug fixes. Where remote stand-alone devices are operated within an RS-485 network, it is advisable that each unit runs the same version of firmware.

The revision of the firmware is often shown on the local display upon power-up of the device. Where this is not the case, a system command to determine the firmware revision is usually provided.

Quality manuals provided with remote stand-alone devices will include a firmware change history, with the revision numbers and brief description of the changes made with each revision. This allows the user to identify problems that are consistent with a previous revision of the software.

7.6 Stand-alone logger/controller software design

The power and flexibility provided by stand-alone logger/controllers has generically resulted in the hardware, and consequently the firmware that controls its operation, becoming necessarily complicated. This however, does not mean that the commands used to instruct stand-alone devices, need to be complicated. In fact, from a programming point of view it is beneficial to keep the basic command and data structure simple and readable. To this end, a simple ASCII-based command structure is commonly implemented. ASCII-based command and data response formats are popular because of their simplicity, especially for stand-alone systems where a serial interface has been added with no major design changes to the existing system. Essentially the additional port is treated like a keypad by the stand-alone device.

Depending on the particular application, some of the tasks stand-alone devices are required to perform are as follows:

- Take measurements from sensors at time intervals determined by the user
- Make the measurements from sensors conditional on certain events or environments
- Adjust the sensor measurement rate so that readings are taken more frequently during conditions of greater interest
- Mathematically combine and manipulate sensor readings
- Apply statistical procedures to reduce the number of readings that need to be stored

- Use different data formats so that the data transmitted to the host PC is suitable for a computer program (i.e. spreadsheet package) or a human operator
- Store sensor measurements within the device or onto memory cards for later transmission to the host PC
- Transmit measurements back to the PC as soon as they are taken. This can be by direct connection over the RS-232 interface, by modem through the telephone network, or by radio modem over a radio link
- Control equipment external to the stand-alone device

This section outlines the basic software protocol (command formats, data formats and error formats) required to program stand-alone logger/controllers as well as detailing the type of commands required to perform some of the tasks outlined above.

It is not our purpose here to detail the exact software structure of a particular stand-alone logger/controller, as this will vary between manufacturers. However, to provide an example, the most important commands, and their use in a program to control the Datataker range of loggers/controllers, by Data Electronics Australia Pty Ltd, is shown in Appendix H.

7.6.1 ASCII-based command formats

Irrespective of whether commands are sent to stand-alone devices via the serial communications interface or using memory cards, the format of the commands is the same. The PC always generates the command sequence. Programming the operation of the remote devices or reading data from them usually requires the user to enter ASCII command strings only. Commands are sent one at time by using a command terminating character after each command. For terminal emulation packages, this is typically the carriage-return (ASCII 0D). Multiple commands can be included on one line by separating them with a delimiter, typically a tab or space character, then terminating the command string with the command-terminating character.

Although the command format is not standardized between manufacturers of different stand-alone devices, several formats are commonly used. These are:

Word commands

These are entered as continuous ASCII text, most commonly in upper case characters and also contain one or more command options that specify the functions the command is required to perform. Although they vary between different manufacturers, command options are typically enclosed in brackets, separated by commas (no spaces), and can be random.

Sometimes command options are referred to as command parameters since the user is required to append parameters that specify particular values associated with a command. These should not be confused with parameter commands.

Switch commands

These are used as an on/off control function that either enables or disables a particular feature or function of a stand-alone device, thereby controlling its operation. The feature 'X' is enabled/disabled by sending the following switches:

/X   feature enabled
/x   feature disabled

Parameter commands

These commands set a value internal to the device that the user may only want to set once, or at least not very often. Parameter commands have a global effect, can be set at any time, take effect immediately and allow the user to set different performance options. A parameter command can often have a wide range of values within the parameter's limits for a particular device.

Where the stand-alone logger/controller is connected as part of a network, all commands must be preceded by the address of the device for which the command is intended.

7.6.2   ASCII-based data formats

All data returned to the host PC from stand-alone devices is in simple ASCII text format. The format of the ASCII string is entirely configurable by the user and will be sent in the format in which the stand-alone device was requested to send it. Data is typically presented in two mathematical formats:

- Floating point format with 'n' (user configurable) significant digits
- Exponential format with 'n' (user configurable) significant digits

In addition, the ASCII string can be made more readable by the addition of the following text:

- Units applicable to the measurement being taken (e.g. mV, Ohms, Hz etc)
- The channel number and type of signal being measured or type of sensor being used (e.g. 1 V, 2 LM35)
- A channel ID string (e.g. boiler temperature # 1)
- The time and date indicating when the reading was taken (e.g. 10:30 12/12/99)

Where the data is to be imported into a spreadsheet software package, it can be formatted so that it contains no additional text except the time (HR:MIN:SEC format) and date (DD/MM/YYYY format), is delimited by commas (ASCII 2C), and each line of data is terminated by carriage return (ASCII 0D).

If the stand-alone logger/controller is connected as part of a network, the data that is returned can be configured to include the address of the device from which the data was returned.

7.6.3   Error reporting

When an error occurs, either in sending a command to the stand-alone device or in performing some function, the error can be reported by returning error messages to the host PC.

Although not differentiated on all makes of stand-alone device, three types of errors are commonly recognized:

- Command errors
- Channel errors
- Operational errors

Command errors

Command errors are reported immediately after a command has been sent to the stand-alone device. They indicate that all or part of the command was either not understood because of incorrect transmission from the host PC, the command contained an incorrect syntax or alternatively the command itself was not actionable. If a command refers to a set of channels, some of which are not used or are an incorrect type, then the appropriate error response is generated but the command is executed for those channels to which it applies.

Channel errors

Channel errors are reported when an error occurs in the measurement from a particular channel. The method of reporting channel errors often varies, depending on the stand-alone device. As well as reporting an error, some devices return a default data error value (i.e. 99999.9) in the logged or displayed data. An example of this type of error is when a channel is read and the analog signal on that channel is out of range.

Operational errors

Operational errors occur when an operational limitation prevents the correct execution of a command. For example, an operational error occurs when the command input buffer becomes full. This can be due to the command being too long or successive commands being sent too quickly. An operational error would also occur when the data memory became full.

Error format

The error message is typically sent in some shorthand ASCII notation, but can be sent in a more descriptive (verbose) ASCII form. In all cases the error format is determined by the configuration in which the remote device was requested to send it. The format is usually set by a command switch. It should be noted that error messages can be turned off – typically by a command switch.

7.6.4 System commands

System commands are used to perform system initialization, hardware initialization, variable initialization, system parameters setting (such as the time, date and password), or return the system status.

7.6.5 Channel commands

The versatility and simplicity of use of stand-alone devices lies in the fact that many different types of sensor can be interfaced to the input channels of the device. In most cases it does not matter to which channel a sensor is connected, provided the device is informed which sensor is connected to each channel. The rate at which each channel is sampled is entirely variable. Some channels can be configured to be sampled continuously, while others take a measurement only when certain operating conditions are met. Once instructed to perform a data acquisition or control task (or many complicated tasks), the remote device will continue to operate by itself, taking measurements, storing data if required, or sending data back to the host PC.

Channel commands allow the user to modify specified channels for:

- Input configuration
- Sensor excitation
- Defining channel constants such as resistive shunts and attenuation factors
- Identifying reference channels for thermocouples, Wheatstone bridges
- Scaling of the channel data by spans, polynomials, factors specifying statistical analysis and histogram extraction
- Specifying progressive difference, rate of change, integral assignment of channel data to temporary storage registers
- Assigning unique names to channels
- Specifying format and resolution of the channel data
- Specifying whether data is returned to the host PC, logged or displayed locally.

Some of these channel commands are discussed in more detail below.

Channel excitation

As we have shown, many sensors require some form of excitation in the form of a voltage supply or constant current source, to enable them to output a signal. Excitation channel options inform the stand-alone device of the excitation that is required for a particular channel.

Statistical channel commands

Channels can be read frequently but produce a statistical summary at longer intervals. This summary is returned, logged or displayed at intervals determined by the pre-defined schedules. Channels that require statistical sampling must include a channel option to indicate the statistical information to report. The following statistical channel options are typically available:

- **Average**
  The sum of all the channel readings divided by the number of readings since the last statistical report. It is very useful in reducing sensor noise by averaging out cyclical noise, such as mains hum.

- **Standard deviation**
  This is the measure of variability of data concerning the average. The variation may be due to noise or process changes.

- **Maximum and minimum**
  This returns the maximum or minimum for the scan period and the time and date that it occurred.

- **Integral**
  This returns the integral (or area under the curve) with respect to the time in seconds using a trapezoidal approximation. The units of integration are those of the original reading multiplied by seconds.

Channel data manipulation

Data manipulation and scaling of data read from a particular channel can be performed automatically before the data is stored, by using some of the following utilities:

- **Channel scaling**
This automatically multiplies the value read by the fixed channel-scaling factor.

- **Intrinsic functions**
These are mathematical functions applied to the data read, for example:

  - Inverse (1/x) of the data
  - Square root ($\sqrt{x}$) of the data
  - Natural logarithm (1n[x]) of the data
  - Base ten logarithm (log[x]) of the data
  - Absolute value [x] of the data
  - Square (x*x)

- **Spans**
These allow sensors with linear calibrations to be converted to engineering units. The end points of the span are defined by the user and the linear calculation of input signal values performed automatically. This is the same as applying a first order polynomial $y = a + bx$ to the input value x, and is particularly suited for 4–20 mA current loops.

- **Polynomial equations**
Linearization of non-linear data can be performed using the '*n*' order polynomial equation shown in Appendix E.

- **Channel variable storage**
Internal variables are used for temporary storage of the readings taken from one or more channel. These readings can then be added to the readings from other channels or mathematical calculations applied before the data is logged, displayed, or returned to the host.

- **Mathematical and logical calculations**
Mathematical expressions containing arithmetic operators (+ = * / %), relational operators (< > < = etc), logical operators (AND, OR, XOR, NOT) and trigonometric functions (SIN, COS, TAN etc) can be applied to the value read from the channel.

7.6.6   Schedules

A schedule is a list that tells the remote device which channel or number of channels data is to read, as well as the method by which the reading of data on each channel is to be triggered. When a trigger event occurs, all channels listed in the schedule are scanned, and depending on the type of schedule, the data logged, displayed or returned to the host PC.

Schedule triggers

Three different schedule triggers are typically defined:

- Trigger by interval
- Trigger on event
- Trigger when condition is true

Trigger by interval

When using this method, the stand-alone device is programmed to take data readings on each channel at a specified scan interval. The format of the scan interval can be every x seconds, minutes, hours or days. Where no interval is defined, scanning will occur as rapidly as possible.

Trigger on event

When using this method the occurrence of an event on a pre-determined digital input or counter input is used to trigger the scanning of all channels in the schedule.

The trigger events commonly used are:

- Trigger on positive voltage transitions (low to high) of a digital input
- Trigger on negative voltage transitions (high to low) of a digital input
- Trigger on both positive and negative transitions of a digital input
- Trigger after 'n' counts of a counter input

When schedule scanning occurs subject to a digital event, the stand-alone device must be constantly checking the required digital inputs or counter inputs for the event to occur. This requires that the device must not go into a low power sleep mode. As there are limitations on the speed at which the device can do a check scan of the required digital or count inputs, care must be taken to ensure that the speed of the trigger event is not faster than the check scan rate, otherwise events may be missed.

Trigger when condition is true

In addition to the various methods of time interval triggering or event triggering a schedule channel scan, it is often possible to enable/disable the trigger event using the state of one or more other digital inputs. This would be very useful, for example, if an alarm limit was reached and this condition was used to enable the trigger event that activated the channel scans.

Types of schedule

Five types of schedule are typically defined:

- Immediate schedules
- Report schedules
- Polled schedule
- Statistical schedules
- Alarm schedules

Immediate schedules

Immediate schedules are used for inspecting and testing input channels and sensors. When this schedule is triggered, either by a command from the host PC or an alarm condition becoming true, the designated channels are scanned only once and data is returned to the host only. The execution of an immediate schedule channel scan does not disrupt any report schedules that may be in progress and have a higher designated priority.

Report schedules

This type of schedule is used for repeated data acquisition from selected input channels and forms the program building blocks of the stand-alone logger/controller program.

The channel scan for this type of schedule can be triggered by

- Time intervals from 1 second to months
- Digital events (conditional on digital state)
- Counter events (conditional on count value)

The data collected from report schedules can be returned to the host, logged or displayed. Stand-alone devices typically define several reporting schedules (e.g. RA, RB, RC, RD, RX), one of which can be used only for host requests (RX).

Polled schedule

A polled schedule that is executed only in response to a special command is assigned a unique identifier. This schedule will only occur when:

- The host sends the host request command
- An alarm condition issues the host request

Statistical sub-schedules

Statistical sub-schedules are used for repeated data acquisition from input channels at short intervals, but produce statistical data such as the average, minimum, maximum etc at longer intervals. One or more statistical sub-schedules, each with a unique schedule identifier, are defined. By default, statistical sub-schedules are scanned at the maximum possible rate unless a user-defined trigger is applied.

The channel scan for this type of schedule can be triggered by

- Time intervals from 1 second to months
- Digital events (conditional on digital state)
- Counter events (conditional on count value)

Channels that are to be statistically sampled must include one or more channel options to indicate the statistic(s) that are required.

Alarm schedules

Alarm schedules determine the rate at which one or more channels will be scanned to check if an alarm condition has been reached (see Alarms, below). Multiple alarm schedules, each with a unique schedule identifier, are defined. By default, alarm schedules are scanned at the maximum possible rate unless a user-defined trigger is applied.

The channel scan for this type of schedule can be triggered by:

- Time intervals from 1 second to months
- Digital events (conditional on digital state)
- Counter events (conditional on count value)

Controlling schedules

When a stand-alone device is connected to a host PC, added flexibility can be achieved by allowing the user to stop, and then resume an individual schedule or all of the schedules, as required. This allows the user to temporarily halt the data acquisition process to check and/or change system parameters or the schedule triggers.

7.6.7   Alarms

Alarms are used to warn of error conditions in an application, allowing the user to make decisions when input channel signals, timers, the time and variables exceed specified alarm limits. Alarms are multi-functional and allow:

- Logical comparisons with set points. The conditional tests available are:
  - less than (<) setpoint (low alarm)
  - greater than (>) setpoint (high alarm)
  - outside the range (<>) of two setpoints (high low alarm)
  - inside the range (><) of two setpoints (in range alarm).
- Control of digital output channels based on the alarm condition, to turn on alarm lights, control relays, etc.
- Issuing of messages to the host PC or local display that may include alarm data (i.e. alarm type) and alarm time and date.
- Issuing of commands to control operation of the device.

The manner in which alarms annunciate the resultant actions to be performed is often configurable, thereby providing flexibility for the user in meeting the requirements of a particular system or process.

Alarms can annunciate actions in the following manner:

- Alarm actions are performed only once when the alarm state first becomes true.
- Alarm actions are performed only once and only when a predefined delay has elapsed after the alarm condition is first met. When a delay period is defined no action is taken unless the delay period has elapsed and the alarm state has not changed during this period. This acts as a filter to prevent nuisance alarms and unnecessary or rapid actions on digital outputs, which may be caused by noise.
- Alarm actions are performed repeatedly while the alarm condition is true.
- Alarm actions are performed repeatedly while the alarm condition is true after a pre-defined delay.

Controlling alarms

When a stand-alone device is connected to a host PC, added flexibility can be achieved by allowing the user to stop, and then resume an individual alarm or all of the alarms, as required. This allows the user to temporarily halt a particular alarm, and to allow sensor, or actuator maintenance, without disrupting an application.

7.6.8   Data logging and retrieval

Storing data

There are two places to store data. The first is the internal memory, the second (where applicable), is the optional higher capacity memory card. The amount of data that can be stored is dependent on the memory capacity and/or the capacity of the memory cards, as well as the format of the data that is stored.

If an empty memory card is inserted into the stand-alone device, then all data in internal memory is transferred to the memory card and logging continues to the end of the card memory. If the memory card is removed, logging continues to internal memory. When a partially full memory card is inserted then logging continues to internal memory.

How data is stored

Data is logged in a fixed non-ASCII format (i.e. 24-bit floating point format) to save space. A fixed length header at the start of each schedule scan is used for identification, time and date. When the data is unloaded, the identification header is used to interpret the data and add the required information for the user. This is why schedules cannot be overwritten when data has been logged. By using encoded headers and fixed length data, the amount of data required is greatly reduced.

In stand-alone devices, memory is a fixed and unchangeable quantity. Two methods of logging data are available:

- **Stop when full mode**
  Logging stops once the memory is full. This retains data in the order it is logged, the latest data being discarded. Where a memory card is used, the internal memory is used only after the memory card is full.

- **Overwrite mode**
  In this mode of logging data, the memory is organized as a circular buffer. The oldest data may be overwritten when the memory is full.

Retrieving data

Data can be unloaded back to the host PC either from internal memory or a memory card, usually using simple commands sent from the host PC. A number of options, defining what logged data is to be unloaded, are usually available.

These are:

- All the data, oldest first
- The most recently logged data only
- Data for a particular schedule
- Data from the point in memory where the last unload finished
- Data logged between certain times.

7.7   Host software

The software running on the host PC completes the data acquisition system utilizing stand-alone logger/controllers, and performs two functions:

- Sends the commands that program the operation of the remote stand-alone devices, including what actions to take at any point in time, where to store the data read from the input channels, the data format, and what data to output on the output channels.
- Acquires data from the remote devices and provides analysis, storage and presentation of the data.

The simplest form of software provided with stand-alone devices is commonly in the form of a ready to use communications package with a graphical user interface. This is typical of proprietary software packages such as 'DeTerminal' provided by Data Electronics.

Stand-alone loggers. More advanced software, either supplied by the manufacturer or developed by the user, provides a higher level of user interface that eliminates the need for the formatting of individual commands and allows the automatic collection of data into files, for graphical presentation or analysis.

Other off-the-shelf software packages that can be used with remote stand-alone devices are:

- Any terminal emulator/communications package
- Spreadsheet packages such as Excel, Lotus 123, Framework, Quattro

In all cases, the format of the commands received and the data sent in return by the remote device is still in simple ASCII format, dictated by the firmware in the remote device.

7.8   Considerations in using stand-alone logger/controllers

Data acquisition and control using stand-alone logger/controllers is an orderly process. When designing a system that utilizes these devices, users should consider the following:

- The first is the number of sensors and control outputs required, and their location in relation to the host PC. This determines how many logger/controllers are required and their location. Where the analog input or digital I/O requirements are greater than can be provided by a single unit, but the increased channel requirements are localized, expansion modules can be used. If the increased I/O capability must be distributed, more than one logger/controller has to be used and can be connected as part of a distributed network. Where the units are remotely located (e.g. at a remote weather station or an electricity sub-station located in the country) and it is not practical for these devices to be connected directly to the host PC, then two options are available. These are: an operator uploads data using a memory card, and the remote device is connected via a modem to the host PC.
- The volume of data that is to be logged. Stand-alone logger/controllers only have a limited amount of memory. How quickly the memory becomes full depends on the number of input channels to be read and their scan rate. If only a limited amount of data is to be stored then the internal memory may be sufficient. Increased capacity can be provided by a memory card. If the internal memory or the memory card becomes full frequently, then the logger/controller will require connection, either directly or via modems.
- How often stored data must be uploaded to the host PC. Equally important as memory considerations, upload of data to a host PC, is largely determined by the type of application. Logging applications in which the data are gathered for analysis over an extended period (e.g. weather monitoring applications), clearly do not require constant uploading. The frequency of uploading would only depend on the memory capacity of the device and the amount of data being logged. Critical applications requiring constant uploading of process data, feedback of alarms and on-line control would need to be directly connected to the host PC.
- The availability of power. Where mains power is not available, the stand-alone logger/controller is powered from a battery supply. The battery life is determined by a number of factors and is not unlimited. In time, the battery will need to be charged or replaced.

7.9 Stand-alone logger/controllers vs internal systems

7.9.1 Advantages

Logger/controllers enjoy the same benefits of distributed I/O in that they are modular and can be located where they are required. Future expansion is easily catered to by increasing the number of devices in the network.

As well as the ability to make decisions remotely, the use of stand-alone logger/controllers increases system reliability. This is because the stand-alone or distributed system can continue to operate even when the host computer is not connected or functional. It also increases overall system performance, by distributing the control decisions, algorithms and other analysis functions to localized processors.

The analog inputs of plug-in data acquisition boards are typically designed to accept voltage signals. Where the signal levels are small, or the sensors do not output voltage signals, or are remotely located from the PC and affected by noise, then some form of external signal conditioning is required. Unlike distributed I/O, logger/controllers typically have more than one channel per unit, and each one of these channels can be configured for operation with a variety of sensors and signals, not just one type of sensor for one channel. Whilst stand-alone devices do not necessarily have inbuilt filters, these are most often not required since the units can be located very close to the signal source and therefore not as susceptible to noise. In addition, statistical sampling methods, such as averaging a number of measurements or integrating over the noise cycle, help to eliminate cyclical noise such as mains hum.

As stand-alone loggers/controllers can be placed near the signal source and do not necessarily have to be connected to the host PC, the requirement for lengthy cabling, which can be affected by noise, is greatly reduced.

A final advantage is that logger/controllers often include a local operator interface, providing feedback on a system or process at the location of the device.

7.9.2 Disadvantages

When connected to the host PC, irrespective of whether there is a single logger/controller or a network of devices, all communications must proceed via the RS-232 interface. The speed at which data can be transmitted to and from the logger/controller(s) is limited by the speed at which data can be transferred across the single RS-232 communications path. This is an important consideration where a number of units send information to be logged by the host PC or there is a large amount of data to be uploaded. In either case, the number of samples that can be taken from each logger/controller is limited by the total amount of information that can be sent via the communications interface.

Unlike specialized high-speed plug-in DAQ boards, stand-alone logger/controllers are not designed for the sampling of high frequency signals. Logger/controllers do not enjoy the benefits of high-speed microprocessors, high-speed data bus and fast memory storage facilities such as DMA that characterize modern PCs. Instead, these devices are powered by local, dedicated microprocessors or microcontrollers. The microprocessor not only performs many tasks associated with the operation of the hardware in the device but possibly multiple data acquisition and control tasks associated with the program they are required to execute. This necessarily means that when operating at full capacity these loggers/controllers can only sample signals at a very low sampling rate and are therefore more suited to applications where the signals change more slowly.

A final limiting factor, especially where the logger/controller is operating in stand-alone mode, is that the device has a limited amount of memory. The greater the number of channels and the faster the sampling rate on each channel, the greater the number of samples that will be taken in a given time period. In time, the memory will become full. Care must be taken to keep the sampling rate of each channel to the minimum necessary, while still obtaining the information required.

Plug-in data acquisition boards can continuously sample data and transfer it directly to the PC's memory. Where even greater storage capacity is required, or the data is to be stored permanently, it can be transferred to the hard disk.

8

IEEE 488 standard

8.1   Introduction

The communications standard now known as the GPIB (general purpose interface bus), was originally developed by Hewlett Packard in 1965 (when it was called the Hewlett Packard interface bus – HPIB), to connect and control their programmable test instruments. Due to its speed, flexibility, and usefulness in connecting programmable instruments to computers, it gained widespread acceptance, and was adopted by different manufacturers for their own programmable instrumentation.

It soon became clear that with the introduction of digital controllers and programmable instruments from different manufacturers, a standard high-speed data communications interface was required so that instruments from different manufacturers could communicate with each other. Following a study by a committee of the IEEE, the interface was published in 1975 as the IEEE Standard 488-1975. The standard was updated, with minor revisions, in 1978, to coincide with the international issue of the standard as IEC-625, the latter designation being more commonly used in Europe. The current version of the same standard is referred to as the IEEE Standard 488.1-1987. This standard greatly simplified the connection of programmable instrumentation by defining the mechanical, electrical and hardware protocol specification of the communications interface. For the first time, instruments from different manufacturers were connected by a standard cable. However, the standard did not address the data formats, status reporting, message exchange protocol, common configuration commands, or device-specific commands, all of which were subsequently implemented differently by different equipment manufacturers.

In 1987, the IEEE Standard 488.2 was introduced to define data formats, status reporting, controller functionality, error handling, and common commands. IEEE 488.2 concentrates on software protocol issues and maintains full compatibility with devices that comply with the hardware-oriented IEEE 488.1 standard.

EOI. In 1990, a group of instrument manufacturers announced a further extension of the standard, known as SCPI (standard commands for programmable instruments), which uses IEEE 488.2 as a basis and defines a common command set that can be used for programming instruments with any hardware link. Figure 8.1 illustrates the relationships among the standards.

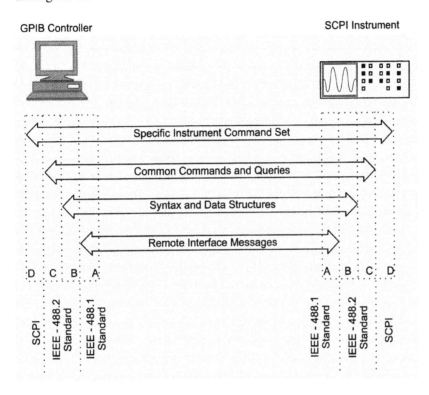

Figure 8.1
IEEE 488.1, IEEE 488.2, and SCPI standards

Apart from the three standards associated with the term IEEE 488 (IEEE 488.1, IEEE 488.2, and the SCPI) the term GPIB is often used interchangeably.

The GPIB is an interface design that allows the simultaneous connection of up to 15 devices or instruments on a common parallel data communications bus. This allows instruments to be controlled or data to be transferred to a controller, printer, or plotter. It defines methods for the orderly transfer of data, addressing of individual units, standard bus management commands, and defines the physical details of the interface. These are discussed in the following sections.

8.2   Electrical and mechanical characteristics

The GPIB bus is carried inside a shielded 24-wire cable with standard connectors at each end. The connector used is the Amphenol MICRORIBBON, Cinch Series 57 MICRORIBBON, or AMP CHAMP type, shown in Figure 8.2, which has both a plug and receptacle (male/female). Adding a new device to the bus is done by connecting a new cable in a star or chain configuration. Screws hold each connector securely to the next one. Since the 24-pin connectors are usually stackable, it is easy to connect or disconnect devices to the bus.

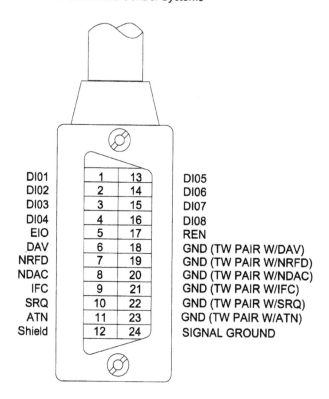

DIO1	1	13	DIO5
DIO2	2	14	DIO6
DIO3	3	15	DIO7
DIO4	4	16	DIO8
EIO	5	17	REN
DAV	6	18	GND (TW PAIR W/DAV)
NRFD	7	19	GND (TW PAIR W/NRFD)
NDAC	8	20	GND (TW PAIR W/NDAC)
IFC	9	21	GND (TW PAIR W/IFC)
SRQ	10	22	GND (TW PAIR W/SRQ)
ATN	11	23	GND (TW PAIR W/ATN)
Shield	12	24	SIGNAL GROUND

Figure 8.2
GPIB connector (IEEE 488) and pin assignments

The 24 lines, in each cable, consist of 8 data lines and 8 pairs (16) of control and bus management lines. The data lines are used exclusively to carry data, in a parallel configuration, one byte at a time, along the bus. The control and bus management lines are used for various bus management tasks that synchronize the flow of data. When data or commands are sent down the bus, the bus management lines distinguish between the two. The GPIB uses binary voltage signals to represent the information that is carried on the lines of the bus. It uses the symbols 'true' and 'false' to represent the two states of voltages on the lines. The GPIB uses the logic convention called 'low-true' or negative logic, where the lower voltage state is 'true' and the higher voltage state is 'false'. Standard TTL voltage levels are used. For example, when a line is true, or asserted, the TTL voltage level is low (≤ 0.8 V), and when the line is 'false', or unasserted, the TTL level is high (≥ 2.0V). The 'low-true' logic means that any device can set a bus control voltage 'low-true', but no line can be at a voltage 'high=false' unless **all the devices** on that line allow it to go 'high=false'.

8.3   Physical connection configurations

The devices on the GPIB can be connected in a star configuration, as shown in Figure 8.3, or in a chain (linear) configuration as shown in Figure 8.4.

A star configuration is one where each instrument is connected, by means of a separate GPIB cable, directly to the controller. The connectors are all connected to the same port at the controller. A drawback to this simple configuration is that all of the devices on the bus must be relatively close to the controller because of the length limitation of each cable.

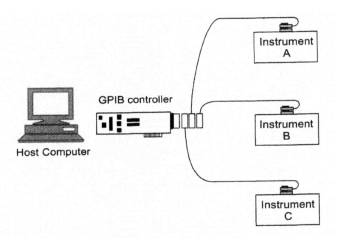

Figure 8.3
The GPIB star configuration

In the chain configuration, each device, including the controller, is connected to the next in a chain. The controller does not have to be the first or last device in the chain, but can be linked in anywhere. It is a controller only in the sense that it co-ordinates the events on the bus. Physically and electrically, it is similar to any other device connected to the GPIB. This configuration is usually the most convenient way to connect equipment. A disadvantage of the chain configuration is that the software may require reconfiguration if a device and its cable are removed.

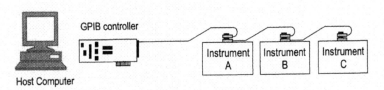

Figure 8.4
The GPIB chain (linear) configuration

Although the star and chain configurations are suggested for GPIB, connections can be made in any other way, provided that the following rules are observed:

- All devices are connected to the bus
- No more than 15 devices, including the controller, may be on the bus, with no less than two thirds powered on
- Cable length between any two devices may not exceed 4 meters with an average separation of 2 meters over the entire bus
- Total cable length may not exceed 20 meters

A single device on the GPIB can transfer data up to 14 other devices. Because the GPIB uses asynchronous (un-clocked) handshaking, the actual data transfer rate is more dependent on the devices themselves rather than on the hardware interface.

8.4   Device types

From a data communications standpoint, there are four different groups of devices that communicate on a GPIB interface:

- **Talkers**

  A talker is a one-way communicating device that can only send data to a listening device. It does not receive data. The talker waits for a signal from the controller and then places its data on the bus. Only one device can talk at a time. Common examples are simple DVMs (digital voltmeters) and some A/D converters.

- **Listeners**

  A listener is a one-way communicating device that can only receive data from another device. It does not send data. It receives data when the controller signals it to read the bus. Common examples are printers, plotters, and recorders.

- **Talkers/Listeners**

  A talker/listener has the combined characteristics of both talkers and listeners. However, it is never a talker and a listener at the same time. A common example is a programmable DVM, which is a listener while its range is being set by the controller and a talker while it sends the results back to the controller. Most modem digital instruments are talker/listeners as this is the most flexible configuration.

- **Controllers**

  A controller manages and controls everything that happens on the GPIB. It is usually an intelligent or programmable device, such as a PC or a microprocessor-controlled device. It determines which devices will send data (talkers) and which will receive data (listeners), and when. To avoid confusion in any GPIB application, there can only be one active controller, called the controller in charge (CIC). The key word is active. There can be several controllers, but to avoid confusion, only one can be active at any time. A controller also has the features of a talker/listener. In some cases, when several PCs are simultaneously connected on a GPIB, one is usually configured as the controller and the others configured as talkers/listeners. The controller needs to be involved in every transfer of data. It needs to address a talker and a listener before the talker can send its message to a Listener. After the message is sent, the controller un-addresses both units. Some GPIB configurations do not require a controller, e.g. when only one talker is connected to one or more listeners. A controller is necessary when the active or addressed talker or listener must be changed. In this case, a device sees its talk address on the bus, and knows that it has to act as a talker and hence required to send data. Conversely, when it sees its listen address on the bus, it knows it is required to act as a Listener and hence receive data.

8.5   Bus structure

The GPIB interface system illustrated in Figure 8.5 consists of 16 signal lines and 8 ground return or shield drain lines.

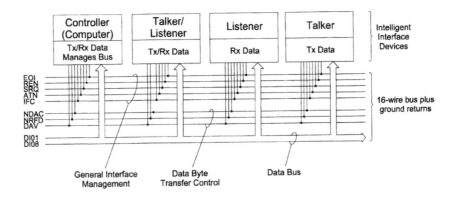

Figure 8.5
The GPIB bus structure

The 16 signal lines consist of data lines (D101–D108) and 8 control lines. Three of the eight control lines are the handshaking lines that coordinate the transfer of data (DAV, NRFD and NDAC), while the remaining five lines are used for bus control and management (ATN, REN, IFC, SRQ and EOI). The 8 'ground' lines provide electronic shielding and prevent bus control signals from interfering with one another or from being influenced by external signals. A summary of the signal lines on the GPIB is as follows:

- Data bus lines   D101   –   D108
- Handshaking lines   DAV   –   Data available
                      NRFD –   Not ready for data
                      NDAC –   No data accepted
- General interface   ATN   –   Attention
- Management lines   FC   –   Interface clear
                      SRQ   –   Service request
                      REN   –   Remote enable
                      EOI   –   End or identify

8.5.1   Data Lines

The eight data lines D101 to D108 carry both data and command messages. All commands and most data use the 7-bit ASCII code, in which case the eighth bit, D108, is either unused or used for parity. The state of the attention (ATN) line determines whether the information is data or commands. Command messages are sent with the ATN line asserted, while data messages are sent with the ATN line unasserted.

8.5.2   Interface management lines

Five signal lines manage the flow of information across the GPIB. These are described below.

- ATN (attention) – The controller asserts the ATN line 'true' when it uses the data lines to send commands. All devices become listeners and participate in the communication. When ATN is unasserted, information on the bus is interpreted as data.
- IFC (interface clear) – This line can only be controlled by the system controller, which drives the IFC line to initialize the bus and become

controller in charge (CIC). The IFC line is the master reset of the GPIB and when asserted all devices return to a known quiescent state.

- REN (remote enable) – The system controller drives the REN line to put devices into a remote state. When the REN line is asserted and a device is addressed to listen, the device is placed into a remote programming state.
- SRQ (service request) – Any device can drive the SRQ line to asynchronously notify the CIC that it needs service. It is the responsibility of the CIC to monitor the SRQ line, poll the device, and determine the type of service the device needs. SRQ will remain asserted until the CIC polls the device requesting service.
- EOI (end or identify) – The EOI line has two purposes. Its first use is when a talker asserts the EOI line to indicate the last byte of data in the message string. A listener stops reading data when EOI is asserted 'true'. A second use for the EOI line is to tell devices to identify their response in a parallel poll.

8.5.3   Handshake lines

Three handshake lines asynchronously control the transfer of message bytes between devices. The GPIB uses a three-wire interlocking handshake scheme that guarantees that message bytes on the data lines are sent and received without error. The handshake lines and their use are discussed below.

- NRFD (not ready for data) – The NRFD handshake line indicates whether a device is ready to receive a message byte or not. When receiving commands, the line is driven by both talkers and listeners, and only by listeners when receiving data messages.
- NDAC (no data accepted) – This line indicates whether a device has or has not accepted a message byte. NDAC is driven by all devices (i.e. talkers and listeners) when receiving commands, and only by listeners when receiving data messages.
- DAV (data valid) – The DAV handshake line indicates whether signals on the data lines are stable and therefore valid and can be accepted by devices. The controller controls DAV when sending commands and the talker controls DAV when sending data.

8.6   GPIB handshaking

Data is transmitted asynchronously on the GPIB parallel interface one byte at a time. The transfer of data is coordinated by the voltage signals on the three bus-control 'handshake' lines (DAV, NDAC and NRFD). The process is called a *three-wire interlocked handshake*. Handshaking ensures that a talker will put a data byte on the bus, only when all listeners are ready and will keep the data on the bus until it has been read by all listeners. It also ensures that listeners will accept data only when a valid byte is available on the bus.

The talker first un-asserts DAV then monitors the NDAC and NRFD lines. The talker must wait for the NRFD line to go high ('false') before any data can be put onto the bus. The NRFD line is controlled by the listeners. Only when NRFD voltage is high ('false') are all listeners ready to receive data. A short delay after NRFD goes high, the talker asserts DAV 'true' (voltage low) to indicate valid data is available on the bus. The delay is determined by the type of drivers the talker uses on the data lines (trio-state requires less delay than open collector). When the listeners detect the low voltage level on DAV, they read the byte on the data lines and immediately assert the NRFD line to indicate that no further data should be sent. As each listener accepts the data, it releases NDAC. After the last listener has accepted the data, the NDAC line voltage goes high ('false'), signaling the Talker that the data has been accepted. Only when the data byte has been accepted by all the listeners, can the talker allow DAV voltage to go high ('false') and remove its data from the bus. The listeners then assert NDAC to signify that the data transfer has ended, in preparation for the next cycle. Figure 8.6 illustrates this handshaking sequence for one message byte.

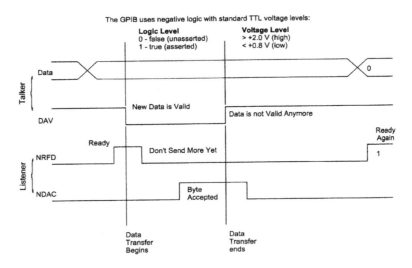

Figure 8.6
GPIB handshaking timing diagram

It should be noted that since a talker waits until all listeners are ready (NRFD is 'false') before sending a message byte and waits for all listeners to accept the message byte (NDAC line is 'false') before transferring any more data, the maximum data transfer rate of the GPIB is determined by the slowest listener on the bus.

8.7 Device communication

GPIB devices communicate with other interconnected GPIB devices by passing device-dependent messages and interface messages one byte at a time through the parallel data communications interface.

- Device-dependent messages contain specific information related to a particular device, including programming instructions, measurement results, device status, and data files. These messages are often called data messages.
- Interface messages manage the operation of the communications bus, performing such tasks as initializing the bus, addressing, and disabling devices, setting modes for remote or local programming. Such messages are usually referred to as command messages. The term 'command message', used here, can sometimes be confused with device-specific commands that are contained in data messages. For example, the identification query command *IDN? that is used to identify a particular device, is a command that a device understands but is sent in what is known as a GPIB data-message.

8.7.1 GPIB addressing

Each device connected to the GPIB has a unique device address and each device must be designed with enough intelligence to identify whether the **data** or **command** sent down the data lines is meant for it or for another device. Device addresses are arbitrary and are set by the user, usually on a DIP switch, located on the back of the device, or by programming the device software. For the controller, each connected device is identified in the software of the controller's program. The only limitation in choosing a device address is that it must be an integer number in the range 0 to 30.

To communicate over the GPIB, the controller must first address the appropriate devices. To address a device and configure it to take part in communications on the bus, the controller sends an address command message to the required device. Address command messages have the format shown in Figure 8.7.

Bit 7	Bit 6	Bit 5	Bit 4	Bit 3	Bit 2	Bit 1	Bit 0
	TA	LA	X	X	X	X	X

Figure 8.7
Address command message format

Bit 0 to bit 4 represent the binary GPIB primary address of the device, to be included in the communication. If bit 5 (LA) is set, the device with the primary address specified is to be configured as a listener, while bit 6 (TA) is set if the device is to be configured as a Talker. Bit 7 is considered a don't-care bit as it is never used, and can assumed to be zero. The command message byte for a device with address of 0x05, to be configured as a listener, would be 0x25 ('%'). The command message byte for the same device configured as a talker would be 0x45 ('E').

8.7.2 Un-addressing devices

Once a communication has taken place between two or more devices, the controller must clear the bus of the current talker and listener(s) before assigning a new talker and listener(s). Sending an unused talk address, or more typically, the un-talk (UNT) command message (0x5F), un-addresses the current Talker. This command is not required since addressing one talker automatically un-addresses all others. The current listener or multiple listeners are automatically un-addressed using the un-listen (UNL) command message (0x3F). A single listener cannot be un-addressed if multiple listeners have been previously addressed.

8.7.3 Terminating data messages

There are three methods for terminating data messages:

- EOI method – When using this method, the EOI bus control line is asserted when the talker places the last byte of data onto the data bus. This line is monitored by all listeners.
- EOS method – This method does not make use of any bus control lines, instead, appending a pre-determined end-of-string byte to the end of the data string and sending this data byte to the listener(s). The EOS byte is usually the new line character (0x0A) or the carriage return character (0x0D)

- Count method – Using the count method requires that the controller stop the talker from sending any more bytes once the specified number of bytes has been sent to the listener. This can be achieved by asserting the NRFD and NDAC bus control lines, but requires the application software for the GPIB controller hardware to control this facility.

8.7.4 Sending and receiving data

Assuming the GPIB has been initialized and configured correctly, the protocols for sending and receiving data are as follows:

Sending data
1 Set ATN 'true' and EOI 'false'.
2 Send the controller talk address.
3 Send the UNL command message.
4 Send the listen address for each device to be configured as a listener.
5 Set ATN 'false'.
6 Send the required data bytes one byte at a time.
7 Send the terminating character or set EOI 'true'.

Receiving data
1 Set ATN 'true' and EOI 'false'.
2 Send the UNL command message.
3 Send the controller listening address.
4 Send the required talker address.
5 Set ATN 'false'.
6 Receive data bytes until terminating character received or EOI 'true'.

8.8 IEEE 488.2

IEEE 488.2 was drafted to correct problems that arose with the original IEEE 488 standard. At the same time, it was designed to ensure compatibility with existing IEEE 488.1 devices. The concept used in expanding the standard was that the manner, in which controllers and instruments would function as talkers, would be precisely defined. This means that fully compatible IEEE 488.2 systems would be reliable and efficient. However, backward compatibility with IEEE 488.1 devices could be ensured by designing the standard so that IEEE-488.2 devices, when acting as listeners, would be able to accept a wide range of commands and data formats from IEEE 488.1 devices.

8.8.1 Requirements of IEEE 488.2 controllers

The IEEE 488.2 standard sets out a number of requirements for a controller, including an exact set of IEEE 488.1 interface capabilities as follows:

- Sensing the state and transitions of the SRQ line
- Setting and detecting EOI
- Pulsing the interface clear line for 100 microseconds
- Setting/asserting the REN line
- Timing out on any I/O transaction

8.8.2   IEEE 488.2 control sequences

One of the additional features that the IEEE 488.2 standard has over the IEEE 488.1 standard is the definition of the exact messages which are sent by the controller, as well as the ordering of those messages if more than one message is sent. IEEE 488.2 defines fifteen required control sequences and four optional control sequences as shown in Table 8.1. These control sequences describe the exact state of the GPIB, how devices should respond to specific messages, and the ordering of command messages for each of the defined operations.

Description	Control sequence	Compliance
Send ATN-'true' commands	Send command	Mandatory
Set address to send data	Send setup	Mandatory
Send ATN-'false' data	Send data bytes	Mandatory
Send a program message	Send	Mandatory
Set address to receive data	Receive setup	Mandatory
Receive ATN-'false' data	Receive response message	Mandatory
Receive a response message	Receive	Mandatory
Pulse IFC line	Send IFC	Mandatory
Place devices in DCAS	Device clear	Mandatory
Place devices in local state	Enable local controls	Mandatory
Place details in remote state	Enable remote	Mandatory
Place devices in remote with local lookout state	Set RWLS	Mandatory
Place devices on local lockout	Send LLO	Mandatory
Read IEEE 488.1 status byte	Read status byte	Mandatory
Send group execution trigger (GET) message	Trigger	Mandatory
Give control to another device	Pass control	Optional
Conduct a parallel poll	Perform parallel poll	Optional
Configure device's parallel poll responses	Parallel poll configure	Optional
Disable device's parallel poll capability	Parallel poll unconfigure	Optional

Table 8.1
IEEE 488 2 required and optional control sequences

8.8.3   IEEE 488.2 protocols

Protocols are high-level functions that combine a number of control sequences to perform common test functions, thereby reducing test program development time. The IEEE 488.2 defines two mandatory protocols and six optional protocols, as shown in Table 8.2.

Keyword	Name	Compliance
RESET	Reset system	Mandatory
FINDRQS	Find device requesting service	Optional
ALLSPOLL	Serial poll all devices	Mandatory
PASSCTL	Pass control	Optional
REQUESTCTL	Request control	Optional
FINDLSTN	Find listeners	Optional
SETADD	Set address	Optional but requires FINDLSTN
TFSTSYS	Self-test system	Optional

Table 8.2
IEEE 488.2 controller protocols

The functions performed by each of the defined protocols are as follows:

RESET	Initializes the GPIB bus, clears and sets all devices to a known state.
FINDRQS	The controller senses the 'FALSE' to 'TRUE' transition of the SRQ line and then services the most critical devices first.
ALLSPOLL	Serial polls each device and returns the status byte of each device.
PASSCTL	Passes bus control to other devices.
REQUESTCTL	Requests bus control.
FINDLSTN	The controller issues this command with a particular listener's address and monitors the NDAC handshake line to determine if a device exists at that address.
SETADD	Used in conjunction with FINDRQS sets a device address.
TESTSYS	The controller instructs each device to perform a self-test and report back whether it has any problems or is OK.

As an example of the usefulness of protocols, consider the serial polling of one or more devices on a GPIB system. Multiple devices can asynchronously request service from the controller by asserting the service request (SRQ) line. It is the responsibility of the controller to determine which device(s) made the request by performing a poll of the known active devices. Serial polling of a particular device can be implemented with the IEEE 488.1 protocol.

The following sequence of events would occur:

- Controller sends the un-listen (UNL) command message, which disables all Listeners from listening.
- Set the listen address of the controller.
- The controller sends the command message containing the serial poll enable (SPE) command to the device. This command directs the device to return its serial poll status byte, by setting the IEEE 488.1 serial poll mode in the device. When the device is addressed to talk, it will return the serial poll status byte.

- The controller temporarily configures the required device as a talker by putting the talk address of the device onto the bus.
- Controller reads a byte (status byte) from the device configured as a talker.
- The controller sends the command message containing the serial poll disable (SPD) command to the device. This command resets the serial poll mode in the device.

The mandatory protocol ALLSPOLL serial polls all devices in the GPIB, whose addresses are provided with the protocol (e.g. ALLSPOLL 1,2,3,4). The steps required to serial poll each device are carried out invisibly to the user.

8.8.4 Device interface capabilities

IEEE 488.2 defines a minimum set of IEEE 488.1 interface capabilities that a device must have. The device interface capabilities, (i.e. the functions that a device can perform) are represented by explicit codes that should be found below or near the interface connector of each device. The minimum functions that a device must be able to perform are shown in Table 8.3.

Capability	Code	Comment
Source handshake	SH1	Full capability
Acceptor handshake	AH1	Full capability
Talker	T(TE)5 or T(TE)6	Basic talker, serial poll or untalk on MLA
Listener	L(LE)3 or L(LE)4	Unlisten on MTA or full capability
Service request	SR1	Full capability
Device clear	DC1	Full capability
Remote local	RL0 or RL1	No or full capability
Parallel poll	PP0 or PP1	No or full capability
Device trigger	DT0 or DT1	No or full capability
Controller	C0 or C4 with C5, C7, C8 or C11	No or full capability with send if message, pass, receive control
Electrical interface	E1 or E2	Open collector or tri-state

Table 8.3
A minimum set of device interface capabilities for IEEE 488.2

For a complete description of what each code represents, and all other codes describing device functionality, the user should refer to the IEEE 488 standard or alternatively the specifications of a particular device.

8.8.5 Status reporting model

One of the problems associated with the IEEE 488.1 was that status reporting from individual devices was unique, with each device reporting its status information using a different set of bits in the status byte. IEEE 488.2 solved this problem by defining a standard status-reporting model as shown in Figure 8.8.

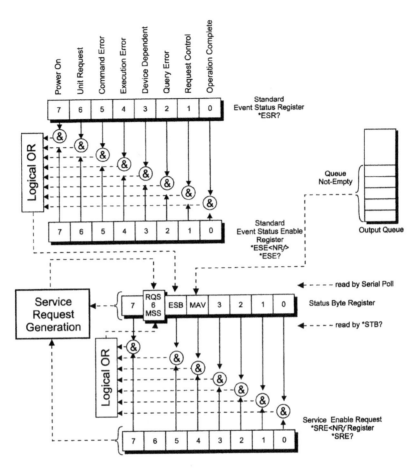

Figure 8.8
IEEE 488.2 status reporting model

This model builds upon the status byte definition of IEEE 488.1, which only had bit 6 defined as the RQS (request service) bit. The RQS bit is set to indicate that a device is requesting service by asserting the service request (SRQ) line. A further two bits are defined, the event status bit ESB) and the message available (MAV) bit. Other bits are defined by the manufacturer. The ESB indicates that one of the standard events, as defined in the standard event status register, has occurred. Only the events for which the user has set the corresponding bit in the standard event status register will cause the ESB to be set in the status byte. The MAV bit is used to indicate if a message is available in the device's output queue.

Whether a device asserts the service request (SRQ) line when one of the bits in its status register becomes set, is enabled or disabled, using the service request enable register. If a bit is set by the user in this register, then the setting of the corresponding bit in the status register will cause the device to assert the SRQ line.

8.8.6 Common command set

IEEE 488.2 also designates a minimum set of standard commands that a device must have. These are not new bus command messages, but are new data messages common to all devices. Table 8.4 lists the mandatory IEEE 488.2 commands, indicating the command group to which each command belongs.

Mnemonic	Group	Description
*IDN?	System data	Identification query
*RST	Internal operations	Reset
*TST?	Internal operations	Self-test query
*OPC	Synchronization	Operation complete
*OPC?	Synchronization	Operation complete query
*WAI	Synchronization	Wait to complete
*CLS	Status and event	Clear status
*ESE	Status and event	Event status enable
*ESE?	Status and event	Event status enable query
*ESR?	Status and event	Event status register query
*SRE	Status and event	Service request enable
*SRE?	Status and event	Service request enable query
* STB?	Status and event	Read status byte query

Table 8.4
IEEE 488.2 commands

The SCPI takes the common command set further by defining specific commands that each instrument class (i.e. multimeters, oscilloscopes, digital voltmeters, etc), from different manufacturers, must obey.

8.9 Standard commands for programmable instruments (SCPI)

A lack of standardization between the command sets for programming instruments developed by different manufacturers led to a group of manufacturers developing the SCPI specification. Using the IEEE 488.2 standard as a basis, SCPI defines a single comprehensive programming command set that can be used with any hardware link and different instruments.

For example, the command ':MEAS:FREQ?' or ':MEAS:VOLT?' for reading of a frequency and voltage measurement respectively, would be applicable to any instrument capable of processing these parameters. The instrument could be a voltmeter, an oscilloscope, or a frequency counter.

IEEE 488.1, IEEE 488.2, and SCPI instruments and controllers can be used in the same system, but the most easily programmed, flexible, fast, and interchangeable systems make use of IEEE 488.2 controllers and SCPI instruments.

8.9.1 IEEE 488.2 common commands required by the SCPI

As a minimum requirement for SCPI, the common and mandatory commands defined in the IEEE 488.2 standard and needed by all SCPI compatible instruments was defined in Table 8.4. This minimum command set does not handle device specific commands (i.e. :MEAS:FREQ?), but consists of program commands and status queries such as device reset, device self test, service request enable reporting, device identification, operation synchronization and standard event status enabling and reporting.

8.9.2   SCPI required commands

To build on the common command set defined in IEEE 488.2, SCPI also defines its own set of required common commands, as shown in Table 8.5.

For example, the status reporting defined by IEEE 488.2 is expanded with the OPERation and QUEStionable status registers, for which there are commands for reading the contents of the EVENt arid CONDition registers, setting the ENABle mask, or reading the ENABle mask.

The command set allowed by the introduction of the SYSTem parent command defines functions that perform general housekeeping duties such as setting the TIME or SECurity access. The subcommand query ERRor? requests the next entry from the error/event queue of the SCPI device.

Command	Description
:SYSTem	Collects functions not related to instrument performance
:ERRor?	Requests the next entry from the instruments queue
:STATus	Controls the SCPI-defined status reporting structures
:OPERation	Selects the operation structure
[:EVENt]?	Returns the contents of the event register
:CONDition?	Returns the contents of the condition register
:ENABle?	Reads the enable mask
:QUEStionable	Selects the questionable structure
[:EVENt]?	Returns the contents of the event register
:CONDition?	Returns the contents of the condition register
:ENABle	Sets the enable mask which allows event reporting
:ENABle?	Reads the enable mask
:PRESet	Enables all required event reporting

Table 8.5
SCPI required commands

8.9.3   The SCPI programming command model

As a means of achieving compatibility and breaking commands up into groups, SCPI defined a model of a programmable instrument, as shown in Figure 8.9, which applies to all the different types of instrumentation.

When using this model it should be noted that not all instruments have the functionality indicated; it is a generalized model of an instrument. For example, an oscilloscope does not have the functionality defined by the signal generation block in the SCPI model, where as a function generator will have. In addition, the function generator will probably not have the functionality described by the measurement function block.

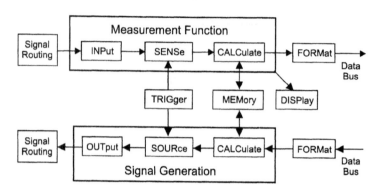

Figure 8.9
SCPI instrument model

The functional components of the SCPI instrument model are described in the following sections:

Measurement function

The measurement function component is used to convert a signal into a preprocessed form. It is divided into three distinct parts:

INPut   Performs signal conditioning, such as filtering, biasing and attenuation on the raw incoming signal.

SENse   Converts the conditioned input signal into a data format that can be manipulated by the user. This function controls parameters such as range, resolution, gate time, and normal mode rejection.

CALCulate   Converts the data into a format that is more useful for a particular application, such as converting the data to engineering units.

The measurement function described by the three functions above is specified by the signal parameters, and not by the instrument's functionality. This usually provides the highest level of compatibility by allowing different instruments that make the same measurements to be interchanged without changing the SCPI command.

Signal generation

The signal generation function converts internal data to real-world signals. It is also divided into three distinct parts:

OUTput   This signal block conditions the output signal by filtering, attenuation and biasing.

SOURce   Generates a signal based on specific characteristics and internal data, specifying such signal parameters as current, voltage, power and frequency.

CALCulate   This functional block converts the data from one set of engineering units to another and also takes into account abnormalities, which occur in the generation of the signal data.

Signal routing

The signal routing function controls the connection of a signal to the internal functions of a particular instrument. The TRIGger component synchronizes instrument options with external conditions, such as an analog or digital signal, internal instrument events involving the instrument's functionality or a software command. The MEMory component is used for internal storage of data; while the FORMat component converts the data from the instrument to a form suitable for transmission across a standard bus. The DISPlay component allows for display of the data on a screen.

8.9.4   SCPI hierarchical command structure

The SCPI instrument model defines the functional blocks for each of the main command categories. Each of these main categories has a hierarchical command structure, referred to as a command tree, comprising sub-commands, parameters and options that provide the specific details of the function to be performed. This is shown in Figure 8.10.

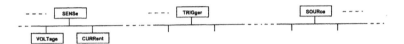

Figure 8.10
SCPI hierarchical command structure

The SENSe partial command tree of Figure 8.11 contains sub-commands that program an instrument to control the conversion of a signal into the required internal data format. This data can be further manipulated by the user.

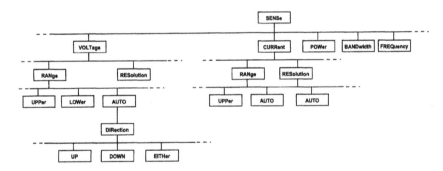

Figure 8.11
Partial command tree for SENSe command structure

SENSe commands control parameters such as resolution and range. Therefore, to program a digital voltmeter to measure a voltage with autoranging, the command would be as follows:
SENS:VOLT:RANG:AUTO:DIR:EITH
Colons are used to separate each command and instruct the instrument's command interpreter to move to next level down in the command tree hierarchy. Note that command words are not broken by any spaces.

Consider the programming of a digital voltmeter to measure voltage with specific upper and lower range values of 10 and 5 respectively. The command would be as follows:
SENS:VOLT:RANG:UPP10:LOW 5
In this command, there are two sub-commands that are on the same level of the hierarchical command tree. When two commands are issued without changing levels, a semicolon is used to separate the commands. Also, note that commands and any related parameters must be separated by white space.

9

Ethernet LAN systems

9.1   Ethernet and fieldbuses for data acquisition

Ethernet, in the past, has been usually thought of as an office networking system. Now many manufacturers are using Ethernet and industrial fieldbuses as communication systems to interconnect data acquisition devices. This can take the form, among others, of connecting computers that are using plug-in data acquisition cards or data loggers that are networked together. Fieldbuses such as Profibus and Foundation Fieldbus are being used to interconnect devices that are doing data acquisition. To this end, it is necessary to define both Ethernet and industrial fieldbuses.

The Ethernet network concept was developed by Xerox Corporation, at its Palo Alto Research Center (PARC), in the mid-seventies. It was based on the work done by researchers at the University of Hawaii, where there were campus sites on the various islands. Its ALOHA network was set up using radio broadcasts to connect the various sites. This was colloquially known as their 'Ethernet' since it used the 'ether' as the transmission medium and created a network 'net' between the sites. The philosophy was quite straightforward. Any station that wanted to broadcast to another station could do so immediately. The receiving stations then had a responsibility to acknowledge the message; thus advising the original transmitting station of successful reception of the original message. This primitive system did not rely on any detection of collisions (two radio stations transmitting at the same time) but rather waited for an acknowledgement back within a predefined time.

The initial system installed by Xerox was so successful that they soon applied the system to their other sites typically connecting office equipment to shared resources such as printers, and large computers acting as repositories of large databases.

In 1980, the Ethernet Consortium consisting of Xerox, Digital Equipment Corporation and Intel (sometimes called the DIX consortium) issued a joint specification based on the Ethernet concepts and known as the Ethernet Blue Book 1 specification. This was later superseded by the Ethernet Blue Book 2 specification, which was offered to the IEEE as a standard. In 1983, the IEEE issued the 802-3 standard for carrier sense; multiple access;

collision detect LANs based on the Ethernet standard, which gave this networking standard even more credibility.

As a result of this, there are three standards in existence. The first – often termed Ethernet Version 1 – can be disregarded as very little equipment based on this standard is still in use. Ethernet Version 2, or 'Blue Book Ethernet' is, however, still in use and there is a potential for incompatibility with the IEEE 802.3 standard. The differences between these two later standards are discussed in section 9.5. While these differences are minor, they are nonetheless significant. Despite the generic term 'Ethernet' being applied to all CSMA/CD networks, it should be reserved for the original DIX standard. This manual will continue with popular use and refer to all the LANs of this type as Ethernet, unless it is important to distinguish between them.

Ethernet uses the CSMA/CD access method discussed in section 9.3. This gives a system, which can operate with little delay, if lightly loaded, but the access mechanism can fail completely, if too heavily loaded. Ethernet is widely used commercially, and the NICs are relatively cheap and produced in vast quantities. Because of its probabilistic access mechanism, there is no guarantee of message transfer and messages cannot be prioritized. It is becoming more widely used industrially despite these disadvantages.

9.2 Physical layer

802.3 standard defines a range of cable types that can be used for a network based on this standard. They include coaxial cable, twisted-pair cable and fiber optic cable. In addition, there are different signaling standards and transmission speeds that can be utilized. These include both baseband and broadband signaling, and speeds of 1 Mbps and 10 Mbps. The standard is continuing to evolve, and this manual will look at 100 Mbps CSMA/CD systems in the next chapter.

The IEEE 802.3 standard documents (ISO 8802.3) support various cable media and transmission rates up to 10 Mb/s as follows:

- **10BASE-2**
  Thin wire coaxial cable (0.25 inch diameter), 10 Mbps, single cable bus
- **10BASE-5**
  Thick wire coaxial cable (0.5 inch diameter), 10 Mbps, single cable bus
- **10BASE-T**
  Unscreened twisted-pair cable (0.4 to 0.6 mm conductor diameter), 10 Mbps, twin cable bus
- **10BASE-F**
  Optical fiber cables, 10 Mbps, twin fiber bus
- **1BASE-5**
  Unscreened twisted-pair cables, 1 Mbps, twin cable bus
- **10BROAD-36**
  Cable television (CATV) type cable, 10 Mbps, broadband

9.2.1 10Base5 systems

This is a coaxial cable system and uses the original cable for Ethernet systems – generically called 'Thicknet'. It is a coaxial cable, of 50 ohm characteristic impedance, and yellow or orange in color. The naming convention for 10Base5: means 10 Mbps; baseband signaling on a cable that will support 500-meter segment lengths. It is difficult to work with, and so cannot normally be taken to the node directly. Instead, it is laid in a cabling tray etc and the transceiver electronics (medium attachment unit, MAU) is installed directly on the cable. From there an intermediate cable, known as an attachment unit interface (AUI) cable is used to connect to the network interconnection card (NIC). This cable, which can be a maximum of 50 meters long, compensates for the lack of flexibility of placement of the segment cable. The AUI cable consists of 5 individually shielded pairs – two each (control and data) for both transmitting and receiving, plus one for power.

The MAU connection to the cable can be made by cutting the cable and inserting an N-connector and a coaxial tee or more commonly by using a 'bee-sting' or a 'vampire' tap. The tee is a mechanical connection that clamps directly over the cable. Electrical connection is made via a probe that connects to the center conductor and sharp teeth, which physically puncture the cable sheath to connect to the braid. These hardware components are shown in Figure 9.1.

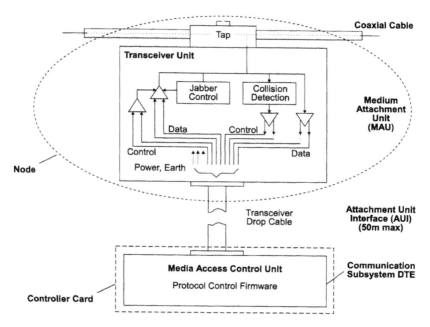

Figure 9.1
10Base5 hardware components

The location of the connection is important to avoid multiple electrical reflections on the cable, and the Thicknet cable is marked every 2.5 meters with a black or brown ring to indicate where a tap should be placed. Fan out boxes can be used if there are a number of nodes for connection, allowing a single tap to feed each node, as though it was individually connected. The connection at either end of the AUI cable is made through a 25 pin D-connector, with a slide latch, often called a DIX connector after the original consortium.

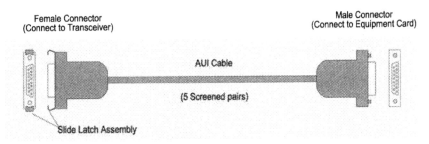

Figure 9.2
AUI cable connectors

There are certain requirements if this cable architecture is used in a network. These include:

- Segments must be less than 500 meters in length to avoid signal attenuation problems
- No more than 100 taps on each segment i.e. not every connection point can support a tap
- Taps must be placed at integer multiples of 2.5 meters
- The cable must be terminated with a 50 ohm terminator at each end
- It must not be bent at a radius exceeding 25.4 cm or 10 inches
- One end of the cable must be earthed

The physical layout of a 10Base5 Ethernet segment is shown in Figure 9.3.

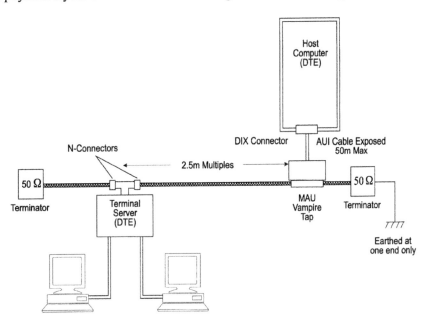

Figure 9.3
10Base5 Ethernet segment

The Thicknet cable was extensively used as a backbone cable until recently (1995) but 10BaseT and fiber are becoming more popular. Note that when a MAU (tap) and AUI cable is used, the on board transceiver on the NIC is not used. Rather, there is a transceiver in the MAU and this is fed with power from the NIC via the AUI cable. Since the transceiver is remote from the NIC, the node needs to be aware that the termination can detect collisions if they occur. This confirmation is performed by a Signal Quality Error (SQE), or heartbeat test function, in the MAU. The SQE signal is sent from the MAU to the node on detecting a collision on the bus. However, on completion of every frame transmission by the MAU, the SQE signal is asserted to ensure that the circuitry remains active, and that collisions can be detected. You should be aware that not all components support SQE test and mixing those that do with those that don't can cause problems. Specifically, if a NIC was to receive a SQE signal after a frame had been sent, and it was not expecting it, the NIC could think it was seeing a collision. In turn, as you will see later in the manual, the NIC will then transmit a jam signal.

9.2.2   10Base2 systems

The other type of coaxial cable Ethernet networks is 10Base2 and often referred to as 'Thinnet' or sometimes 'thinwire Ethernet'. It uses type RG-58 A/U or C/U with a 50 Ω characteristic impedance and of 5 mm diameter. The cable is normally connected to the NICs in the nodes by means of a BNC T-piece connector, and represents a daisy chain approach to cabling. Connectivity requirements include:

- Termination at each end with a 50 ohm terminator
- The maximum length of a cable segment is 185 meters and NOT 200 meters!
- No more than 30 transceivers can be connected to any one segment
- There must be a minimum spacing of 0.5 meters between nodes
- It may not be used as a link segment between two 'Thicknet' segments
- The minimum bend radius is 5 cm

The physical layout of a 10Base2 Ethernet segment is shown in Figure 9.4.

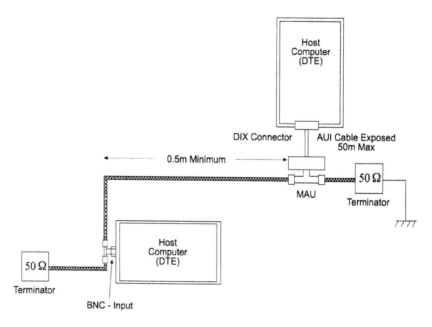

Figure 9.4
10Base2 Ethernet segment

The use of Thinnet cable was, and remains, very popular as a cheap and relatively easy way to set up a network. However, there are disadvantages with this approach. A cable fault can bring the whole system down very quickly. To avoid such a problem, the cable is often taken to wall connectors with a make-break connector incorporated. The connection to the node can then be made by 'fly leads' of the same cable type. It is important to take the length of these fly leads into consideration in any calculation on cable length! There is also provision for remote MAUs in this system, with AUI cables making the node connection, in a manner similar to the Thicknet connection.

9.2.3 10BaseT

The 10BaseT standard for Ethernet networks uses AWG24 unshielded twisted pair (UTP) cable for connection to the node. The physical topology of the standard is a star, with nodes connected to a wiring hub, or concentrator. Concentrators can then be connected to a backbone cable that may be coax or fiber optic. The node cable can be category 3 or category 4, although you would be well advised to consider category 5 for all new installations. This will allow an upgrade path as higher speed networks become more common, and given the small proportion of cable cost to total cabling cost, will be a worthwhile investment. The node cable has a maximum length of 100 meters; consists of two pairs for receiving and transmitting and is connected via RJ45 plugs. The wiring hub can be considered as a local bus internally, and so the topology is still considered as a logical bus topology. Figure 9.5 shows schematically how the hub interconnects the 10BaseT nodes.

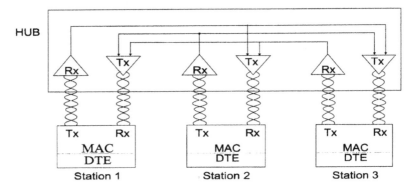

Figure 9.5
Schematic 10BaseT system

Collisions are detected by the NIC and so an input signal must be retransmitted by the hub on all output pairs. The electronics in the hub must ensure that the stronger retransmitted signal does not interfere with the weaker input signal. The effect is known as far end crosstalk (FEXT), and is handled by special adaptive crosstalk echo cancellation circuits

The standard has become increasingly popular for new networks, although there are some disadvantages that should be recognized:

- The cable is not very resistant to electrical noise, and may not be suitable for industrial environments.
- While the cable is inexpensive, there is the additional cost of the associated wiring hubs to be considered.
- The node cable is limited to 100 m.

Advantages of the system include:

- Intelligent hubs are available that can determine which spurs from the hub receive information. This improves on the security of the network – a feature that has often been lacking in a broadcast, common media network such as Ethernet.
- Flood wiring can be installed in a new building, providing many more wiring points than are initially needed, but giving greater flexibility for future expansion. When this is done, patch panels – or punch down blocks – are often installed for even greater flexibility.

9.2.4   10BaseF

This standard, like the 10BaseT standard, is based on a star topology using wiring hubs. The actual standard had been delayed by development work in other areas, and was ratified in September 1993. It consists of three architectures. These are:

- **10BaseFL**
  This is the fiber link segment standard that is basically a 2-km upgrade to the existing fiber optic inter repeater link (FOIRL) standard. The original FOIRL as specified in the 802.3 standard was limited to a 1km fiber link between two repeaters, with a maximum length of 2.5 km if there are 5 segments in the link. Note that this is a link between two repeaters in a network, and cannot have any nodes connected to it.

- **10BaseFP**
  This is a star topology network based on the use of a passive fiber optic star coupler. Up to 33 ports are available per star, and each segment has a maximum length of 500 m. The passive hub is immune to external noise and is an excellent choice for noisy industrial environments.

- **10BaseFB**
  This is the fiber backbone link segment in which data is transmitted synchronously. It is designed only for connecting repeaters, and for repeaters to use this standard, they must include a built-in transceiver. This reduces the time taken to transfer a frame across the repeater hub. The maximum link length is 2 km, although up to 15 repeaters can be cascaded, giving greater flexibility in network design.

9.2.5   100 Base-T (100 Base-TX, T4, FX, T2)

This is the preferred approach to 100 Mbps transmission that uses the existing Ethernet MAC layer with various enhanced physical media dependent (PMD) layers to improve the speed. These are described in the IEEE 802.3u and 802.3y standards as follows:
IEEE 802.3u defines three different versions based on the physical media:

- 100Base-TX which uses two pairs of category 5 UTP or STP
- 100Base-T4 which uses four pairs of wires of category 3,4 or 5 UTP
- 100Base-FX which uses multimode or single-mode fiber optic cable
  IEEE 802.3y:
- 100Base-T2, which uses two pairs of wires of category 3,4 or 5 UTP.

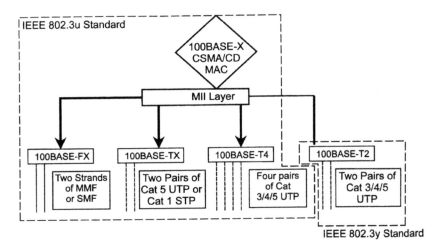

Figure 9.6
Summary of 100Base-T standards

This approach is possible because the original 802.3 specification defines the MAC layer independently of the various physical PMD layers it supports. As you will recall, the MAC layer defines the format of the Ethernet frame and defines the operation of the CSMA/CD access mechanism. The time dependent parameters are defined in the 802.3 specification in terms of bit-time intervals and so are speed-independent. The 10 Mbps Ethernet inter-frame gap is actually defined as an absolute time interval of 9.60 microseconds, equivalent to 96 bit times; while the 100 Mbps system reduces this by ten times to 960 nanoseconds.

One of the limitations of the 100Base-T system is the size of the collision domain, which is 250 m. This is the maximum sized network in which collisions can be detected; being one tenth of the size of the maximum 10 Mbps network. This limits the distance between our workstation and hub to 100 m, the same as for 10Base-T, but usually only one hub is allowed in a collision domain. This means that networks larger than 200 m must be logically connected together by store and forward type devices such as bridges, routers or switches. However, this is not a bad thing since it segregates the traffic within each collision domain, and hence reducing the number of collisions on the network. The use of bridges and routers for traffic segregation, in this manner, is often done on industrial CSMA/CD networks.

The dominant 100Base-T system is 100Base-TX, which accounts for about 95% of all fast Ethernet shipments. The 100Base-T4 systems were developed to use four pairs of category 3 cable; however few users had the spare pairs available and T4 systems are not capable of full-duplex operation, so this system has not been widely used. The 100Base-T2 system has not been marketed at this stage, however its underlying technology using digital signal processing (DSP) techniques is used for the 1000Base-T systems on two category 5 pairs. With category 3 cable diminishing in importance, it is not expected that the 100Base-T2 systems will become significant.

Ethernet signals are encoded using the Manchester encoding scheme. This method allows a clock to be extracted at the receiver end to synchronize the transmission/ reception process. The encoding is performed by performing an 'exclusive-or' between a 20 MHz clock signal and the data stream. In the resulting signal, a 0 is represented by a high to low change at the center of the bit cell, whilst a 1 is represented by a low to high change at the center of the bit cell. There may or may not be transitions at the beginning of a cell as well, but these are ignored by the receiver. The transitions in every cell allow the clock to be extracted, and synchronized with the transmitter.

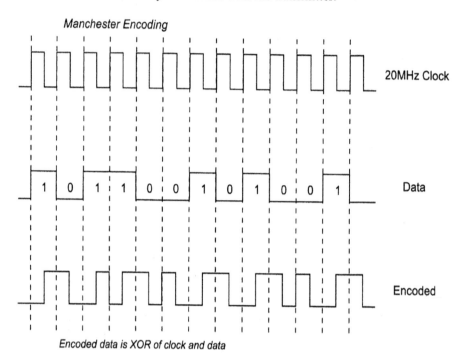

Encoded data is XOR of clock and data

Figure 9.7
Manchester encoding

The voltage swings were from −0.225 to −1.825 volts in the original Ethernet specification. In the 802.3 standard, voltages on coax cables are specified to swing between 0 and −2.05 volts with a rise and fall time of 25 ns at 10 Mbps.

9.3   Medium access control

Essentially, the method used is one of contention. As was described in the first section on this architecture, each node has a connection via a transceiver to the common bus. As a transceiver, it can both transmit and receive at the same time. Each node can be in any one of three states at any time. These states are:

- Idle, or listen
- Transmit
- Contention

In the idle state, the node merely listens to the bus, monitoring all traffic that passes. If a node then wishes to transmit information, it will defer while there is any activity on the bus, since this is the 'carrier sense' component of the architecture. At some stage, the bus will become silent, and the node, sensing this, will then commence its transmission. It is now in the transmit mode, and will both transmit and listen at the same time. This is because there is no guarantee that another node at some other point on the bus has not also started transmitting having recognized the absence of traffic. After a short delay as the two signals propagate towards each other on the cable, there will be a collision of signals. Quite obviously, the two transmissions cannot coexist on the common bus, since there is no mechanism for the mixed analog signals to be 'unscrambled'. The transceiver quickly detects this collision, since it is monitoring both its input and output and recognizes the difference. The node now goes into the third state of contention. The node will continue to transmit the jam signal for a short time, to ensure that the other transmitting node detects the contention, and then performs a back-off algorithm to determine when it should again attempt to transmit its waiting frames.

When a frame is to be transmitted, the medium-access-control monitors the bus and defers to any passing traffic. After a period of 96 bit times, known as the inter-frame gap, to allow the passing frame to be received and processed by the destination node, the transmission process commences. Since there is a finite time for this transmission to propagate to the ends of the bus cable, and thus ensure that all nodes recognize that the medium is busy, the transceiver turns on a collision detect circuit whilst the transmission takes place. In fact, once a certain number of bits (576 bits in a 10 Mbps system) have been transmitted, provided that the network cable segment specifications have been complied with, the collision detection circuitry can be disabled. If a collision should take place after this, it will be the responsibility of higher protocols to request retransmission – a far slower process than the hardware collision detection process. Here is a good reason to comply with cable segment specifications! This initial 'danger' period is known as the collision window, and is effectively twice the time interval for the first bit of any transmission, to propagate to all parts of the network. The slot time for the network is then defined as the worst-case time delay that a node must wait, before it can reliably know that a collision has occurred. It is defined as:

Slot time = 2 * (transmission path delay) + safety margin
For a 10 Mbps system, the slot time is FIXED as 512 bits or 64 octets.

The transceiver of each node is constantly monitoring the bus for a transmission signal. As soon as one is recognized, the NIC activates a carrier sense signal to indicate that transmissions cannot be made. The first bits of the MAC frame are a preamble and consist of 56 bits of 1010 etc. On recognizing these, the receiver synchronizes its clock, and converts the Manchester encoded signal back into binary form. The eighth octet is a start of frame delimiter, and this is used to indicate to the receiver that it should strip off the first eight octets and commence determining whether this frame is for its node by reading the destination address. If the address is recognized, the data is loaded into a frame buffer within the NIC. Further processing then takes place, including the calculation and comparison of the frame CRC with the transmitted CRC, checking that the frame contains an integral number of octets and is neither too short nor too long. Provided all is correct, the frame is passed to the LLC layer for further processing.

Collisions are a normal part of a CSMA/CD network. The monitoring and detection of collisions is the method by which a node ensures unique access to the shared medium. It is only a problem when there are excessive collisions. This reduces the available bandwidth of the cable and slows the system down while re-transmission attempts occur. There are many reasons for excessive collisions and these will be investigated shortly. The principle of collision cause and detection is shown in the diagram below.

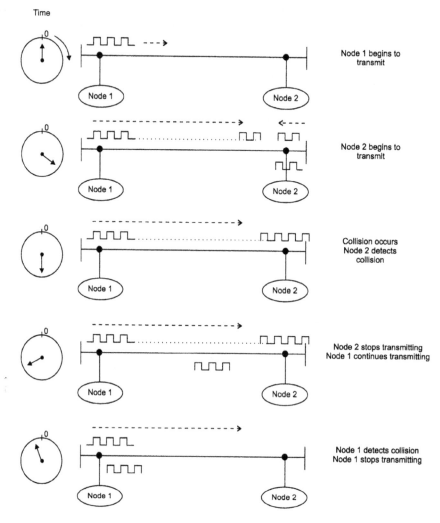

Figure 9.8
CSMA/CD collisions

Assume that both node 1 and node 2 are in listen mode and node 1 has frames queued to transmit. All previous traffic on the medium has ceased i.e. there is no carrier, and the inter-frame gap from the last transmission has timed out. Node 1 now commences to transmit its preamble signal, which immediately commences to propagate both left and right on the cable. At the left end, the transmission is absorbed by the termination resistance, but the signal continues to propagate to the right. However, the MAC sub-layer in node 2 has also been given a frame to transmit from the LLC sub-layer, and since the node 'sees' a free cable, it too commences to transmit its preamble. Again, the signals propagate on to the cable, and some short time later, they 'collide'. Almost immediately, node 2's transceiver recognizes that the signals on the cable are corrupted, and the logic incorporated on the NIC asserts a collision detect signal. This causes node 2 to send a jam signal of 32 bits of random data, and then stop transmitting. In fact, the standard allows any data to be sent as long as, by design, it is not the value of the CRC field of the frame. It appears that most nodes will send the next 32 bits of the data frame as a jam, since it is instantly available.

This jam signal continues to propagate along the cable, as a contention signal, since it is 'mixed' with the signal still being transmitted from node 1. Eventually, node 1 recognizes the collision, and goes through the same jam process as node 2. You can see from this that the frame from node 1 must be at least twice the end-to-end propagation delay of the network, or else the collision detection will not work correctly. The jam signal from node 1 will continue to propagate across the network until absorbed at the far end terminator, meaning that the system vulnerable period is three times the end to end propagation delay.

After the jam sequence has been sent, the transmission is halted. The node then schedules a retransmission attempt after a random delay controlled by a process known as the truncated binary exponential back-off algorithm. The length of the delay is chosen so that it is a compromise between reducing the probability of another collision and delaying the retransmission for an unacceptable length of time. The delay is always an integer multiple of the slot time. In the first attempt, the node will choose, at random, either 1 or 0 slot-times delay. If another collision occurs, the delay will be chosen at random from 0, 1, 2 or 3 slot-times, thus reducing the probability that a further collision will occur. This process can continue for up to 10 attempts, with a doubling of the range of slot times available for the node to delay transmission at each attempt. After ten attempts, the node will attempt 6 more retries, but the slot times available for the delay period will remain as they were at the tenth attempt. After 16 attempts, it is likely that there is a problem on the network and the node will cease attempting to retransmit.

9.4   MAC frame format

The basic frame format for an 802.3 network is shown below. There are eight fields in each frame, and they will be described in detail.

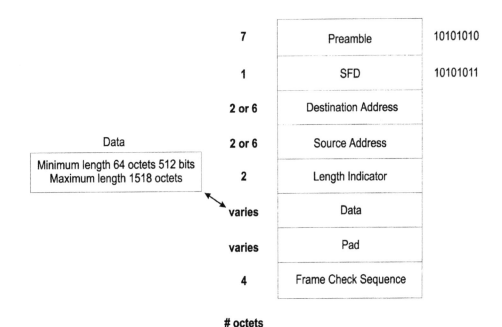

Figure 9.9
MAC frame format

- **Preamble**

  This field consists of 7 octets of the data pattern 10101010. It is used by the receiver to synchronize its clock to the transmitter.

- **Start frame delimiter**

  This single octet field consists of the data 10101011. It enables the receiver to recognize the commencement of the address fields.

- **Source and destination address**

  These are the physical addresses of both the source and destination nodes. The fields can be 2 or 6 octets long, although the six-octet standard is the most common. The six-octet field is split into two three-octet blocks. The first three octets describe the block number to which all NICs of this type belong. This number is the license number and all cards made by this company have the same number. The second block refers to the device identifier, and each card will have a unique address under the terms of the license to manufacture. This means there are 2^{48} unique addresses for Ethernet cards.

There are three addressing modes that are available. These are:

1. Broadcast – the destination address is set to all 1s or FFFFFFFFFFFF.
2. Multicast – the first bit of the destination address is set to a 1. It provides group restricted communications.
3. Individual, or point to point – first bit of the address set to 0, and the rest set according to the target destination node.

- **Length**

  This is defined as a two-octet field that contains the length of the data field. This is necessary since there is no end delimiter in the frame.

- **Information**

  This is defined as the information that has been handed down from the LLC sub-layer.

- **Pad**

  Since there is a minimum length of frame of 64 octets (512 bits or 576 bits if the preamble is included) that must be transmitted to ensure that the collision mechanism works, the pad field will pad out any frame that does not meet this minimum specification. This pad, if incorporated, is normally random data. The CRC is calculated over the data in the pad field. Once the CRC checks OK, the receiving node discards the pad data, which it recognizes by the value in the length field.

- **FCS**

  A 32-bit CRC value that is computed in hardware at the transmitter and appended to the frame. It is the same algorithm used in the 802.4 and 802.5 standard.

9.5   Difference between 802.3 and Ethernet

As has already been discussed, there is a difference between an 802.3 network and a Blue Book Ethernet network. These differences are primarily in the frame structure and are tabulated below.

802.3 Network	Ethernet network
Star topology supported using UTP, fiber etc	Only supports bus topology
Baseband and broadband signaling	Baseband only
Data link layer divided into LLC and MAC	No subdivision of DLL
7 octets of preamble plus SFD	8 bytes of preamble with no separate SFD
Length field in data frame	Field used to indicate the higher level protocol using the data link service
SQE can be used as network management device	SQE can only be used in version 2.0

Table 9.1
Differences between IEEE 802.3 and Blue Book Ethernet (V2)

The significant difference in the frame is the length-field in 802.3 vs the higher protocol field in Ethernet. Since an 802.3 frame cannot be longer than 1500 bytes, the values in the protocol type field of the Ethernet frame commences at 1500. This allows protocol analyzers to recognize one type of frame vs the other.

9.6 Reducing collisions

The main reasons for collision rates on an Ethernet network are:

- The number of packets per second
- The signal propagation delay between transmitting nodes
- The number of stations initiating packets
- The bandwidth utilization
  A few suggestions on reducing collisions in an Ethernet network are:
- Keep all cables as short as possible
- Keep all high activity sources and their destinations as close as possible. Possibly isolate these nodes from the main network backbone with bridges/routers to reduce backbone traffic
- Use buffered repeaters rather than bit repeaters
- Check for unnecessary broadcast packets which are aimed at non existent nodes
- Remember that the monitoring equipment to check out network traffic can contribute to the traffic (and the collision rate)

9.7 Ethernet design rules

The following design rules on length of cable segment, node placement and hardware usage should be strictly observed.

9.7.1 Length of the cable segment

It is important to maintain the overall Ethernet requirements as far as length of the cable is concerned. Each segment has a particular maximum length allowable. For example, 10BASE-2 allows 200 m maximum length. The recommended maximum length is 80%

of this figure. Some manufacturers advise that you can disregard this limit with their equipment. This can be a risky strategy and should be carefully considered.

System	Maximum	Recommended
10Base5	500m	400m
10Base2	185m	150m
10BaseT	100m	80m
1Base5	500m	400m

Table 9.2

Cable segments need not be made from a single homogenous length of cable, and may comprise of multiple lengths joined by coaxial connectors (two male plugs and a connector barrel). Although ThickNet (10BASE-5) and ThinNet (10BASE-2) cables have the same nominal 50 ohm impedance they can only be mixed within the same 10BASE-2 cable segment to achieve greater segment length.

To achieve maximum performance on 10BASE-5 cable segments, it is preferable that the total segment be made from one length of cable or from sections off the same drum of cable. If multiple sections of cable from different manufacturers are used, then these should be standard lengths of 23.4 m, 70.2 m, or 117 m ($\pm$ 0.5 m), which are odd multiples of 23.4 m (half wavelength in the cable at 5 MHz). These lengths ensure that reflections from the cable-to-cable impedance-discontinuities are unlikely to be added in phases. Using these lengths exclusively, a mix of cable sections should be able to make up the full 500 m segment length.

If the cable is from different manufacturers and one suspects potential mismatch problems, then one should check that signal reflections due to impedance mismatches do not exceed 7% of the incident wave.

9.7.2   Maximum transceiver cable length

In 10BASE-5 systems the maximum length, of transceiver cables, is 50 m but it should be noted that this applies to specified IEEE 802.3 compliant cables only. Other AUI cables using ribbon or office grade cables can only be used for short distances (less than 12.5 m) so check the manufacturer's specifications for these!

9.7.3   Node placement rules

Connection of the transceiver media access units (MAU) to the cable causes signal reflections due to their bridging impedance. Placement of the MAUs must therefore, be controlled to ensure that reflections from them do not significantly add in phases.

In 10BASE-5 systems the MAUs are spaced at 2.5 m multiples, coinciding with the cable markings.

In 10BASE-2 systems, the minimum MAU spacing is 0.5 m.

9.7.4   Maximum transmission path

The maximum transmission path is made of five segments connected by four repeaters. The total number of segments can be made up of a maximum of three coax segments containing station nodes and two link segments, having no intermediate nodes. This is summarized as the 5-4-3-2 rule.

5 segments 4 repeaters 3 coax segments 2 link segments	OR	5 segments 4 repeaters 3 link segments 2 coax segments

Table 9.3
5-4-3-2 rule

It is important to verify that the above transmission rules are met by all paths between any two nodes on the network.

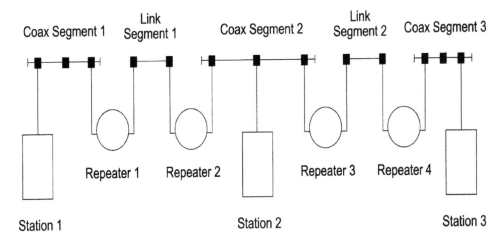

Figure 9.10
Maximum transmission path

Note that the maximum sized network, of four repeaters, supported by IEEE 802.3, can be susceptible to timing problems. The maximum configuration is limited by propagation delay.

Note that 10BASE-2 segments should not be used to link 10BASE-5 segments.

9.7.5   Maximum network size

- 10BASE-5   =   2800 m node to node (5 × 500 m segments + 4 repeater cables + 2 AUI)
- 10BASE-2   =   925 m node to node (5 × 185 m segments)
- 10BASE-T   =   100 m node to hub

9.7.6   Repeater rules

Repeaters are connected to transceivers that count as one node on the segments. Special transceivers are used to connect repeaters and these do not implement the signal quality error test (SQE).

Fiber optic repeaters are available giving up to 3000 m links at 10 Mbps. Check the vendor's specifications for adherence with IEEE 802.3 repeater-performance and compliance with the fiber optic inter repeater link (FOIRL) standard.

9.7.7   Cable system grounding

Grounding has safety and noise connotations. 8802.3 states that the shield conductor of each coaxial cable shall make electrical contact with an effective earth reference at one point only.

The single point earth reference for an Ethernet system is usually located at one of the terminators. Most terminators for Ethernet have a screw terminal to which a ground lug can be attached using a braided cable preferably to ensure good earthing.

Ensure that all other splices, taps, or terminators are jacketed so that no contact can be made with any metal objects. Insulating boots or sleeves should be used on all inline coaxial connectors to avoid unintended earth contacts.

9.8   Fieldbuses

There are currently several hybrid analogue and digital standards available for communication between field devices and also between field devices and a master system. But only a fully compatible digital communication standard will provide the maximum benefits to end users and one such standard, currently being proposed, is Foundation FieldBus

It would be helpful to ask, why is there considerable effort, time and money being invested in searching for a 'perfect' digital communication network? Why are there several approaches and not just one unified effort? Aren't there enough standards and what is wrong with the one's we have?

The current approach to cabling of a typical control system is shown in Figure 9.10. Note that each instrument and actuator is connected back to the instrument room (to a controller) with an individual pair of wires.

The strategy espoused today is to replace this with a communication cable, which connects all the instruments and actuators together, and has several significant advantages listed below. Each instrument and actuator now becomes an 'intelligent device'. An intelligent device can be considered a computer-controlled device that takes analog data (e.g. flow meter), performs an operation on it (e.g. square root extraction), and sends this up a communication network to another device(s), which requires this data. Similarly, an intelligent actuator can control a valve to a specific position with a data value sent down the communications network from another device.

The answer then lies in the potential benefits, and to a lesser degree, in the realization that the older standards were developed for specific reasons that may not be applicable to the demands placed on present day communications systems.

Benefits

There are real benefits to be gained from emerging networks, including:

- Greatly reduced wiring costs
- Reduced installation and start-up time
- Improved on-line monitoring and diagnostics
- Easier change-out and expansion of devices
- Improved local intelligence in the devices
- Improved interoperability between manufacturers

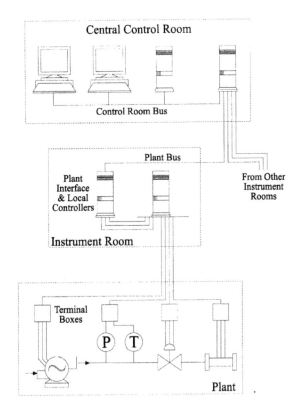

Figure 9.11
Current approach to cabling of a typical control system

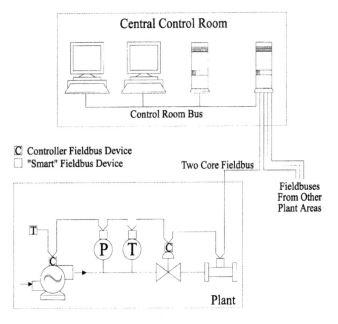

Figure 9.12
Fieldbus approach to cabling of a typical control system

Different systems

Just as with any technology, each of the different approaches being undertaken has benefits and drawbacks. It would at first seem that a single digital bus system would be beneficial to all users, but this is not the case. Very simple field devices (proximity switches, limit switches, and basic actuators), for example, only require a few bits of digital information to communicate 'off or on' states, and are usually associated with real time control applications where update times of a few milliseconds are required. The associated electronics necessary to communicate with these systems can be simple, compact and inexpensive enough to be integrated in the device itself.

Alternatively, complicated devices like PLCs, operator stations (sometimes referred to as man-machine-interfaces or MMIs) and DCSs (distributed control systems), require multi-byte length messages (up to 256 bytes in some systems) and may only require update times of 10–100ms depending on the application. The associated electronics are more expensive and require larger packaging that restricts the integration of these systems into very small devices.

The solution is to select the digital communication network that is best suited to the application, and integrate information up through the higher speed networks as required. Toward this end, several approaches in digital networks have been developed over the last few years, each with a different target application, speed, and technology.

These different approaches are being generically referred to as device networks or fieldbuses and are typically categorized by the length of the 'message' required by the devices to adequately convey information to the host or network.

This method of categorization allows these networks to be placed in one of the following three network oriented classes:

- Bit (sensor)
- Byte (device)
- Message (field)

For example, the bit-oriented systems are used with simple binary type devices such as proximity sensors, contact closures, (pressure switches, float switches, etc), simple pushbutton stations, and pneumatic actuators. These types of networks are also known as 'sensor bus' networks due to the nature of the devices (sensors and actuators) typically used. Excellent examples of this type of system are ASI (actuator sensor interface).

Byte oriented systems are used in much broader applications such as motor starters, bar-code readers, temperature and pressure transmitters, chromatographs and variable-speed drives, due to their larger addressing capability and the larger information content of the several byte length message format. These networks are also referred to as 'device bus' systems or networks. An excellent example of this approach is CANBus or DeviceNet.

Message oriented systems, which are those systems containing over 16 bytes per message, are finding application in interconnecting more intelligent systems like PCs, PLCs, operator terminals, and engineering workstations where uploading and down-loading system or device configurations are required, or in linking the above-mentioned networks together. These systems represent the higher end of the new pure digital networks for industrial and automation environments and are referred to in general as 'FieldBus' systems or networks. Examples of these systems are Interbus-S, Profibus and Foundation Fieldbus.

10

The universal serial bus (USB)

10.1 Introduction

In September 23, 1998 Microsoft, Intel, Compaq, and NEC developed Revision 1.1 of the universal serial bus. The objective was to standardize the input/output connections on the IBM PC for devices like printers, mice, keyboards and speakers. Data acquisition (DAQ) devices were not envisioned to be connected to the USB system. But that does not mean that the USB cannot be used for DAQ. In many ways, the USB is well suited for DAQ systems in the laboratory or other small-scale systems.

Small-scale DAQ systems have traditionally suffered from the need of an easy to use and standardized bus system for connecting smart DAQ devices. The nearest thing was the IEEE 488 GPIB system. The GPIB system is expensive and uses very old technology. There is need for an easy-to-operate, inexpensive, and standardized bus system to connect small-scale DAQ devices. The USB can fill those needs. With its plug and play ability, it is extremely easy to implement and use, and it is now standard on all IBM compatible PCs. Although it is not in any way as cheap as say an RS 232 connection, it is affordable.

The USB is limited by its very nature, for its application, to DAQ systems. The biggest problem is the maximum cable distance. The low speed version is limited to 3 meters and the high-speed version is limited to 5 meters in total cable length. This requirement reduces the ability of USB to be used in a large, factory or plant, environment. Typically in these industries, DAQ systems need distances up to 1 kilometer. Due to the timing requirements of the USB, the length of the cable cannot be increased with repeaters. This limits the use of the USB to laboratories and or bench top systems.

10.2 USB – overall structure

The USB is a master/slave, half-duplex, timed communication bus system designed to connect close peripherals and hubs to an IBM compatible PC. It runs at either 1.5 Mbps (low speed) or 12 Mbps (high speed). The PC's software program using device drivers create packets of information that are going to be sent to devices connected on the USB bus. The USB drivers in the computer allocate a certain time within a frame for the information. The packet is then placed in this 1 ms frame that can contain many packets. One frame might contain information for many devices or may contain information for only one device. The frame is then sent to the physical layer via USB drivers and then on to the bus.

The device receives its part of the packet and if necessary formulates a response. It then places this response on the bus. The USB drivers in the PC hear the response on the bus and verify that the frame is correctly using CRC. If the CRC indicates that the frame is correct, the software in the PC accepts the response.

The devices connected to the USB bus can also be powered off the bus cable. Devices can use no more than 500 mA. This works well for small-scale DAQ devices; larger DAQ devices usually use external power supplies. Both power and communications are on the same cable and connector.

There are many parts in the USB system that make the communication possible. These include:

- Host hubs
- External hubs
- Type A connector
- Type B connector
- Low speed cables
- High speed cables
- USB devices
- Host hub controller hardware and driver
- USB software driver
- Device drivers

10.2.1   Topology

The USB uses a pyramid shaped topology with everything starting at the host hub. The host hub usually consists of two USB ports on the back of the PC. These ports are basically in parallel with each other. Each port is a four-pin socket with two pins reserved for power and two for communications. The cables from eternal hubs or USB devices are plugged into host hub ports. One or both of the ports can be used. It doesn't matter which one is used if only one connection is being made. If the external device or hub has a removable cable then a type 'A to type B' cable is used to make the connection. The A plug goes into the back of a PC (host hub) and the B plug goes into the device or external hub. If the external hub or device has an in-built cable then the A plug is plugged into the host hub port. The socket on the host hub is keyed so the plug will only go in one way. B plugs will not go into A sockets and vice-versa.

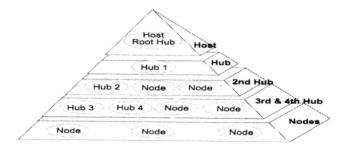

Figure 10.1
USB topology

Cable lengths are very important (and short) for the USB system. All cables, even if they come out of repeater hubs, must be counted in the total length of the cables.

10.2.2   Host hubs

The controller chips for the host hub usually reside on the motherboard inside the PC, (although the hub could be a PCB in a PCI slot). The host controller does the parallel to serial and serial to parallel conversion from the PCI bus to the USB connectors. Sometimes a pre-processor is used to improve efficiency of the USB system. This host controller and connector combination is called the root hub or host hub. The host Hubs' function is to pass the information to and from the PCI bus to the data lines (+D and –D) on the USB socket. The host controller can control the speed at which the USB operates. It also connects power lines (+5 V and ground) to a USB device via the USB cable. The external USB device may be another USB hub or a USB type device like a printer.

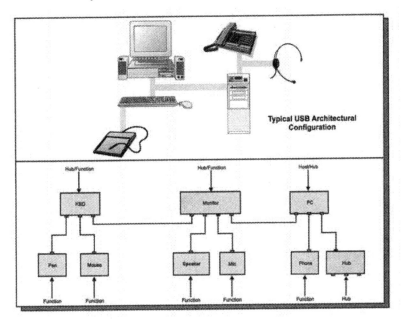

Figure 10.2
Host hub block diagram

The root hub has complete control over the USB ports. This control would include:

- Initialization and configuration
- Enabling and disabling the ports
- Recognizing the speed of devices
- Recognizing that a device has been connected
- Getting information from the application software
- Creating a packet and then frame
- Sending the information on to the bus
- Waiting and recognizing a response
- Error correction
- Recognizing that a device has been disconnected
- Using the port as a repeater

10.2.3   The connectors (type A and B)

There are two types of connectors, type A and type B. The reason there are two types is that some devices have built in cables while others have removable cables. If the cables were the same it would be possible to connect a host hub port to another host hub port. Because of the polarity of the connectors, the +5 volts would be connected to ground. To keep this from happening the hub's output ports use type A connectors and the device input ports use type B. This means that it is impossible to connect one hub port to another hub port. On an external hub, the input to the hub is a type B connector unless the cable on the hub is permanently connected (no connector).

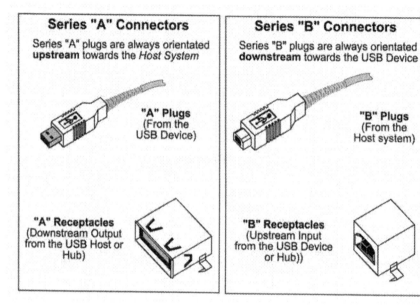

Figure 10.3
USB connectors

10.2.4   Low-speed cables and high-speed cables

The USB standard states that the USB will run at either 1.5 Mbps (slow speed) or 12 Mbps (fast speed). The USB must have low-speed cables and high-speed cables. This is due to the impedance difference caused by the different frequencies of data transfer. Low speed cables use untwisted and unshielded. The data pair is 28 AWG and the power pair is 20–28 AWG. The low-speed cable is used on devices like keyboards and mice. The maximum distance for low-speed cabling is 3 meters. High-speed cables are twisted and shielded. The data pair is 28 AWG and the power pair is 20–28 AWG. The maximum propagation delay must be less than 30 ns. The maximum distance for high speed USB is 5 meters.

10.2.5   External hubs

External hubs are used to increase the amount of devices connected to the system. Usually they have four USB output ports and either one type B input connector or a dedicated cable. This cable has a type A plug on it. It is usually connected to a host hub, but could also be connected to the output socket (type A) of another external hub. Even though the external hub is a repeater it cannot extend the overall length of the system. This is because of the timing requirements of the USB standard.

The external hub is an intelligent device that can control communication lines and power lines on its USB ports. It is a bi-directional repeater for information coming from the host hub and from USB devices. It talks to and even acts like an external USB device to the host hub. It plays an integral part in the configuration of devices at startup. There is no physical limit to the number of hubs.

10.2.6 USB devices

The USB system supports every peripheral that can be currently connected to a PC. It has also been adapted to devices that are not usually considered peripherals. This would include data acquisition devices such as digital I/O modules and analog input and output modules. All USB devices must be intelligent devices. Smart devices obviously cost more than the old dumb RS-232 and RS-485 connected devices. With this cost the user gets more functions, ease of use and the ability to connect more devices to the PC. With the old non-USB system the computer was limited to a few devices. The USB system allows 127 devices to be connected to the PC at the same time. There are two types of USB devices, low-speed and high-speed.

Low-speed devices

Low-speed devices are not only limited in their speed but also in features. These devices could be keyboards, mice and digital joysticks. Since these devices put out small amounts of information they are polled less and are slower than other devices. When the USB bus is being accessed by high-speed devices, the low-speed device communication is disabled. Turning off the low-speed device ports at the root or external hubs, disables the low-speed devices. The hubs re-enable the low-speed ports after receiving a special preamble packet.

High-speed devices

High-speed devices like printers, CD-ROMs and speakers need the speed of the 12 Mbps bus, to transfer the large amount of data required for these devices. All high-speed devices see all traffic on the bus. They are never disabled like the low-speed ones. When a device like a microphone is 'connected' to the speakers, most of the traffic, and therefore packets, will be used by the audio system. Other traffic like keyboard and mouse functions will have to wait. The host hub controller driver decides who has to wait and how long.

10.2.7 Host hub controller hardware and driver

The host hub controller hardware and software drivers, control all transactions. The host hub controller hardware does the physical connections from the PCI bus to the USB connectors. It enables and initializes the host ports one at a time and determines the speed and direction of data transfer on both host ports. The host controller in conjunction with the host hub software driver determines the frame contents, prioritization of the devices, and how many frames are needed for a particular transfer.

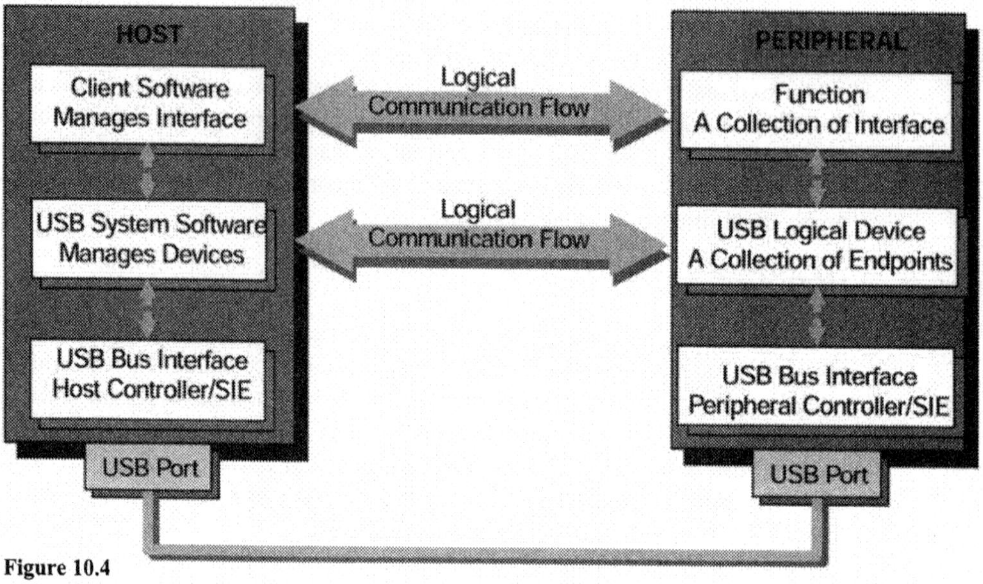

Figure 10.4
Host hub controller diagram

10.2.8   USB software driver

The USB software driver handles the interface between the USB devices, the device drivers and the host hub driver. When it receives a request from a device driver in the PC to access a certain device, it organizes the request with other device requests from the application software, in the PC. It works with the host hub controller driver to prioritize packets, before they are loaded into a frame. The USB software driver gets information from the USB devices during device configuration. It uses this information to tell the host hub controller how to communicate to the device.

10.2.9   Device drivers

Each USB device must have a device driver loaded into the PC. This device driver is a software interface between the external USB device and the application software, the USB software driver, and the host hub controller driver. It has information about that particular device's needs for the other drivers. This information is used to determine things like the type, speed (although that information can be determined physically by the hub ports), priority, function of the device, and the size of the packet needed for data transfer.

10.2.10   Communication flow

As mentioned before, the USB system is a master/slave, half duplex, timed communication bus system, designed to connect peripherals and external hubs. This means that the peripherals cannot initiate a communication on the USB bus. The master or host hub has complete control over the transaction. It initiates all communications with hubs and devices. The USB is timed because all frames are sent within a 1 ms time slot. More than one device can have a packet of information inside that 1 ms frame. The host hub driver, in conjunction with the USB software driver, determines the size of the packet, and how much time each device gets in one frame.

If the application's software wants to send or receive some information from a device, it initiates a transfer via the device driver. This device driver is supplied by either the manufacture of the device, or comes with the operating system. The USB driver software then takes this request and places it in a memory location with other requests from other device drivers. Working together, the USB driver, host hub driver and the host hub controller place the request, data and packets from the device drivers into a 1 ms wide frame. The host controller then transfers the data serially to the host hub ports. Since all the devices are in parallel on the USB bus, all of the devices hear the information, (except low-speed devices, unless it is a low-speed transfer. Low-speed devices are turned off when they are not being polled). The host then waits for a response (if necessary). The remote USB device then responds with an appropriate packet of information. If a device does not see any bus activity for 3 ms then it will go into the suspend mode.

8 bits	1023 bytes	16 bits
PID	DATA	CRC

Figure 10.5
Example of a USB packet

There are four types of IN packets (reading information from a device) and three types of OUT packets (sending information out to a device). Certain devices like mice and keyboards need to be polled (IN packets), but not too often. The USB software driver knows about these devices and schedules a regular poll for them. Included in the response are three levels of error correction. This type of transfer is very reliable. These peripherals are usually low-speed devices and therefore need a distinct low-speed packet to enable them. This packet is called a preamble packet. The preamble packet is sent out before each poll. The low-speed devices are disabled until they receive this preamble packet. Once they are enabled, they hear the poll and respond. Only one device can be polled at a time and therefore only one device will respond. USB has no provisions for multiple responses from devices.

On the other hand there are devices that need constant attention but polling is not possible. These would be devices like microphones (IN packets), speakers (OUT packets) and CD-ROMs (both types of packets IN and OUT). The transfer rate is very important to these devices. Obviously they would use the high-speed transfer rate. They would also use a large portion of the frame (up to 90%). The receiving device does **NOT** respond to the data transfer. This transfer is a one-way data transfer or 'simplex'. This means that error correction is effectively turned off for these types of transfers.

10.3   The physical layer

The physical layer of the universal serial bus is based on a differential +/− 3 V DC communication system. It is in some ways very similar to the RS-485 voltage standard. Unfortunately it does not have the range of 485. This is not because of the type of wire used or because of the USB voltage standard itself. It is because of the timing requirements of the USB protocol. In order to fit in all of the things the peripherals do on a USB bus, it was necessary to put very strict time requirements on the USB. The USB physical standard has a lot of benefits to the user. It is fast, 12 MHz. It is very resistant to noise and is very reliable, as long as the user follows the cabling rules. With standardized cables and connectors it is very hard for the user to get things wrong when cabling the USB system.

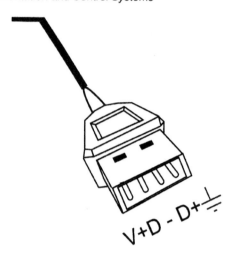

Figure 10.6
USB connector pins

There is a story that one day Bill Gates was watching some installers put in his new computer. When he saw all the wires coming out of the back of the computer he called the CEO at Intel and said, 'We have to get rid of this mess of cables and connectors'. And as they say, the rest is history.

10.3.1   Connectors

The plugs and sockets on the USB have two wires for data communication and two wires for power. Using bus-powered devices is optional. The pins on the plug are not the same length. The power pins are 7.41 mm long and the communication pins are 6.41mm long. This means that if a cable is plugged in 'hot' the power will be applied to the device before the communications lines. More importantly it also means that when a cable is unplugged the data communications lines will be disconnected before the power. This would reduce the possibility of back EMF voltages damaging the equipment. There are two types of connectors for the USB, type A and type B.

Type A connector is a flat semi-rectangular keyed connector that is used on the host ports, external hubs and devices. The type B keyed connector is half-round and smaller than the type A connector. Note that both type A and B plugs have the USB symbol on the **top** of the connector. This is for orientation purposes.

The hubs and devices all have female sockets, while the cables have a type A male plug, on one end, and a type B on the other end. This is because if there were a type A on both ends it would be possible to connect two host hub sockets or external hub sockets together. Cables that are not removable from the device or external hub, have only the type A plug on one end.

10.3.2   Cables

The cables for the USB are specified as either low- or high-speed cables. Both the low- and high-speed cables can use type A connectors, but only a high-speed device can use type B connectors. Detachable cables are therefore always high-speed cables.

Due to that fact that the impedance of a cable is determined in part by the frequency of the signal, the two speeds need two different cables. External hubs are always high-speed units, but they will accept low- and high-speed cables. Low-speed devices like keyboards will only connect to other low-speed devices using low-speed cables. The ports on the hub can detect the speed of the device on the other end. If the D+ line is pulled high (+3.0 V DC to +3.6 V DC) then the device is considered high-speed. If the D– line is pulled high then the device is considered low-speed.

The low-speed (1.5 Mbps) cable uses unshielded, and untwisted data cables. The communication pair is 28 AWG gauge, but due to the lack of shielding and twisting, the overall diameter of the cable is smaller than that of a high-speed cable. The maximum distance for the low-speed cable is 3 meters. This includes all host hub ports to external hub, as well as the external hub to device cables. Usually on data communication systems slower data speeds mean longer distances. In this case, the cable is unprotected against noise, and because of the FCC restrictions on 1 to 16 Mbps communication, its length is severely limited.

The high-speed (12 Mbps) cable uses a shielded and twisted-pair 28 AWG gauge wire. The maximum distance for high-speed cables is 5 meters. Again this includes all hub to hub and hub to device connections. The shield is internally connected to chassis-ground at both ends. Usually on data communication systems, the ground is connected at only one end, but because the distances are short, this is not a problem.

NOTE: It might be wise though, to measure the chassis to chassis ground difference between both devices before making the connection.

The power pair on both low- and high-speed cables is 20 to 28 AWG gauge. The power pair supplies between 500 and 100 mA to external devices at +5 V DC. Every port on a hub will provide this power to the devices if enabled by the hub. All hubs can decide if a port has power applied to the connector. If an external hub is itself powered by the bus then it will divide the 500 mA up into 100 mA or so per port.

10.3.3　Signaling

When a device is plugged in to a hub, the port on the hub immediately determines the speed of the device. The port looks at the voltage on the D+ and D– lines. If the D+ line goes positive, the port knows that the device is a high-speed device. If the D– line goes positive, the port knows that the device is a low-speed device. If both D+ and D- voltages fall below 0.8 V DC for more than 2.5 microseconds, the hub sees this as the device having been disconnected. If the voltage on either line is raised above 2 V DC for more than 2.5 microseconds, the port sees this as the device having been plugged-in.

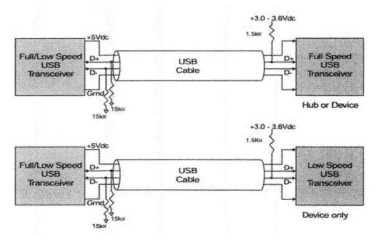

Figure 10.7
USB wiring diagram

The idle states for low- and high-speed devices are opposite of each other. For the low-speed device the idle state is the D+ line at 0 volts, and the D– a positive voltage. The idle state for a high-speed device is such that the D+ is a positive voltage and the D– is 0 volts. In most data communications, a positive voltage indicates a zero (0) condition, and a one (1), a minus voltage. In the USB system it is not possible to say this because it uses an encoding system called NRZI.

The voltages used for differential balanced signaling are:

- Maximum voltage transmitted +3.6 V DC
- Minimum voltage transmitted +2.8 V DC
- Minimum voltage needed to sense a transition +/– 2 V DC
- Typical line voltage as seen from the receiver +/– 3 V DC

10.3.4   NRZI and bit stuffing

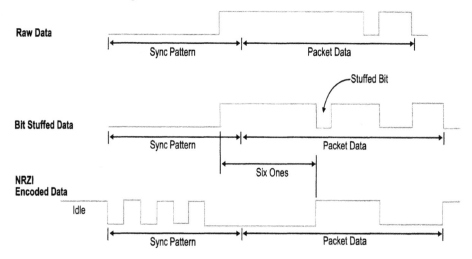

Figure 10.8
NRZI example

The USB uses non-return to zero inverted (NRZI) encoding scheme. In NRZI, a '1' is defined as 'no change' or 'transition' of voltage whereas a '0' is a change or transition of voltage. A string of '0s' would cause a clock-like data stream. The USB signaling system uses the transition from one voltage to another to synchronize the receivers. A stream of ones therefore would mean no transitions. This would cause the receiver to lose synchronization. To overcome this problem the USB system uses a 6 of 7 bit stuffing technique. If six or more ones are to be transmitted in a row the transmitter stuffs in a zero (a transition). If the receiver sees six ones in a row, it knows that the next transition (zero) is to be ignored.

10.3.5   Power distribution

Devices like keyboards and mice need power to operate. This power is supplied by the USB system through the cables and hubs. External hubs can be either self-powered or powered off the bus. The voltage supplied by a USB hub is +5 V DC. The hubs must be able to supply a minimum of 100 mA and a maximum of 500 mA, through each port. If an external hub with four ports is powered off the bus, then it will divide the 500 mA supplied between the ports. Four times 100 mA equals 400 mA. This leaves 100 mA to run the hub. It is not possible to connect two bus-powered hubs together unless the devices connected to the last hub were self-powered. If the external hub is self powered, that is mains powered, it should be able to supply 500 mA to each of the ports.

10.4 Data link layer

The data link layer within the USB specification defines the USB as a master/slave, half duplex, timed communication bus system, designed to connect close peripherals and external hubs. The hardware and software devices such as the host hub controller hardware and driver, USB software driver and device drivers all contribute to the data link layer of the USB.

All these devices working together accomplish the following:

- Collects data off the PCI bus via the device drivers
- Processes the information or data
- Verifies, determines and processes the different transfer types
- Calculates and checks for errors in the packets and frames
- Puts the different packets into a 1 ms frame
- Checks for start of frame delimiters
- Sends the packets to the physical layer
- Receives packets from the physical layer

Contact number	Signal name	Cable color
1	VCC	Red
2	–Data	White
3	+Data	Green
4	Ground	Black

Table 10.1
USB data link layer block diagram

10.4.1 Transfer types

A good place to start when looking at the data link layer of the USB is at the four different transfer types. The wide range of devices that the USB has to deal with require that there be multiple transfer types.

These are:

- Interrupt transfer
- Isochronous transfer
- Control transfers
- Bulk transfers

As stated before, there are two speeds that can be used in the USB system. For the most part the data link layer is the same, but there are some differences. The low speed devices do not support bulk and isochronous transfers. The reason for this will become apparent in the following transfer descriptions.

Interrupt transfer

The interrupt transfer type is used for devices that traditionally used IRQ lines. Devices like keyboards, mice and DAQ cards use the IRQ lines to tell the computer that they need servicing. The USB does not support devices that initiate requests to the computer. To overcome this problem the USB driver initiates a poll of those devices that it knows need periodical attention. This poll must be frequent enough so that data does not get lost but not too frequent as to use up much needed bandwidth. When installed, the device determines its minimum requirements for polling. Devices that need to be polled are rarely polled on every frame. The keyboard is typically polled only every 100th frame.

Isochronous transfer

Isochronous transfer is used when the devices need to be written to or read from, at a constant rate. This would include devices like microphones and speakers. The transfer can be done in an asynchronous, synchronous, or device specific manner, depending on the device in use. This constant attention requires that the bulk of the bandwidth of the frame be allocated to one or two devices. If too many of these transfers take place at the same time data could be lost. This type of transfer is not data quality critical. There is no error correction and lost or data that is in error, is ignored. Low-speed devices cannot use isochronous transfer because of the small amounts of data being transferred. It is not possible to move data fast enough using low-speed devices. In an isochronous transfer, the maximum amount of data that can be placed in one packet is 1023 bytes. There is no maximum number of packets that can be sent.

Control transfer

Control transfers are used to transfer specific requests and information to specific devices. This method is used mostly during the configuration and initialization cycles. These transfers are very data critical and require a response or acknowledgement from the device. Full error correction is in force for this type of transfer. All devices use this type of transfer at one time or another. These transfers use very little bandwidth but because the device must respond back to the host hub the frames are dedicated to this one transfer.

Bulk transfer

Bulk transfers are used to transfer large blocks of data to devices that are not time dependent but where data quality is important. A typical device that would use the bulk transfer method would be a writeable CD and a printer. These devices need large amounts of data but there is no time constraint like that in a speaker. If data gets there in the first 10 ms or in the next, is not a problem. But they do need correct data, so this type of transfer includes handshaking and full error correction.

10.4.2   Packets and frames

The USB protocol can and often does use a multi-packet-frame format. The USB frame is made up of up to three parts. One frame equals one transaction.
The packets are of the following type:

- The token packet
- The data packet
- The handshaking

Every frame starts with a token packet. The token packet includes other smaller packets. These are, the synchronization pattern, packet type I.D. and token packet type.

There are four types of token packets – start of frame, in packets, out packets and setup packets. The start of frame/token packet indicates the start of the packet. This tells the receiver that this is the beginning of the 1 ms frame. The in packets are packets that will transfer data in from the devices to the PC. The out packets are packets that will transfer data out from the PC to the device. The setup packet is used to ask the devices or hubs for startup information and have information for the devices or hubs.

A special packet is only used on low-speed transfers. It is called the preamble packet. It is a shorter packet than the high-speed frame. It only holds up to 64 bytes of data and always uses handshaking. It only has three variations – in packet, out packet and setup packet.

At the end of all packets, except the isochronous-frames, there is an error correction packet. On high-speed frames, it is a 16-bit CRC. Low-speed devices use a 5-bit CRC, because of their smaller packets. If a device or host hub sees an end of frame message, it checks the CRC. If the CRC is correct then it assumes that this is the end of the message. If the CRC is not correct and the time-out limit has not been reached, the receiver waits. If the CRC is not correct and the time has been reached, the receiver will then assume that the frame is not correct.

10.5 Application layer (user layer)

The application layer can be divided into two sub-layers, the operating system (such as WIN 2000) and the device application software (such as a modem application program). The application layer of the USB standard is in reality a User layer. This is because the USB standard does not really define a true application layer. What it does define is a user layer that can be used (by an application programmer) to build an application layer.

The operating system user layer includes:

- Commands
- Software drivers
- Hub configuration
- Bandwidth allocation

Device applications would use

- Commands
- Device drivers
- Device configuration

Specific user layer information can be found in the universal serial bus specifications at the USB implementers forum web page at http://www.usb.org

10.6 Conclusion

Designed as a peripheral connection system for the PC, the universal serial bus can be adapted to be used on data acquisition systems. Now that the DAQ industry is developing more and more intelligent data acquisition and control systems, the USB is easily adaptable to modern DAQ. The devices can either be low- or high-speed devices, and are quickly and easily connected to a PC. There are many devices in the market now and it is bound to grow in the future. Engineers are usually more interested in getting the job done than in spending a lot of time and trouble getting the DAQ system up and running. With the plug and play system incorporated in USB the user does not have to spend hours or even days configuring a DAQ system. This time saved will often offset the added cost of the devices.

The target speed of USB 2.0 is 480 Mbps, as announced by the USB 2.0 Promoter Group, consisting of Compaq, Hewlett-Packard, Intel, Lucent, Microsoft, NEC and Philips. The target speed announcement coincides with the release of the USB 2.0 specification draft to industry developers.

10.6.1 Acknowledgements

Included in the document above are the following information sources.
Universal serial bus specification – USB implementers forum web page at:
http://www.usb.org
Universal serial bus system architecture – MindShare, Inc by Don Anderson:
http://www.amazon.com
Intel USB product specifications Intel 8×930 and 8×931 USB peripheral controllers at:
http://www.intel.com/design/usb/prodbref/29776501.htm
Other web sites used
http://www.lucent.com/micro/suite/usb.html
http://www-us.semiconductors.philips.com/usb/

11

Specific techniques

11.1   Open and closed loop control

By definition, a data acquisition and control system is not only required to acquire data from a system or process, but also to act on it. In an industrial environment the methods and techniques used to calculate and perform the appropriate actions at any given time, are often extremely critical. Large or incorrect control actions can adversely affect the performance of the system, and can indeed prove to be extremely costly. One well used method of controlling a system or process, in which the current state of the system is fed back to the controller (i.e. the PC), is closed loop control. This method, and the use of a PID control algorithm to implement it, are discussed in the following section.

11.1.1   Definitions

Control systems are classified as open loop or closed loop systems. The distinction is determined by the control action, which is the mechanism responsible for activating the system to produce the output. An open loop control system is one in which the control action is independent of the output. In this type of control system, there is no feedback from the process on the results of a given control action taking place.

Two important features of open loop control are:

- Their ability to perform accurately is determined by their inherent accuracy and their calibration. Calibration is the re-establishing of an input–output relationship, to obtain the desired system accuracy.
- They are not generally troubled with problems of instability.

A closed loop control system is one in which the control action is dependent on the output. In this type of control system there is continuous feedback from the process on the results of a given control action taking place.

The most important features of a closed loop control system are:

- Increased accuracy.

- The sensitivity of the output/input relationship (transfer characteristic) to variations in system characteristics being reduced.
- Reduced effects of non-linearities.
- Increased bandwidth.
- Tendency towards oscillation and instability.

11.1.2 Fluid level closed loop control system

Consider the simple closed loop control system shown in Figure 11.1, in which the fluid in a tank is being used for an industrial process. The process requires that the fluid in the tank must be maintained at a certain level.

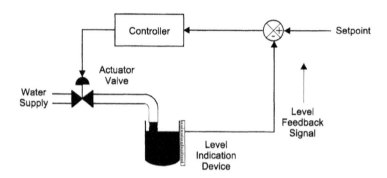

Figure 11.1
Fluid level closed loop control system

The required fluid level is called the reference, or SetPoint, and is the input $\{s(t)\}$ to the system. Depending on the fluid level requirements, the SetPoint may vary with time. The actual fluid level is the output of this system $\{l(t)\}$ and will vary in time according to the use of water in the tank.

The input to the controller is the error difference $\{e(t)\}$ between the required level $\{s(t)\}$ and the output level $\{l(t)\}$.

The output of the controller $\{m(t)\}$ sets the valve of the actuator to supply more or less fluid flow to the tank, depending on the level of water in the tank.

If the level of the tank is lower than the SetPoint, the value of the error difference is positive. A positive signal is sent to the valve to open up and allow more fluid to flow into the tank. Conversely, if the fluid level in the tank is greater than the SetPoint, the value of the error difference is negative. A negative signal is sent to the valve to close up and restrict the flow of fluid into the tank.

Where the output is subtracted from the reference input, the system is known as having negative feedback.

11.1.3 PID control algorithms

The closed loop control process, described above, can be represented by the block diagram shown below in Figure 11.2.

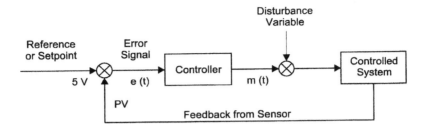

Figure 11.2
Block diagram of a closed loop control system

One effective method of calculating the required controller output $m(t)$ for a given control process, is the PID (proportional, integral and derivative) control algorithm, which is the sum of four terms. This is shown in the following two equations for both the real time continuous and discrete time processes:

$$m(t) = K_p \, e(t) + K_i \int e(t) dt + K_d \, \frac{de(t)}{dt} + Bias$$

Where:

$m(t)$ is the output
K_p is the proportional gain constant (1/sec)
K_i is the integral gain constant (1/sec)
K_d is the derivative gain constant (sec)
$e(t)$ is (SP–PV) [set point – process variable]
'*Bias*' is a constant determined from knowledge of the system

$$m(i) = K_p \, e(i) + T \times K_i \sum_{k=0}^{k=i} e(k) + K_d \times \frac{[e(i) - e(i-1)]}{T} + Bias$$

Where:

$m(i)$ is the output at time of the ith sample ($= i * T$)
K_p is the proportional gain constant
K_i is the integral gain constant (1/sec)
K_D is the derivative gain constant (sec)
T is the time interval for sampling
i is the number of samples
$e(i)$ is the error at ith sampling interval
$e(i-1)$ is the error at $(i-1)$th previous sampling interval
'*Bias*' is the feed-forward or constant-bias
$e(i)$ is the SetPoint (i) – process variable (i) (measured at the ith sample)

The first term (proportional term) of these equations is directly proportional to the current process error. The value of the proportional constant (K_p) determines how hard the system reacts to differences between the SetPoint and the actual process variable.

Simple proportional control cannot take into account load changes in the process under control. This is handled by the integral term of the PID equation, which sums up the long-term error (m) in the system and adds a correctional value to the controller output, proportional to the integral constant (K_i).

The rate of change of the process error is compensated for by the derivative term. This results in a much faster process response. The derivative term results in a much harder control response, when the error term is going in the wrong direction and a dampening effect when the error term is going in the right direction.

This can be described in another way. If the error term is getting larger, the derivative term will contribute a positive correction to the output; the size of the correction being proportional to the speed at which the error term is getting larger. Conversely, when the error term is getting smaller, the derivative term is negative. If the rate at which the derivative term is getting smaller is too quick, the output from the controller will be reduced, thereby dampening the output.

The bias term is quite simply the value of the controller output that is required to maintain the output at the SetPoint reference.

11.1.4   Transient performance – step response

The response of a closed loop system to a step change in the input reference is known as the step response of the system. This is illustrated in Figure 11.3. The step response provides an insight into the transient response of the system, in particular its speed of response and relative stability.

The **overshoot** is the maximum difference between the transient and steady state responses of the control system. It is a measure of the relative stability.

The **rise time** is defined as the time required for the output response to a unit-step function input to rise from 10% to 90% of its final value.

The **settling time** is defined as the time required for the response to a unit-step input to reach and remain within a specified percentage of its final value (steady state value).

The values of rise time and settling time indicate the speed of response of the control system.

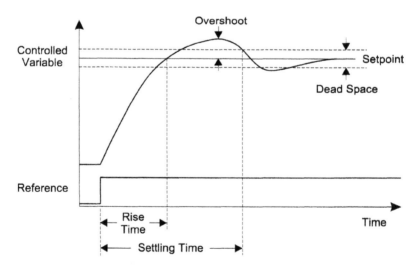

Figure 11.3
Step response of a closed loop system

The values of K_p, K_i and K_d affect the characteristics of the step response. This is shown in Figure 11.4.

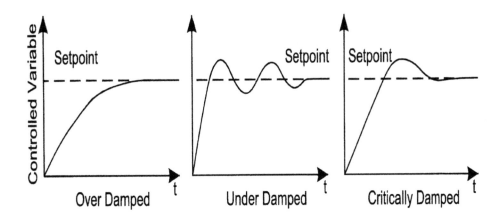

Figure 11.4
Effect of damping on the step response of a closed loop system

11.1.5   Deadband

The natural tendency of closed loop systems to oscillate around the required output value can be seen from the step response. In addition to this, there are many practical control systems in which it is almost impossible to entirely eliminate the error.

Such systems allow for a zero crossing deadband. This adjustable deadband allows the user to select an error range above and below the SetPoint where the output will not change.

This deadband is useful in ensuring that the output does not oscillate even though there is a small error in the system.

11.1.6   Output limiting

A feature many controllers incorporate, is output limiting (using an anti-reset windup), whereby the software acts to limit the output from the PID equation from exceeding a certain value.

In terms of the PID control algorithm, the integral term is excluded from further calculations until the output returns to a value within the correct operating range.

11.1.7   Manual control – bumpless transfer

Where a control system allows for manual user control of the output, a return to automatic control could cause a 'bump' in the controller output, and subsequently in the system output. Bumpless transfer allows the system to transfer from the manual mode to the automatic mode (where the PID equation determines the output), without the output bumping up or down. This is achieved in software by calculating a required integral term in the PID equation for automatic mode, so that no immediate 'bump' is caused to the output of the controller. The system then slowly adjusts back to the reference output under automatic control.

11.2   Capturing high speed transient data

Transient signals are by their nature very fast. In the frequency domain, a transient pulse contains many high frequency components – the narrower the pulse, the wider the range of frequencies over which the pulse can be represented. Theoretically, an infinite impulse is represented by all frequencies across the frequency spectrum. Intuitively, it is obvious that the narrower the pulse, the higher the rate at which it must be sampled to be accurately represented. The following sections discuss the special data acquisition hardware requirements for capturing high-speed transient data as well as the special triggering techniques used.

11.2.1   A/D board operation and memory requirements

Consider a system, with a sampling rate of 10 MHz (i.e. a sampling period of 100 ns), producing 10 million samples/second. Apart from the speed limitations that could prevent the storage of such data to the computer's memory, there is the obvious question of the amount of data being stored, especially when the transient pulse to be captured may only be of 5 µs duration.

Therefore high-speed data acquisition systems used to capture transient data, consist of an A/D converter followed by fast digital memory, which stores the sampled values sequentially, in a circular buffer. A circular buffer is used so that no matter how long it takes to get a trigger event, the system never stops converting the incoming signal. If a trigger event never happens, the A/D system should keep on storing data in the buffer indefinitely; continually overwriting the old data with the new.

When a trigger occurs, the circular buffer information can be saved, thus capturing the latest '*n*' seconds of data for display, analysis or permanent storage in the computer's memory. The amount of memory required is determined by the speed of the fastest transient that will be recorded (affects the sampling rate) and the amount of samples before and after a triggerable event that needs to be stored.

11.2.2   Trigger modes (pre- and post-triggering)

Old style oscilloscopes only allowed the viewing of a transient event, after the trigger event (i.e. post-triggered). In high-speed A/D systems, where data is continuously acquired and stored in a circular buffer, it is possible to capture and view what happened before a transient event. This is known as pre-triggering.

Depending on the equipment being used several trigger modes are usually available:

- Post-trigger – collect N samples following the trigger.
- Pre-trigger – collects data into the circular buffer, terminating in the trigger.
- Pre/post trigger – collects data into the buffer and N additional samples following trigger.
- Delay trigger mode – collects N samples a certain delay after the trigger

11.2.3   Trigger source and level

A number of trigger sources and programmable trigger levels are available on high-speed boards for triggering the acquisition.

Analog trigger mode

An analog trigger on a single channel of the board or from an external analog trigger source starts the acquisition.

The threshold level and slope at which the trigger begins the acquisition is commonly programmable. A high resolution DAC output generates the programmed voltage threshold, which is then compared to the analog voltage level from the trigger input. When the two voltage levels are equal and the slope polarity of the trigger is correct, the acquisition begins.

Where a trigger capability is specified as above or below level, only the value of the analog input trigger and its level with regard to a programmable threshold, is considered. The trigger slope is ignored.

Digital trigger mode

An external digital trigger input, TTL compatible and programmable as active on the rising or falling edge, triggers the acquisition process. When using the digital trigger mode, some boards specify a minimum pulse width – *be wary of this!*

Software trigger mode

The data acquisition process is started by a call from software.

Multiple trigger modes

Some boards have dual-trigger capability, which allows triggering to occur on a combination of trigger inputs. The data acquisition will not occur unless both trigger inputs reach their programmed threshold levels.

Logic analyzers and digital storage oscilloscopes, which allow multi-event, multi-level or sequential trigger modes, are examples of equipment that require more complex triggering capabilities.

12

The PCMCIA Card

Introduction

The PCMCIA (PCMCIA stands for Personal Computer Memory Card International Association) standard was developed to standardize the many different types of memory cards so that they could interface to a wide variety of personal computers. In 1995 the name was changed to simply the PC Card. The PC Card can be considered the same as a standard plug-in card and is used for many different types of devices including flash memory, analog to digital conversion, GPIB, digital I/O, networks, and modems. Special drivers for these cards are supplied by the PC Card manufacturers. All settings, including base address, IRQ and DMA, if available, are set up by the user through the installation software. Temporary stay resident programs (or TSRs) are used to allow constant access to the card by the user. The PC Card consists of miniaturized electronics inside a small metal box with a connector at one end. The connector is an industry standard 68 dual in-line socket that fits into the host computer's PC Card adapter. The other end of the PC Card may or may not have a connector, depending on the type of device.

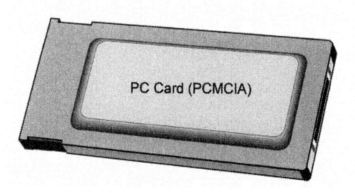

Figure 12.1
PCMCIA Card

12.1   History

In 1988, the Personal Computer Memory Card International Association was formed in America, by representatives from memory card and chip manufacturers. In Japan, the Japan Electronics Industry Development Association (JEIDA) had been working on a memory card standard since 1985.   In 1989 the PCMCIA adopted the 68-pin socket from JEIDA. Since then JEIDA and PCMCIA have been working very closely to keep their standards compatible. By 1990, version 1.0 of the PC Card standard was released. This standard specified memory cards and a socket for virtual drives. The idea was that the memory card would replace the floppy disk. Initially the specifications were developed for the IBM PC, but soon other types of computers, such as the Apple Macs, were incorporated into the standard.

With the release of version 2.0 in November 1992, the PCMCIA standard now includes provisions for input and output devices such as modems, hard disk, and data acquisition cards. It was about this time that the association realized that for full compatibility they needed to include software in the standard. This software included card services and socket services to interface the PC Card to the computer. In 1995 the Personal Computer Memory Card Association decided that the acronym PCMCIA was too confusing so they put out a notice that the PCMCIA card would now be known as the PC Card.   In conjunction with this, they released a new standard with no version number. This new standard is called the PC Card Standard. Below is a list of release dates for the hardware and software standards.

Nov 1990   V 1.0 type 1 memory card
Aug 1991   V 1.0 socket services
Sept 1991   V 1.01 type I, II and III added
Dec 1991   V 1.0 (draft) card services
Nov 1992   V 2.0 I/O devices
July 1993   V 2.1 plug and play
1995   PCMCIA changed to the PC Card (DMA)
1996   CARDBUS (32-bit)

As with most industry driven standards it is often hard to say who started the PC Card. The list of manufacturers is long, but it can be said that Intel had no small part, in so much as they were one of the first to push their memory chip sets, to encourage the manufacture of PC Cards. Soon afterwards, companies such as Cirrus and Vadem picked up the ball as it were, and ran with it. Also, it must be mentioned that the United States government supply office's decision to require all their computers be supplied with PC Card interfaces, has helped ensure the future of the PC Card.

12.2   Features

12.2.1   Size and versatility

The PC Card's main feature from the start was that it was small and portable. The PC Cards are small enough to fit in a shirt pocket. It is no surprise then that the portable computer market was the first to include PCMCIA host bus adapters as a standard item. The PC Card can be used on virtually any PC or Macintosh computer. It is a 16-bit bus device that plugs into one of two slots in a LAPTOP or a full size computer. The host card adapter can either be supplied with the computer or added latter. The host card adapter on a full size computer has a cable with an edge connector that plugs into the bus slots on the motherboard. These bus adapters can be purchased as third party devices. Usually the adapters can hold two type I, two type II or one type III card.

12.2.2 16-bit

All PC Cards, including the old PCMCIA types, are 16 bit. There are twenty-six address lines that allow up to 64 megabytes of addressable memory space. The CARDBUS is a 32-bit device that addresses up to 4.2 billion addresses. This allows (when the prices are low enough) for a four-gigabyte, type III memory card.

12.2.3 Direct memory access (DMA)

DMA is included in the PC CARD standard but not on the PCMCIA versions 1, 2, or 2.1 cards. It is also included in the CARDBUS part of the standard. DMA is a very useful tool in transferring information from data acquisition devices such as analog to digital converter cards. DAQ devices need DMA to transfer information that is sampled at speeds greater than 40 kHz. PCMCIA cards of version 2.1 and earlier can only transfer information using polling or interrupt methods.

12.2.4 Multi-functional and transparent

One of the greatest advantages of the PC Card is that many different types of devices can be built into one card. Manufacturers are producing combination PC Cards that include a cellular-ready 28.8 k baud modem, fax and Ethernet. In the future, it may be possible to have every type of device that can be plugged into a computer on one card. This multifunction ability also allows different formats, such as DOS and Unix files, to be placed on the same card. Once the PC Card is installed and the correct software has been loaded the card becomes transparent to the system.

The PC Card then responds to the PC as though it was a standard card plugged into the bus. This makes the card problem-free from the users' point of view. Generic and third party software can then access the card as though it was an old standard plug-in card. This eliminates the need for buying new software for the PC Card.

12.2.5 Low voltage

New technology in the PC industry has seen the development of low voltage microprocessors. The new PC Card supports 5.0, 3.3 volt, and what the PC Card people call 'x.x' volts. This 'x.x' volt is placed in the standard for future development by computer chip manufacturers. There are three different voltage types of host bus adapters (i.e. the computer adapters): 5 volt only, 3.3 volt only, and a combination 5 volt and 3.3 volt adapter. This combination adapter can sense which voltage the card needs, and automatically configure the power supplied to the PC Card. This combination of voltages relates to plug and play.

12.2.6 Plug and play

Plug and play is the ability of a PC to automatically recognize a PC Card. This recognition would determine if the card has been inserted, the type of card, and the configuration of the software needed to run the card. The software allows the PC Card slot to adjust its voltage for the level needed for the card. True plug and play has the ability to boot up and run off the embedded auto-executable batch files, by looking at the configuration of the programs stored on the PC Card. For plug and play to be fully realized, the PC, BIOS, PC Card, card and services software, and application software must all follow the same PC Card standard. The problem is that PC Cards, PCs, and software packages are made by different companies. Getting them to follow a voluntary standard can be extremely difficult.

12.2.7 Execute in place

Execute in place means that a PC Card such as a flash memory device can run a program from the card instead of downloading the program from memory. A floppy disk can do this, but because of the seek-time on the drive, it takes too long. Flash memories have very low seek-time and therefore it would be the same as having the program in the memory of the computer. Auto-execution of programs when the PC Card is inserted into the computer has its own problems that are not within the scope of this book.

12.2.8 Problems

The downside of the PC Cards is two fold. One is that the retailers have problems with them being stolen because they are so small. If you wish to buy a PC Card, you usually have to ask the retailer to open up a locked case where they are kept. The other problem is that the external connectors are usually small. This means that they can easily break when slightly tugged. It is best when PC Cards are used in a portable PC to mount the PC in a docking station or temporary jig so that the PC doesn't move. Cables attached to the PC Card should be so placed that they are not snagged by accident.

12.3 Products

In the beginning, the PC Card cards were just memory cards, but it soon became evident that this new format was perfect for all kinds of other devices. The manufacturers found that it was easy to include many functions on one card. The following is a short list of devices that are produced. Many of these devices can be combined on one PC Card.

12.3.1 Memory cards

The memory cards include Flash memory, RAM, and ROM devices. Flash memory is used to download data, increase memory and load programs in products like data loggers. RAM cards are used as virtual floppy disk drives and ROM cards are used as fixed program cards.

12.3.2 Disk drives

There are two types of disk drives for the PC Card. One is the all-electronic type. This card uses RAM to save the information. The other uses a modified version of the mechanical spinning magnetic disk. Both types of hard disk have their advantages and disadvantages. The RAM based drive is faster and has no moving parts, but is very expensive at the moment. The mechanical drive is cheap and at the moment can hold larger amounts of memory, but it is slow (12 ms seek-time) and prone to breakage due to the presence of moving parts. Eventually the price of memory cards will fall and the mechanical hard disk will go the way of the punch card.

12.3.3   Pagers

Pager cards are used in offices and factories on portable PCs to allow the user to receive or transmit messages to either a central location in the office or around the world. This feature can be included with other functions and increases the versatility of personal and business communications.

12.3.4   Local area networks

The local area network market is one of the fastest growing electronic markets. PC Cards that include Ethernet or token ring devices on the card are very popular. The newest PC Card is the wireless LAN that can remotely connect to a network at distances of 300 to 500 meters without a repeater. These PC Cards function as modems, faxes, and LANs all on the same card. 10 BASE 2, 10 BASE T, 100 BASE T, token ring and wireless Ethernet are just some of the local area networks that are available on PC Cards.

12.3.5   Modems

There are three types of modems available on the market, 14.4 K bps, 28.8 K bps, and ISDN in the PC Card format. The 14.4 K bps model is the cheapest and some would argue that this is enough for current communications. However, those with vision and a local web server find that 28.8K bps works best for them. ISDN technology is growing daily and will soon be available in all areas. ISDN provides a large increase in speed over standard telephone lines. Most PC Card modems now come with FAX capabilities. Some cards can also be plugged into a cellular telephone. The cable for connecting to the cellular phone is ordered after the modem is purchased from the PC Card manufacturer because there are many different types of mobile phones. The main features to look for in a PC Card modem are speed, which countries it can be used in, and the durability of the cable connection.

12.3.6   Cellular telephone

As of this writing, manufacturers of digital cellular telephones are developing a PC Card telephone. This will give the user total portable communications anywhere in the world. Features of this card could include modem, fax, voice mail and voice communications.

12.3.7   Data acquisition

The data acquisition market has also increased tremendously in the last few years. Many manufacturers have transferred their products from standard plug-in cards to PC Cards. With the advent of the PC Card standard, it's now possible to purchase a multi-function DAQ card that completely replaces the standard plug-in card. The cost has been coming down and now it is often cheaper to buy PC Cards than full-size cards. Almost all DAQ cards are now available in the PC Card format. DAQ PC Cards include analog to digital converters, digital input and output, counter timers, digital to analog converters and 488 GPIB systems.

12.3.8   Digital multimeter

Some manufacturers are producing dedicated devices on PC Cards. One such device is the digital multimeter. The PC Card voltmeter has fixed wires on one end that the user uses to read voltages on electronic equipment. Windows software graphically displays the voltmeter on the screen. Now you can turn your $5,000 laptop into a $29 voltmeter!

12.3.9 GPS systems

Recently a few manufacturers have developed a global positioning system that works on a PC Card. This device will tell you your location and the time anywhere in the world in a matter of seconds. The price of these units will soon rival stand-alone units. Additional features that are in the pipeline are multi-functional units with, faxes to download weather maps from satellites, and voice mail.

12.3.10 Pocket organizer

The pocket organizer is becoming more powerful every day. The users are demanding greater versatility and usability. This has pushed the manufacturers to include a PC Card slot on their pocket organizers. The UK HP OMNIGO 100 is just one product that has a PC Card slot on board. This slot gives the user access to many features that other organizers lack. It also will insure the future usability of the pocket organizer as PC Card technology expands.

12.3.11 Stand-alone products

Although the PC Card was developed as a PC interface, its uses have not gone unnoticed by other product manufacturers. Some products have been using memory cards for a long time. Data logger manufacturers use memory cards in their products to increase the memory size, download data and programs. Banks are now starting to use the PC Card to replace credit and debit cards. JVC is producing an overhead LCD projection viewer that has a PC Card slot. This will let the user carry only the viewer and a PC Card containing the graphics and text for the lecture. It is said that some day almost all electronic products will be PC Card compatible.

12.3.12 Full size computers

At the moment portable computers make up the biggest users of PC Cards, but this will soon change. Already the US government requires all PCs to be equipped with a PC Card interface. PC Card drives can be purchased for about the cost of a floppy drive. These interfaces have a cable attached that plugs into the standard bus sockets inside the computer. Computer manufacturers such as IBM (Optiva model) are including the adapters as standard equipment. The trend is that the PC Card will replace the floppy drive and eventually the hard disk on all full-size computers.

12.4 Construction

The construction of the PC Card is very simple from the user's point of view, but because it is so small it is very difficult from a manufacturing point of view. All of the electronic components are flat pack instead of the usual dual inline pin type. The PCB has to be very thin and this causes problems when soldering the part to the board. Special high temperature boards have been specially developed for PC Cards. Normal PCBs are either too thick or too flexible. The large number of parts that need to be put on the board causes its own problems. Manufacturers are constantly searching for ways to overcome these problems.

12.4.1   Size and types

Size 85.6 mm × 54.0 mm
Type I 3.3 mm thick
Type II 5.0 mm thick
Type III 10.5 mm thick
Type I (extended) 3.3 mm thick
Type II (extended) 5.0 mm thick

12.4.2   Extended types

The standard allows PC Cards to extend past the host bus adapter for any length, the manufacturer wishes, on type I and II cards. This provides for larger cards and therefore more electronics to be incorporated into the PC Card. The only restriction on this is that the card must not change thickness until it is 10 mm past the host bus adapter.

12.5   Hardware

All PC Cards are 85.6 mm long by 54.0 mm wide. The electronics is usually encased in a metal box with a 68-socket connector at one end. This socket plugs into the host bus adapter on the computer with pins of different length. The ground and power pins (GND and VCC) are 2.5 mm long. This allows the card to be powered up before the I/O section. The general interface pins are 2.1 mm long and the card detection pins are 1.5 mm long. This completely powers up the memory and I/O before the computer knows that it has been installed. PC Cards will work in an environment of at least −20 degrees to 60 degrees C and 95% humidity. They should, according to the standard, take up to 10,000 insertions. The contacts of the sockets should pass up to 0.5 amps.

The standard also includes thermal shock, electrostatic discharge, x-ray exposure, EMI, vibration, shock, bend test, warp test and even a drop test. The only other test that could be added is a cable pull test for externally connected cables such as the cable that goes to the telephone jack on a modem. These external cable connections to the card seem to be the weakest point of a PC Card. The PC Card can work (depending on the type of card) at 5 volts, 3.3 volts or some other future voltage. This means that there are three different types of cards that can be inserted into a host bus adapter. Three different keying structures are defined in the PC Card standard – a standard card, 5 volts only, a low voltage card, 3.3 or less and a double key that allows both types of card to be inserted.

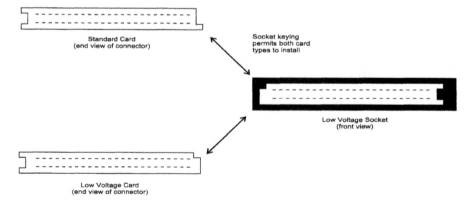

Figure 12.2
Standard and low voltage keying

12.5.1 Power

Power usage on the PC Card and HBA can vary widely. Some host bus adapters use power even when a card is not inserted. An external PC Card hard disk can use up to 20% of the battery power when in use. It is therefore prudent to inquire about the power usage when purchasing a PC Card.

12.5.2 Pin assignments

The obvious physical attributes of the PC Card vary little between the different classifications of cards. There are five basic types of PC Cards:

- Memory only cards (usually type I)
- Memory or I/O cards (usually type II)
- I/O with DMA (usually type II)
- ATA interface (usually type III)
- AIMS interface (usually type III)

12.5.3 Memory only cards

There are two types of memory space within a PC Card: attribute memory and common memory. Attribute memory is where the configuration information is stored. This includes the configuration registers and the CIS (card information structure). The configuration registers hold information such as software reset, bus size, power down control, audio enable and interrupt pending. The CIS contains information about timing, memory addresses, device type, device speed, and common memory size. The common memory is the working address space that is used to map the memory arrays that store data and/or programs. Memory only sockets have the following pin outs.

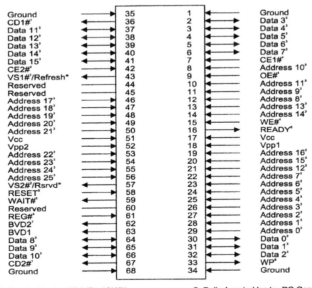

Ground		35	1		Ground
CD1#'		36	2		Data 3'
Data 11'		37	3		Data 4'
Data 12'		38	4		Data 5'
Data 13'		39	5		Data 6'
Data 14'		40	6		Data 7'
Data 15'		41	7		CE1#'
CE2#'		42	8		Address 10'
VS1#'/Refresh*		43	9		OE#'
Reserved		44	10		Address 11'
Reserved		45	11		Address 9'
Address 17'		46	12		Address 8'
Address 18'		47	13		Address 13'
Address 19'		48	14		Address 14'
Address 20'		49	15		WE#'
Address 21'		50	16		READY*
Vcc		51	17		Vcc
Vpp2		52	18		Vpp1
Address 22'		53	19		Address 16'
Address 23'		54	20		Address 15'
Address 24'		55	21		Address 12'
Address 25'		56	22		Address 7'
VS2#'/Rsrvd*		57	23		Address 6'
RESET*		58	24		Address 5'
WAIT#'		59	25		Address 4'
Reserved		60	26		Address 3'
REG#*		61	27		Address 2'
BVD2'		62	28		Address 1'
BVD1		63	29		Address 0'
Data 8'		64	30		Data 0'
Data 9'		65	31		Data 1'
Data 10'		66	32		Data 2'
CD2#'		67	33		WP*
Ground		68	34		Ground

1. Pulled-up to Vcc by HBA($R \geq 10K\Omega$)
2. Pulled-up to Vcc by HBA ($R = 10K\Omega - 100K\Omega$)
3. Pulled-down by PC Card ($R \geq 100K\Omega$)
4. Pulled-up to Vcc by PC Card ($R \geq 10K\Omega$)
5. Pulled-up to Vcc by PC Card ($R \geq 100K\Omega$)
6. Pulled-up to Vcc by HBA ($R \geq 10K\Omega$)
7. Pulled-up to Vcc by HBA

Figure 12.3
Memory only pin outs

12.5.4   I/O Cards

The memory only interface is converted to an I/O interface after the HBA detects that an I/O card has been inserted. Pins that are used by the memory only interface are converted to the appropriate I/O pins. I/O devices would include modems, data acquisition, and LAN PC Cards. Below is a list of pins that are added or changed.

INPACK# PIN 60 ADDED
IORD# PIN 44 ADDED
IOWR# PIN 45 ADDED
IREQ# PIN 16 CHANGED FROM READY
IOIA16# PIN 33 CHANGED FROM WP
SPKR# PIN 62 CHANGED FROM BVD2
STSCHG# PIN 63 CHANGED FROM BVD1

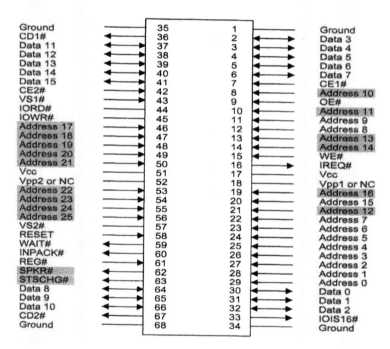

Figure 12.4
I/O interface pin outs

12.5.5   I/O with direct memory access

With the addition of DMA, six pins are reassigned by the HBA. These pins are used by the DMA mechanism to transfer control of the DMA system.

The pins that have been changed are:

DACK PIN 61 CHANGED FROM INPACK#
TC(WRITE) PIN 9 CHANGED FROM OE#
TC(READ) PIN 15 CHANGED FROM WE#. CAN BE ASSIGNED TO ANY ONE OF THREE PINS
DREQ PIN 60 CHANGED (POSSIBLY) FROM INPACK#
DREQ PIN 61 CHANGED (POSSIBLY) FROM REG#
DREQ PIN 33 CHANGED (POSSIBLY) FROM IOIS16#

The DMA interface pin changes are shown below:

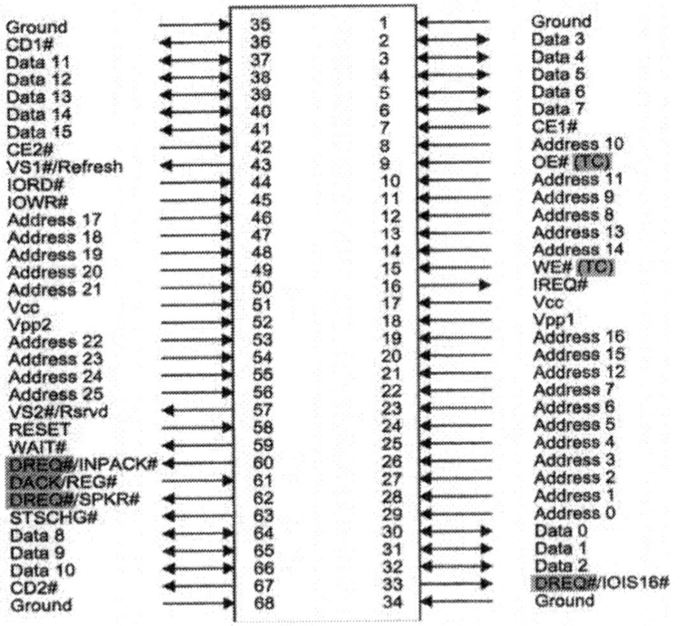

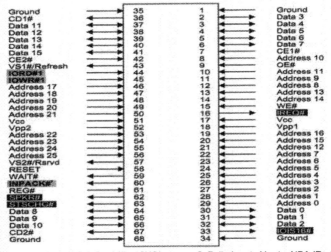

Figure 12.5
The direct memory access interface pin outs

12.5.6   ATA interface (AT attachment)

The ATA is the interface that connects the hard disks and floppy drives to the computer's data bus. This type of PC Card interface usually used a Type III PC Card.

The ATA pin out requirements are shown below:

1. Pulled-up to Vcc by PC Card (R≥10KΩ)   2. Pulled-up to Vcc by HBA (R≥10KΩ)

Figure 12.6
ATA Interface pin outs

12.5.7   AIMS interface (auto-indexing mass storage)

The AIMS interface supports large data structures such as multimedia with text and pictures. It is also used to store still or moving video pictures for electronic cameras. AIMS cards require the ability to transfer large amounts of data using block transfer methods. To do this the AIMS interface requires 8-bit writes to load the entire 32-bit address register.

The figure below shows the minimum signals required for the AIMS interface.

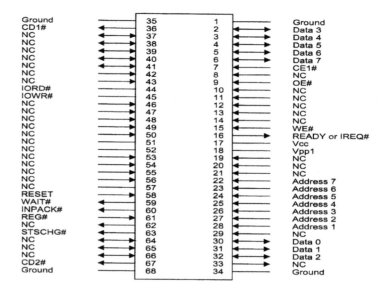

Figure 12.7
AIMS interface pin outs

12.6   Software

The figure below shows the block diagram of the relationship of software and hardware in the operation of PC Cards.

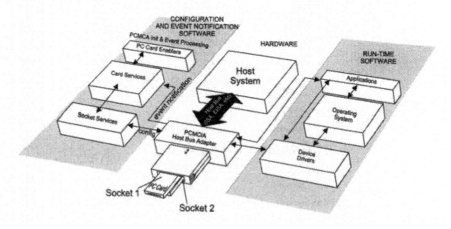

Figure 12.8
PC Card software block diagram

12.6.1   PC Card environment

The PC Card is designed to interface with many different brands of PCs. Various standard software packages are used to guarantee that each card can talk to the different makes of PCs.

Five different types of software are used to accomplish this feat. PC Card enablers, which include:

- Card services
- Socket services
- Test software
- Driver software

In addition, other support software and firmware is used to enable the PC Card to function correctly. These include:

- CIS (card information structures)
- TSRs (temporary stay resident programs)
- Operating systems (such as Windows 95)
- Autoexec.bat files
- Config.sys files
- DOS

12.7   PC Card enablers and support software

C&SS – card and socket services Software

The card and socket services software comes with the particular board when purchased. The software forms an interface between the PC Card controller in the computer and the PC Card. This software may do testing, setup, and control functions.

Socket services

The socket services software is used to interface between the card services and the host bus adapter. It allows any type of card to talk to the card services software. The socket services software interrogates the PC Card and passes that information on to the card services software. It interprets the PC Card's configuration requirements. It determines if the resources needed by the card are available. It checks for changes such as the removal of the card. It also releases resources in the HBA if the card is removed.

Card services

The card services software interfaces the socket services software to the PC Card enabler. The card services software job is to notify the enabler that a card has been removed or inserted, manage the resources of the PC Card, and to run PC Card utilities. It also handles bulk transfers to and from memory cards.

Test software

Test software is used to test the PC Card. These tests may include availability of the card, configuration, and/or setup. The test software may come with the card when purchased, or from a third party.

Driver software

Driver software is provided by the manufacturer of the card, host bus adapter or by a third party. This software interfaces the application software to the PC Card.

Card information structure (CIS)

The card information structure is the software inside the PC Card that tells the computer what type of card is inserted in the slot.

12.8 Future

The PC Card has the potential to replace the floppy disk, hard disk and CD ROM. We may find one day that the PC Card may even be called just 'the card'. It has the ability to consolidate many devices on one small device. In the future, more and more devices will be combined and miniaturized until the whole computer, display, I/O, phone, modem, fax, GPS, video, audio and memory are on one PC Card.

12.8.1 Magazine list and PCMCIA address

- PC Laptop Computer Magazine

- PC Magazine

- PC Card System Architecture - MindShare INC. Don Anderson

12.8.2 Personal Computer Memory Card International Association

PC Card Headquarters
1030G East Duane Ave
Sunnyvale CA 94086 USA
Tel: (408) 433 2273

Appendix A

Glossary

This glossary explains some common data communications terms. Also included are some general computer terms, used in this book. Terms in *italics* are explained elsewhere in the glossary.

24-hour time-of-day mode	**Counter/timer circuit that may be used as real-time clock. Able to control triggering based on the time of day.**
Absolute addressing	**A mode of addressing, containing both the instruction and location (address) of data.**
Accuracy	**Closeness, of indicated or displayed value, to the ideal measured value.**
Acknowledge	**A handshake line that is used by the receiving device to indicate that it has read the transmitted data.**
Active device	**Device capable of supplying current for the loop.**
Active filter	**A combination of active circuit devices (usually amplifiers), with passive circuit elements (resistors and capacitors) that have characteristics that more closely match ideal filters than do passive filters.**
Actuator	**Control element or device used to modulate (or vary) a process parameter.**
A/D	**Analog to digital conversion.**

A/D conversion time	This is the length of time a board requires to convert an analog signal into a digital value. The theoretical maximum speed (conversions/second) is the inverse of this value. See *Speed/Typical Throughput.*
Address	Designator for the location of data in a storage device; allows the retrieval of data by reading the contents of a specific location. Also, the identity of a peripheral device. These (normally unique) allow individual devices on a single communications line to recognize and respond to messages directed at them.
Address register	A register that holds the address of a location containing a data item called for by an instruction.
AFB	Access frame buffer.
Algorithm	Normally used as a basis for writing a computer program. This is a set of rules with a finite number of steps for solving a problem.
Alias frequency	A false lower frequency component that appears in data reconstructed from original data acquired at an insufficient sampling rate (less than two times the maximum frequency of the original data).
ALU	Arithmetic logic unit.
Analog	A continuous real-time phenomenon in which the information values are represented in a variable and continuous waveform.
Analog input board	Printed circuit board converting incoming analog signals to digital values.
Analog output board	Printed circuit board converting outgoing digital values to analog signals.
Analog slope channel	Sampling can be triggered at a user-selectable point on an incoming analog slope. Triggering can be set to occur at a specific threshold level, including selected modes of ± slope level high, and level low.
Analog trigger channel	An auxiliary analog channel, in addition to the analog input channels to be measured, which is used solely for an analog trigger signal.
ANSI	American National Standards Institute. The principal organization developing standards in the USA. Many of the ANSI standards are of great interest to the engineering and software fields.

AOI	Area of interest.
AppleTalk	A proprietary computer networking standard initiated by Apple Computer for use in connecting the Macintosh range of computers and peripherals (including LaserWriter printers). This standard operates at 230 kilobits/second.
Application layer	The highest layer of the 7-layer *OSI* model structure, containing all user or application programs.
Application program	A sequence of instructions written to solve a specific problem facing organizational management. These programs are normally written in a high-level language and draw on resources of the operating system and the computer hardware in executing its tasks.
ARP	Address resolution protocol. A transmission control protocol/internet protocol *(TCP/IP)* process that maps an ID address to an Ethernet address; required by TCP/IP for use with Ethernet.
ARQ	Automatic request for transmission. A request by the receiver for the transmitter to retransmit a block or a frame because of errors detected in the originally received message.
ASCII	American Standard Code for Information Interchange. A universal standard for encoding alphanumeric characters into 7 or 8 binary bits. Drawn up by ANSI to ensure compatibility between different computer systems.
Asynchronous	Communications in which characters can be transmitted at an arbitrary, unsynchronized time, and where the time intervals between transmitted characters may be of varying lengths. Communication is controlled by start and stop bits at the beginning and end of each character.
Attenuation	The decrease in signal magnitude or strength between two points.
Auto-ranging	An auto-ranging board can be set to monitor the incoming signal and automatically select an appropriate gain level based on the previous incoming signals.
Background program	An *application program* that can be executed whenever the facilities of the system are not needed by a higher priority program.

Back-plane	A panel containing sockets into which circuit boards (such as I/O cards, memory boards and power supplies) can be plugged.
Band pass filter	A filter that allows only a fixed range of frequencies to pass through. All other frequencies outside this range (or band) are sharply reduced in magnitude.
Bandwidth	The range of frequencies available, expressed as the difference between the highest and lowest frequencies, in hertz (cycles per second, abbreviated Hz).
Barcode symbol	An array of rectangular parallel bars and spaces of various widths designed for the labeling of objects with unique identifications. A bar code symbol contains a leading quiet zone, a start character, one or more data characters including, in some cases, a check character, a stop character, and a trailing quiet zone.
Base address	A memory address that serves as the reference point. All other points are located by offsetting in relation to the base address.
Baud	Unit of signaling speed derived from the number of events per second (normally bits per second). However, if each event has more than one bit associated with it, the baud rate and bits per second are not equal.
Baudot	Data transmission code in which five bits represent one character. Sixty-four alphanumeric characters can be represented. This code is used in many teleprinter systems with one start bit and 1.42 stop bits added.
Bell 212	An AT&T specification of full duplex, asynchronous or synchronous 1200 bps data transmissions for use on the public telephone network.
BERT/BLERT	Bit error rate/block error rate testing. An error checking technique that compares a received data pattern with a known transmitted data pattern to determine transmission line quality.
Binary coded decimal (BCD)	A code used for representing decimal digits in a binary code.
BIOS	The basic input/output system for the computer, usually firmware-based. This program handles the interface with the PC hardware and isolates the operating software (OS) from the low-level

activities of the hardware. As a result, application software becomes more independent of the particular specifications of the hardware on which it runs, and hence more portable.

Bipolar range, bipolar inputs	A signal range that includes both positive and negative values. Bipolar inputs are designed to accept both positive and negative voltages. Example: ±5 V.
Bisynchronous transmission	See *BSC.*
Bits & bytes	One bit is one binary digit, either a binary 0 or 1. One byte is the amount of memory needed to store each character of information (text or numbers). There are eight bits to one byte (or character), and there are 1024 bytes to one kilobyte (KB). There are 1024 kilobytes to one megabyte (MB). Data acquisition boards typically take 2-byte samples; a board acquiring data at a 20 kHz sample rate is actually gathering 40,000 bytes of data per second.
Block	In block-structured programming languages, a section of programming languages or a section of program coding treated as a unit.
Bloom	The bleeding of one color over another on a video screen. This is often the result of the side-by-side presentation of strongly and weakly saturated colors.
Border	On a VGA display the border represents the area all around the visible image. It is usually black.
bps	Bits per second. Unit of data transmission rate.
Broad band	A communications channel that has greater bandwidth than a voice grade line and is potentially capable of greater transmission rates.
BSC	Bisynchronous transmission. A byte- or character-oriented communication protocol that has become the industry standard (created by IBM). It uses a defined set of control characters for synchronized transmission of binary coded data between stations in a data communications system.
Bubble memory	Describes a method of storing data in memory where data is represented as magnetized spots called magnetic domains that rest on a thin film of semiconductor material. Used in high-vibration, high-temperature, or harsh industrial locations.

Buffer	An intermediate temporary storage device used to compensate for a difference in data rate and data flow between two devices (also called a spooler for interfacing a computer and a printer).
Bus	A data path shared by many devices, with one or more conductors for transmitting signals, data or power. Also, the expansion connector built into a computer. Boards are inserted into this connector, and all communication between the computer and the board occurs through the computer's bus. There are several different expansion buses available, including the XT, AT & Micro-Channel buses for IBM-compatible PCs and the NuBus for the Apple Macintosh II PC line.
BW	Bandwidth.
Cache memory	A fast buffer memory that fits between the CPU and the slower main memory to speed up CPU requests for data.
CCD	Charge-coupled device (camera).
CCIR	Comité Consultatif Internationale des Radiocommunications.
CCITT	Comite' Consultatif Internationale de Télégraph et Téléphone. An international association that sets worldwide standards (for example, V.21, V.22, V.22 bis).
CGA	Color graphics adapter. A computer standard utilizing digital signals offering a resolution of 320 by 200 pixels, a palette of 16 colors.
Character	Letter, numeral, punctuation, control figure or any other symbol contained in a message.
Chroma	The amount and relative brightness of color (or hue) as measured in a video signal.
Chrominance	The color component of a video signal.
Clock	The source of timing signals for sequencing electronic events such as synchronous data transfer or CPU operation in a PC.
Clock pulse	A rising edge, then a falling edge (in that order), such as applied to the clock input of an 8254 timer/counter.
CMRR	Common mode rejection ratio – A board's ability to measure only the voltage difference between the

leads of a transducer, rejecting what the leads have in common. The higher the CMRR, the better the accuracy.

Cold-junction compensation	Thermocouple measurements can easily be affected by the interface the thermocouples are connected to. Cold-junction compensation circuitry compensates for inaccuracies introduced in the conversion process. The *STT* screw terminal panels feature a heavy isothermal plate for high-accuracy cold-junction compensation.
Cold-junction compensation channel	This is an additional data acquisition input channel used exclusively for cold-junction compensation, leaving all of the standard input channels free to be used for data acquisition. The ACPC boards include this channel for increased accuracy and performance.
Collector	The voltage source in a transistor with the base as the control source and the emitter as the controlled output.
Common carrier	A private data communications utility company that furnishes communications services to the general public.
Common mode rejection ratio	A measure of the ability of an instrument to reject interference caused by a voltage common to its input terminals relative to ground, expressed in dB.
Compiler	A program to convert high-level source code (such as BASIC) to machine code-executable form, suitable for the CPU.
Composite	A video signal that contains all the intensity, color and timing information necessary for a video product.
Composite link	The line or circuit connecting a pair of multiplexers or concentrators; the circuit carrying multiplexed data.
Contention	The facility provided by the dial network or a data PABX that allows multiple terminals to compete on a first-come, first-served basis for a smaller number of computer ports.
Control system	A system in which a series of measured values are used to make a decision on manipulating various parameters in the system to achieve a desired value of the original measured values.
Convolution	An image enhancement technique in which each

pixel is subjected to a mathematical operation that groups it with its nearest neighbors and calculates its value accordingly.

Counter data register

The 8-bit register of a (8254 chip) timer/counter that corresponds to one of the two bytes in the counter's output latch for read operations and count register for write operations.

Counter loading

The transfer of a count from an 8254 timer/counter's count register to its counting element.

Counter/timer trigger

On-board counter/timer circuitry can be set to trigger data acquisition at a user-selectable rate and for a particular length of time.

Counter/timers

User-accessible circuitry built into many of the DAS boards that can be used for event counting or frequency measurement.

CRC

Cyclic redundancy check. A basic error checking mechanism using a polynomial algorithm based on the content of the frame and then matched with the result that is performed by the transmitter and included in a field appended to the frame. Also referred to as CRC-16 or CRC-CCITT

Crossed pinning

Wiring configuration that allows two DTE or DCE devices to communicate. Essentially it involves connecting pin 2 to pin 3 of the two devices.

Crossover

In communications, a conductor that runs through the cable and connects to a different pin number at each end.

Crosstalk

A situation where a signal from a communication's channel interferes with an associated channel's signals.

CSMA/CD

Carrier sense multiple access/collision detection. When two stations transmit at the same time on a local area network, they both cease transmission and signal that a collision has occurred. Each then retries again after waiting for a random time period.

Current inputs

A board rated for current inputs can accept and convert analog current levels directly, without conversion to voltage.

Current loop

A communication method that allows data to be transmitted over a longer distance with a higher noise immunity level than with the standard RS-

232C voltage method. A mark (a binary 1) is represented by current; and a space (or binary 0) is represented by the absence of current.

Current sink	This is the amount of current the board can supply for digital output signals. With 10–12 mA or more of current sink capability, a board can turn relays on and off. Digital I/O boards with less than 10–12 mA of sink capability are designed for data transfer only, not for hardware power relay switching.

D/A	Digital to analog.
DAS	Data acquisition system.
Data integrity	A performance measure based on the rate of undetected errors.
Data reduction	The process of analyzing a large quantity of data in order to extract some statistical summary of the underlying pattern.
DCE	Data communications equipment. Devices that provide the functions required in establishing, maintaining and terminating a data transmission connection. Normally it refers to a modem.
Decibel	A logarithmic measure of the ratio of two signal levels where $dB = 20\log_{10} V1/V2$. Being a ratio, it has no units of measure.
Default	A value or setup condition assigned automatically unless another is specified.
Deviation	A movement away from a required value.
DFB	Display frame buffer.
Diagnostic program	A utility program used to identify hardware and firmware defects related to the PC.
Differential	See *Number of channels.*
Digital	A signal that has definite states (normally two).
Digitize	The transformation of an analog signal to a digital signal.
DILUT	Double-input look-up table.
DIP	Acronym for dual in-line packages referring to integrated circuits and switches.
DMA	Direct memory access. A technique of transferring

	data between the computer memory and a device on the computer bus, without the intervention of the microprocessor.
DR	Dynamic range. The ratio of the full-scale range (FSR) of a data converter to the smallest difference it can resolve. $DR = 2^n$, where n is the resolution in bits.
Drift	A gradual movement away from the defined input/output condition over a period of time.
Driver software	A program that acts as the interface between a higher level coding structure and the lower level hardware/firmware component of a computer.
DSR	Data set ready. An RS-232 modem interface control signal, which indicates that the terminal is ready for transmission.
DTE	Data terminal equipment. Devices acting as data source, data sink, or both.
Dual-ported RAM	Allows acquired data to be transferred from on-board memory to the computer's memory while data acquisition is occurring.
Duplex	The ability to send and receive data simultaneously over the same communications line.
DRAM	Dynamic random access memory. See *RAM*.
EBCDIC	Extended binary coded decimal interchange code. An 8-bit character code used primarily in IBM equipment. The code allows for 256 different bit patterns.
EEPROM	Electrically erasable programmable read-only memory. This memory unit can be erased by applying an electrical signal to the EEPROM and then reprogrammed.
EGA	Enhanced graphics adapter. A computer display standard that provides a resolution of 640 by 350 pixels, a palette of 64 colors, and the ability to display as many as 16 colors at one time.
EIA	Electronic Industries Association. An organization in the USA specializing in the electrical and functional characteristics of interface equipment.

EMI/RFI	Electro-magnetic interference or radio frequency interference. Background 'noise' capable of modifying or destroying data transmission.
Emulation	The imitation of a computer system performed by a combination of hardware and software that allows programs to run between incompatible systems.
EPROM	Read-only non-volatile semiconductor memory that is erasable in an ultra-violet light and is reprogrammable.
Error	The difference between the SetPoint and the measured value.
Even parity	A data verification method normally implemented in hardware in which each character must have an even number of ONE bits.
External pulse trigger	Many of the A/D boards allow sampling to be triggered by a voltage pulse from an external source.
Fan in	The load placed on a signal line by a logic circuit input.
Fan out	The measure of drive capability of a logic circuit output.
FCC	Federal Communications Commission (USA).
FDM	Frequency division multiplexer. A device that divides the available transmission frequency range in narrower bands, each of which is used for a separate channel.
Field	One half of a video image (frame) consisting of 312.5 lines (for PAL). There are two fields in a frame. Each is shown alternately every 1/25 of a second (for PAL).
Firmware	A computer program or software stored permanently in FROM or ROM or semi-permanently in EPROM.
Floating ground	A device that is not attached to a ground and is considered as floating in a voltage sense and whose purpose is to avoid common mode problems.
Flow control	The procedure for regulating the flow of data between two devices, preventing the loss of data once a device's buffer has reached its capacity.

Frame	A full video image comprising two fields. A PAL frame has a total of 625 lines (an NTSC frame has 525 lines).
Frame grabber	An image processing peripheral that samples, digitizes and stores a television camera frame in computer memory.
Fringing	The unwanted bordering of an object or character with weak colors when there should be a clearly delineated edge.
Full duplex	Simultaneous 2-way independent transmission in both directions (4-wire).
Gain	Amplification; applied to an incoming signal, gain acts as a multiplication factor on the signal, enabling a board to use signals that would otherwise be too weak. For example, when set to a gain to 10, a board with a range of ±5 V can use raw input signals as low as ±0.5 V (±500 mV); with a gain of 20, the range extends down to ±250 mV.
Genlock	This is the process of synchronizing one video signal to a master reference, ensuring that all signals will be compatible or related to one another.
GPIB	General purpose interface bus. A standard bus used for controlling electronic instrumentation with a computer. Also designated *IEEE488*.
Graphics mode	In graphics mode each pixel on a display screen is addressable, and each pixel has a horizontal (or X) and a vertical (or Y) coordinate.
Gray scale	In image processing, the range of available gray levels. In an 8-bit system, the gray scale contains values from 0 to 255.
Ground	An electrically neutral circuit having the same potential as the earth. A reference point for an electrical system also intended for safety purposes.
Half duplex	Transmissions in either direction, but not simultaneously.
Handshake lines	Dedicated signals that allow two different devices to exchange data under asynchronous hardware control.
Handshaking	Exchange of predetermined signals between two devices establishing a connection.

Harmonic	**An oscillation of a periodic quantity whose frequency is an integral multiple of the fundamental frequency. The fundamental frequency and the harmonics together form a Fourier series of the original waveform.**
HDLC	**High-level data link control. The international standard communication protocol defined by ISO.**
Hexadecimal number system	**A base 16 number system commonly used with personal computers.**
High pass filter	**See *HPF***
Histogram	**A graphic representation of a distribution function, such as frequency, by means of rectangles whose widths represent the intervals into which the range of observed values is divided and whose heights represent the number of observations occurring in each interval.**
HPF	**High pass filter. A filter processing one transmission band that extends from a cutoff frequency (other than zero) to infinity.**
HPIB	**Hewlett-Packard Interface Bus; trade name used by Hewlett-Packard for its implementation of the IEEE 488 standard.**
I/O Address	**A method that allows the CPU to distinguish between different boards in a system. All boards must have different addresses.**
IEEE	**Institute of Electrical and Electronic Engineers. A US-based international professional society that issues its own standards and, which is a member of ANSI and ISO.**
Illumination component	**An amount of source light incident on the object being viewed.**
ILUT	**Input look-up table.**
Individual gain per channel	**A system allowing an individual gain level for each input channel, thereby allowing a much wider range of input levels and types without sacrificing accuracy on low-level signals.**
Interface	**A shared boundary defined by common physical interconnection characteristics, signal characteristics, and measuring of interchanged-signals.**
Interlace	**This is the display of two fields alternately with one field filling in the blank lines of the other field**

	so that they interlock. The PAL standard displays 25 video frames per second.
Interlaced	Interlaced – describing the standard television method of *raster* scanning, in which the image is the product of two fields, each of which is a series of successively scanned lines separated by the equivalent of one line. Thus adjacent lines belong to different fields.
Interrupt	An external event indicating that the CPU should suspend its current task to service a designated activity.
Interrupt handler	The section of the program that performs the necessary operation to service an interrupt when it occurs.
ISA	**Instrument Society of America.**
ISO	**International Standards Organization.**
Isolation	Electrical separation of two circuits. For example, optical isolation allows a high-voltage signal to be transferred to a low-voltage input without electrical interactions.
ISR	Interrupt service routine. See *Interrupt handler.*
Jumper	A wire that connects a number of pins at one end of a cable.
k	Abbreviation of 'kilo', the SI prefix for '1000'. See also *K*.
K	In computer terminology, a K is $2^{10} = 1024$. This distinguishes it from the SI unit *k* (kilo), which is 1000.
LAN	Local area network. A data communications system confined to a limited geographic area, typically about 10 km, with moderate to high data rates (100 kbps to 50 Mbps). Some type of switching technology is used; but common carrier circuits are not used.
LCD	Liquid crystal display. A low-power display used on many portable PCs and instruments.
LDM	Limited distance modem. A signal converter that conditions and boosts a digital signal so that it may be transmitted further than a standard RS-232 signal.

Leased line (or private line)	A private telephone line without inter-exchange switching arrangements.
LED	Light emitting diode. A semi-conductor light source that emits visible light or infrared radiation.
Line driver	A signal converter that conditions a signal to ensure reliable transmission over an extended distance.
Line turnaround	The reversal of transmission direction from transmitter to receiver or vice versa when a half duplex circuit is used.
Linearity	A relationship where the output is directly proportional to the input.
Link layer	Layer 2 of the OSI reference model; also known as the data link layer.
Listener	A device on the GPIB bus that receives information from the bus.
Loaded line	A telephone line equipped with loading coils to add inductance in order to minimize amplitude distortion.
Loopback	Type of diagnostic test in which the transmitted signal is returned to the sending device after passing through all, or a portion, of a data communication link or network. A loopback test permits the comparison of a returned signal with the transmitted signal.
Low pass filter	See *LPF*
LPF	Low pass filter. A filter processing one transmission band, extending from zero to a specific cutoff frequency.
LSB	Least significant byte or least significant bit.
Luminance	The black and white portion of a video signal that supplies brightness and detail for the picture.
LUT	Look-up table. This refers to the memory that stores the values for the point processes. Input pixel values are those for the original image, whilst the output values are those displayed on the monitor as altered by the chosen point processes.
Lux	SI unit of luminous incidence of luminance, equal to one lumen per square meter.

Lux-second	SI unit of light exposure.
Manchester encoding	**Digital technique (specified for the IEEE 802.3 Ethernet baseband network standard) in which each bit period is divided into two complementary halves. A negative to positive voltage transition in the middle of the bit period designates a binary 1, whilst a positive to negative transition represents a 0. The encoding technique also allows the receiving device to recover the transmitted clock from the incoming data stream (self-clocking).**
MAP	**Manufacturing automation protocol. A suite of networking protocols originated by General Motors which track the seven layers of the OSI model. A reduced implementation is referred to as a miniMAP.**
Mark	**This is equivalent to a binary 1.**
Mask	**A structure covering certain portions of a photosensitive medium during photographic processing.**
Masking	**Programming technique for suppressing the use of certain bits in a register.**
Masking	**Setting portions of an image at a constant value, either black or white. Also the process of outlining an image and then matching it to test images.**
Modem	**Modulator-demodulator. A device used to convert serial digital data from a transmitting terminal to a signal suitable for transmission over a telephone channel or to reconvert the transmitted signal to serial digital data for the receiving terminal.**
Modem eliminator	**A device used to connect a local terminal and a computer port, in lieu of the pair of modems to which they would ordinarily connect, allowing DTE and DTE data and control signal connections, otherwise not easily achieved by standard cables or connections.**
Morphology	**The study of a structure/form of object in an image.**
MSB	**Most significant byte or most significant bit.**
Multidrop	**A single communication line or bus used to connect three or more points.**

Multiplexer	A technique in which multiple signals are combined into one channel. They can then be de-multiplexed back into the original components. With the addition of multiplexing panels, 64, 256, sometimes more, inputs can be fed to a single 16-channel board. This results in a slower sample rate (throughput), but allows very large data acquisition systems to be constructed economically.
Multiplexer (mux)	A device used for division of a communication link into two or more channels, either by using frequency division or time division.
Negative true logic	The inversion of the normal logic where the negative state is considered to be TRUE (or 1) and the positive voltage state is considered to be FALSE (or 0).
Network	An interconnected group of nodes or stations.
Network architecture	A set of design principles, including the organization of functions and the description of data formats and procedures, used as the basis for the design and implementation of a network (ISO).
Network layer	Layer 3 in the OSI model; the logical network entity that services the transport layer responsible for ensuring that data passed to it from the transport layer is routed and delivered throughout the network.
Network topology	The physical and logical relationship of nodes in a network; the schematic arrangement of the links and nodes of a network typically in the form of a star, ring or bus topology.
NMRR	Normal mode rejection ratio – The ability of a board to filter out noise from external sources, such as AC power lines. NMRR filtering compensates for transient changes in the incoming signal to provide greater accuracy. The higher the NMRR, the better the filtering of incoming data will be.
Node	A point of interconnection to a network.
Noise	Undesirable interference superimposed upon a useful signal reducing its information content.
NRZ	Non-return to zero. Pulses in alternating directions for successive 1 bits; no change from existing signal voltage for 0 bits.

NTSC	**National Television System Committee (USA). A television standard specifying 525 lines and 60 fields per second.**
Null modem	**A device that connects two DTE devices directly by emulating the physical connections of a DCE device.**
Number of channels	**This is the number of input lines a board can sample. Single-ended inputs share the same ground connection, while differential inputs have individual two-wire inputs for each incoming signal, allowing greater accuracy and signal isolation. See also *multiplexer*.**
Nyquist sampling theorem	**In order to recover all the information about a specified signal, it must be sampled at least at twice the maximum frequency component of the specified signal.**

OCR	**Optical character recognition, optical character reader.**
OLUT	**Output look-up table.**
On-board memory	**Incoming data is stored in on-board memory before being dumped into the PC's memory. On a high-speed board, data is acquired at a much higher rate than can be written into PC memory, so it is stored in the on-board buffer memory.**
Optical isolation	**A means of connecting two networks without electrical continuity; uses optoelectronic transmitters and receivers.**
OR	**Outside radius.**
OSI	**Open systems interconnection. A set of defined protocol layers with a standardized interface that allows equipment from different manufacturers to be connected.**
Output	**An analog or digital output control type signal from the PC to the external 'real world'.**
Overlay	**One video signal superimposed on another, as in the case of computer-generated text over a video picture.**

Packet	**A group of bits (including data and call control signals) transmitted as a whole on a packet switching network. Usually smaller than a transmission block.**

PAD	Packet access device. An interface between a terminal or computer and a packet switching network.
PAL	Phase alternating lines. This is the television standard used in Europe and Australia. The PAL standard is 25 frames per second with 625 lines.
Parallel transmission	The transmission model where multiple data bits are sent simultaneously over separate parallel lines. Accurate synchronization is achieved by using a timing (strobe) signal. Parallel transmission is usually unidirectional; an example would be the Centronics interface to a printer.
Parity bit	A bit that is set to a 0 or 1 to ensure that the total number of 1 bits in the data field is even or odd.
Parity check	The addition of non-information bits that make up a transmission block to ensure that the total number of ONE bits is always even (even parity) or odd (odd parity) so that transmission errors can be detected.
Passive device	Device that must draw its power from connected equipment.
Passive filter	A circuit using only passive electronic components such as resistors, capacitors and inductors.
Peripherals	The input/output and data storage devices attached to a computer, such as disk drives, printers, keyboards, display, communication boards, etc.
Phase modulation	The sine wave or carrier has its phase changed in accordance with the information to be transmitted.
PIA	Peripheral interface adapter. Also referred to as PPI (programmable peripheral interface).
Pixel	One element of a digitized image, sometimes called picture element, or pel.
Point-to-point	A connection between only two items of equipment.
Polling	A means of controlling I/O devices on a multipoint line in which CPU queries ('polls') the devices at regular intervals to check for data awaiting transfer (to the CPU). Slower and less efficient than *interrupt*-driven I/O operations.

Port	**A place of access to a device or network, used for input/output of digital and analog signals.**
PPI	**See *PIA*.**
Pre-trigger	**Boards with 'pre-trigger' capability keep a continuous buffer filled with data, so when the trigger conditions are met, the sample includes the data leading up to the trigger condition.**
Program I/O	**The standard method of memory access, where each piece of data is assigned to a variable and stored individually by the PC's processor.**
Programmable gain	**Using an amplifier chip on an A/D board, the incoming analog signal is increased by the gain multiplication factor. For example, if the input signal is in the range of -250 mV to $+250$ mV, the voltage after the amplifier chip set to a gain of 10 would be -2.5 V to $+2.5$ V.**
PROM	**Programmable read only memory. This is programmed by the manufacturer as a fixed data or program that cannot easily be changed by the user.**
Protocol	**A formal set of conventions governing the formatting and relative timing of message exchange between two communicating systems.**
Public switched network	**Any switching communications system – such as Telex and public telephone networks – that provides circuit switching to many customers.**
Pulse input	**A square wave input from a real world device such as a flow meter, which sends pulses proportional to the flow rate.**
RAM	**Random access memory. Semiconductor read/write volatile memory. There are two main types of semi-conductor-based memory, generally speaking: RAM (also called 'volatile memory') and *ROM* (also called 'non-volatile memory'). RAM is divided into two further types: The first of these is 'static' RAM, or SRAM, which effectively consists of a series of flip-flop devices which can be set to a 1 or a 0 state. Static RAM chips retain the stored values as long as the RAM chip remains powered up. The second type of RAM is 'dynamic RAM' or DRAM. A DRAM is effectively a series of capacitors which, depending on their charge, store a 0 or a 1. As capacitors lose their charge over time (within a few milliseconds in the case of a DRAM), each memory location needs to be refreshed every few milliseconds by**

dedicated electronic circuitry. In spite of this, DRAM chips are cheaper and consume less power than their static counterparts, and they are therefore used more extensively in modern PCs.

RAMDAC | **Random access memory digital-to-analog converter.**

Range | **The difference between the upper and lower limits of the measured value.**

Range select | **The full-scale range a board uses is selected by one of three methods: through the DAS software, by a hardware jumper on the board, or through the use of an external reference voltage.**

Range select | **The feature on a board (either via software, or by hardware DIP switch or jumper) to change the incoming analog or output analog range voltage range.**

Raster | **The pattern of lines traced by rectilinear scanning in display systems.**

Real-time | **A system is capable of operating in real-time when it is fast enough to react to the real-world events.**

Reflectance component | **The amount of light reflected by an object in the scene being viewed.**

Refresh rate | **The speed at which information is updated on a computer display (CRT).**

Resolution | **The number of bits in which a digitized value will be stored. This represents the number of divisions into which the full-scale range will be divided; for example, a 0–10 V range with a 12–bit resolution will have 4096 (2^{12}) divisions of 2.44 mV each (10 V/2^{12} or 10 V/4096).**

Resolution | **The number of pixels that may be displayed on a monitor screen.**

Response time | **The elapsed time between the generation of the last character of a message at a terminal and the receipt of the first character of the reply. It includes terminal delay and network delay.**

RGB | **Red/green/blue. An RGB signal has four separate elements; red/green/blue and *sync*. This results in a cleaner image than with composite signals due to the lower level of distortion and interference.**

RLE | **Run length encoder. A digital image method**

whereby the first gray level of each sequential point-by-point sample and its position in the succession of gray levels is encoded. It is used where there is a tendency for long runs of repeated digitized gray levels to occur.

ROI   Region of interest.

ROM   Read-only memory. Computer memory in which data can be routinely read, but written to only once using special means when the ROM is manufactured. The ROM is used for storing data or programs on a permanent basis.

RS   Recommended standard, for example, RS-232C. More recent designations use EIA, for example, EIA-232C.

RS-232C   Interface between DTE and DCE, employing serial binary data exchange. Typical maximum specifications are 50 feet at 19200 baud.

RS-422   Interface between DTE and DCE, employing the electrical characteristics of balanced voltage interface circuits.

RS-423   Interface between DTE and DCE, employing the electrical characteristics of unbalanced voltage digital interface circuits.

RS-449   General purpose 37-pin and 9-pin interface for DCE and DTE employing serial binary interchange.

RS-485   The recommended standard of the EIA that specifies the electrical characteristics of drivers and receivers for use in balanced digital multipoint systems.

RTSI bus   The real-time system integration bus is an additional connector present on some DAS boards, allowing two or more of these boards to be connected together. It allows the boards to share data, timing and interrupt information, at DMA transfer rates of up to 2.4 MB per second, leaving the PC bus free for other bus operations.

S-Video   The luminance and chrominance elements of a video signal are isolated from each other, resulting in a far cleaner image with greater resolution.

Simultaneous sampling The ability to acquire and store multiple signals at exactly the same moment. Sample-to-sample inaccuracy is typically measured in nanoseconds. The PC-30DS board simultaneously samples 16 signals to within ±20 ns.

SDLC	**Synchronous data link control. IBM standard protocol superseding bisynchronous.**
Self-calibrating	**A self-calibrating board has an extremely stable on-board reference that is used to calibrate A/D and D/A circuits for higher accuracy.**
Self-diagnostics	**On-board diagnostic routine, which tests most, if not all, of a board's functions at power-up or on request.**
Serial transmission	**The most common transmission mode, in which information bits are sent sequentially over a single data channel.**
Shielding	**The process of protecting an instrument or cable from external noise (or sometimes protecting the surrounding environment of the cable from signals within the cable).**
Short haul modem	**A signal converter which conditions a digital signal to ensure reliable transmission over DC continuous private line metallic circuits, without interfering with adjacent pairs of wires in the same telephone cables.**
Shutter	**A mechanical or electronic device used to control the amount of time a light-sensitive material is exposed to radiation.**
SI	**International metric system of units (Systéme Internationale).**
Signal conditioning	**Pre-processing of a signal to bring it up to an acceptable quality level for further processing by a more general purpose analog input system.**
Signal-to-noise ratio	**The ratio of signal strength to the level of noise.**
Signal-to-noise ratio	**The ratio of the signal amplitude to the noise amplitude (generally both signal and noise voltages are expressed in volts).**
Simplex transmission	**Data transmission in one direction only.**
Single-ended	**See *Number of channels*.**
Smart sensors	**A transducer (or sensor) with an on-board microprocessor to pre-process input signals to the transducer. It also has the capability of communicating digitally back to a central control station.**
Software drivers	**Typically a set of programs or subroutines**

allowing the user to control basic board functions, such as setup and data acquisition. These can be incorporated into user-written programs to create a simple but functional DAS system. Many boards come with drivers supplied.

Software trigger
Software control of data acquisition triggering. Most boards are designed for software control.

Space
Absence of signal. This is equivalent to a binary zero.

Spatial filtering
In image processing, the enhancement of an image by increasing or decreasing its spatial frequencies.

Spatial resolution
A measure of the level of detail a vision system can display. The value, expressed in mils or inches per *pixel*, is derived by dividing the linear dimensions of the field of view (x and y, as measured in the image plane), by the number of pixels in the x and y dimensions of the system's imaging array or image digitizer.

Speed/typical throughput
The maximum rate at which the board can sample and convert incoming samples. The typical throughput is divided by the number of channels being sampled, to arrive at the samples/second on each channel. To avoid false readings, the samples per second on each channel need to be greater than twice the frequency of the analog signal being measured.

Statistical multiplexer
Multiplexer in which data loading from multiple devices occurs randomly throughout time, in contrast to standard multiplexers where data loading occurs at regular predictable intervals.

Straight through pinning
RS-232 and RS-422 configuration that match DTE to DCE (pin 1 with pin 1, pin 2 with pin 2, etc).

Strobe
A handshaking line used to signal to a receiving device that there is data to be read.

Switched line
A communication link for which the physical path may vary with each use, such as the public telephone network.

Sync
A sync, pulse ensures that the monitor displaying the information is synchronized at regular intervals with the device supplying the data, thus displaying the data at the right location. E.g. a sync pulse would be used between a camera and a display device to reset the image to the top of the frame for the beginning of the image.

Synchronization
The co-ordination of the activities of several

circuit elements.

Synchronous transmission	**Transmission in which data bits are sent at a fixed rate, with the transmitter and receiver synchronized. Synchronized transmission eliminates the need for start and stop bits.**

Talker	**A device on the GPIB bus that simply sends information onto the bus without actually controlling the bus.**
TCP/IP	**Transmission control protocol/internet protocol. The collective term for the suite of layered protocols that ensures reliable data transmission in an internet (a network of packet-switching networks functioning as a single large network). Originally developed by the US Department of Defense in an effort to create a network that could withstand an enemy attack.**
TDM	**Time division multiplexer. A device that accepts multiple channels on a single transmission line by connecting terminals, one at a time, at regular intervals, interleaving bits (bit TDM) or characters (character TDM) from each terminal.**
Text mode	**Signals from the hardware to the display device are only interpreted as text characters, leading to a maximum resolution defined by the number of characters across a screen by the number of vertical lines. Text mode can be faster than graphics mode, but the resolution and the type of graphics that can be displayed are limited by the text character set.**
Thresholding	**The process of defining a specific intensity level for determining which of two values will be assigned to each pixel in binary processing. If the pixel's brightness is above the threshold level, it will appear in white in the image, if it is below the threshold level, it will appear black.**
Time sharing	**A method of computer operation that allows several interactive terminals to use one computer.**
Transducer	**Any device that generates an electrical signal from real-world physical measurements. Examples are LVDTs, strain gauges, thermocouples and RTDs. A generic term for sensors and their supporting circuitry.**
Transient	**An abrupt change in voltage of short duration.**
Trigger	**A rising edge at an 8254 timer/counter's gate**

input.

Trunk	A single circuit between two points, both of which are switching centers or individual distribution points. A trunk usually handles many channels simultaneously.
UART	Universal asynchronous receiver transmitter. An electronic circuit that translates the data format between a parallel representation within the computer and the serial method of transmitting data over a communication line.
Unipolar inputs	When set to accept a unipolar signal, the channel detects and converts only positive voltages. (Example: 0 to +10 V).
Unloaded line	A line without loaded coils that may therefore suffer line loss at audio frequencies.
V.35	CCITT standard governing the transmission at 48 Kbps over 60 to 108 kHz group band circuits.
VGA	Video graphics array. This standard utilizes analog signals only (between 0 and 1 V) offering a resolution of 640 by 480 *pixels*, a palette of 256 colors out of 256000 colors and the ability to display 16 colors at the same time.
Vidicon	A small television tube originally developed for closed-circuit television. It is about one inch (2.54 cm) in diameter and five inches (12.7 cm) long. Its controls are relatively simple and can be operated by unskilled personnel. The Vidicon is widely used in broadcast service.
Volatile memory	A storage medium that loses all data when power is removed.
VRAM	Volatile random access memory. See RAM
Wedge filter	An optical filter so constructed that the density increases progressively from one end to the other, or angularly around a circular disk.
Word	The standard number of bits that a processor or memory manipulates at one time. Typical words are 16 bits.
X-ON / X-OFF	Control characters used for flow control, instructing a terminal to start transmission (X-ON) and end transmission (X-OFF).
X.21	CCITT standard governing interface between

DTE and DCE devices for synchronous operation on public data networks.

X.25   CCITT standard governing interface between DTE and DCE device for terminals operating in the packet mode on public data networks.

X.25 Pad   A device that permits communication between non-X.25 devices and the devices in an X.25 network.

Appendix B

IBM PC bus specifications

The information in this section is useful for programmers and for users who need to troubleshoot their system. The following specifications are included in this section:

- Hardware interrupts
- DMA channels
- 8237 DMA controller
- 8259 Interrupt controller
- 8253/8254 Counter/timer
- Address information
- Bus signal information
- Card dimensions
- Centronics interface standard

B.I   Hardware interrupts

Name	Int#	Description
NMI	-	Parity*
0	8	Timer*
1	9	Keyboard*
IRQ2	A	Reserved (XT), Int.8-15(AT)
IRQ3	B	COM or SDLC
1RQ4	C	COM or SDLC
IRQ5	D	Hard disk (XT), LPT (AT)

IRQ6	E	Floppy disk
IRQ7	F	LPT
(The following are AT only)		
IRQ8	70	Real-time clock
IRQ9	71	Re-directed to IRQ2
IRQ10	72	Unassigned
IRQ11	73	Unassigned
IRQ12	74	Unassigned
IRQ13	75	80287 Co-processor
IRQ14	76	Hard disk
IRQ15	77	Unassigned

Interrupts marked with * exist on the system board and are not available on the bus

Table B.1
Hardware interrupts

B.2　DMA channels

Channel	Usage
0	Memory refresh
1	SDLC
2	Floppy disk
3	Unassigned
4	Unassigned
5	Unassigned
6	Unassigned
7	Unassigned

Table B.2
DMA channels

B.3　8237 DMA channels

Page register	I/O address
Channel 0 (AT)	087
Channel 1	083
Channel 2	081
Channel 3	082

Table B.3
Controller 1: 8-bit (ports 000–00F)

Page register	I/O address
Channel 5	08B
Channel 6	089
Channel 7	08A

Table B.4
Controller 2: 16-bit (AT only – ports 0C0–0DF)

Refresh (AT) 08F

Controller registers:

Controller address		Command codes
1	2	
000	0C0	Channel 0/4 base & current address
001	0C2	Channel 0/4 base & current word count
002	0C0	Channel 0/4 base & current address
003	0C6	Channel 0/4 base & current word count
004	0C8	Channel 0/4 base & current address
005	0CA	Channel 0/4 base & current word count
006	0CC	Channel 0/4 base & current address
007	0CE	Channel 0/4 base & current word count
008	0D0	Read status/write command register
009	0D2	Write request register
00A	0D4	Write single mask register bit
00B	0D6	Write mode register
00C	0D8	Clear byte pointer flip/flop
00D	0DA	Read temp register/write master clear
00E	0DC	Clear mask register
00F	0DE	Write all mask register bits

Table B.5
DMA controllers

B.4   8259 Interrupt controller

The 8259 programmable interrupt controller accesses an internal register set through two I/O ports. The 8259 is initialized by loading up to a 4-byte configuration sequence. It then responds in an operation mode. Configuration bytes may vary somewhat, due to hardware implementations. The port usage shown below is in the operation mode.

Interrupt controller #1 ports are 20–21

Interrupts are positive-edge sense.

Port 20 is used to acknowledge and re-enable the 8259

To send non-specific end-of-interrupt code:

> mov al,mask
>
> out 20h,al

Port 21 is used to set/clear the masking register

A mask bit = 0=> enable, 1 => disable a specific IRQ

To read interrupt mask register

> in al,21h   bit 7 – 0 = IRQ 7 – 0

To write interrupt mask register

> mov al,mask
>
> out 21h,al

For 8259 #2 (AT only)

Interrupt controller #1 ports are A0–A1

Interrupts are positive-edge sense.

Port A0 is used to acknowledge and re-enable the 8259

To send non-specific end-of-interrupt code:

> mov al,20h
>
> out A0h,al

Port A1 is used to set/clear the masking register

A mask bit = 0 = > enable, 1=> disable a specific IRQ

To read interrupt mask register

> in al,A1h   bit 7 - 0 = IRQ 15 - 8

To write interrupt mask register

> mov al,mask
>
> out A1,al

B.5   8253/8254 Counter/timer

Port	Section	Description
40	Counter 0	Real time clock tick count = 0FFFFh Output to IRQ0 every 53 ms (18.2 ticks per second)
41	Counter 1	DRAM refresh count = 0012h Refresh logic (AT) (Refresh approx every 15 usec)
42	Counter 2	Speaker oscillator
43	Control register	
	Bit 7, 6	Select counter 0, 1, 2
	Bit 5, 4	Latch, LSB, MSB, LSB-MSB
	Bit 3, 2, 1	Mode
	0 – Interrupt on count	
	1 – One-shot	
	2 – Rate generator	
	3 – Square wave generator	
	4 – Triggered strobe (s/w)	
	5 – Triggered strobe (h/w)	
	Bit 0	Binary/BCD counting

Table B.6
8253/8254 Counter/timer

Register	Port	Reg #	7	6	5	4	3	2	1	0
DATA	BA+0	0	Bit 7	Bit 6	Bit 5	Bit 4	Bit 3	Bit2	Bit 1	Bit 0
DLL	BA+0	0	Baud rate	Gen LSB	Divide	Count (DLAB=1)				
DLH	BA+1	1	Baud rate	Gen MSB	Divide	Count (DLAB=1)				
IER	BA+1	1	0	0	0	–	Modem int enab	RX line int enab		
IIR	BA+2	2	0	0	0	0	0	Active interrupt bit 1	Active interrupt bit 1	Interrupt pending
LCR	BA+3	3	DLA B divisor latch bit	Set break	Parity mode bit 2	Parity mode bit 1	Parity mode bit 0	Stop bit length bit 0	Char length bit 1	Char length bit 0
MCR	BA+4	4	0	0	0	Loop back	–OUT2	0	–RTS	–DTR
LSR	BA+5	5	0	TEMT	THRE	Break	Framing	Parity	Overrun	RxRdy
MSR	BA+6	6	DCD	RI	DSR	CTS	DDCD	TERI	DDSR	DCTS
SCR	BA+7	7								

BA = Base address   **COM1 BA = 3F8 hex**   **COM2 = 2F8 hex**

COM3 BA = 3E8   **COM4 BA = 2E8**

Table B.7
8250 registers

FFFFFF 100000	AT extended memory (15 M)	
FFFFF F0000	ROM	
EFFFF E0000	OPEN in PC/XT (64 K)	
DFFFF D0000	Recommended location for 'LIM' Expanded memory (64 K)	
CCFFF C8000	Fixed disk, XT only (20 K) (See Table B.7)	
C7FFF	ROM expansion (26 K) (See Table B.6)	
C3FFF	OPEN (16 K)	EGA screen buffers and ROM
AFFFF A0000	OPEN (64 K)	EGA screen buffers and ROM
AFFFF A0000	OPEN (64 K)	EGA screen buffers and ROM
9FFFF 80000	128 K RAM expansion area	
7FFFF	512 K RAM expansion area	
00400	DOS (See Table B.4) BIOS (See Table B.3)	
003FF	Interrupt vectors (See Figure B.2)	

Table B.8
Memory map for PC/XT/AT

00000–00003	=	Interrupt 0, divide-by-zero error
00004–00007	=	Interrupt 1, single-step operation 386 Debug
Exception		
00008–0000B	=	Interrupt 2, non-maskable interrupt
0000C–0000F	=	Interrupt 3, break-point
00010–00013	=	Interrupt 4, arithmetic overflow
00014–00017	=	Interrupt 5, BIOS print-screen routine
00018–0001B	=	Interrupt 6, reserved
0001C–0001F	=	Interrupt 7, reserved
00020–00023	=	Interrupt 8, hardware timer 18.2/sec
00024–00027	=	Interrupt 9, keyboard
00028–0002B	=	Interrupt A, reserved
0002C–0002F	=	Interrupt B, communications
00030–00033	=	Interrupt C, communications
00024–00037	=	Interrupt D, alternate printer
00038–0003B	=	Interrupt E, floppy disk atten signal
0003C–0003F	=	Interrupt F, printer control
00040–00043	=	Interrupt 10, invokes BIOS video I/O service routine
00044–00047	=	Interrupt 11, invokes BIOS equipment configuration clock
00048–0004B	=	Interrupt 12, invokes BIOS memory-size check
0004C–0004F	=	Interrupt 13, invokes BIOS disk I/O service routine routines
00050–00053	=	Interrupt 14, invokes BIOS RS-232 I/O routines
00054–00057	=	Interrupt 15, invokes BIOS cassette, I/O, extended AT service routines
00058–0005B	=	Interrupt 16, invokes BIOS keyboard I/O routine
0005C–0005F	=	Interrupt 17, invokes BIOS printer I/O
00060–00063	=	Interrupt 18, ROM BASIC
00064–00067	=	Interrupt 19, invokes BIOS boot-strap start-up routine
00068–0006B	=	Interrupt 1A, invokes BIOS time-of-day routines
0006C–0006F	=	Interrupt 1B, BIOS ctrl-break control
00070–00073	=	Interrupt 1C, gen at timer clock tick
00074–00077	=	Interrupt 1D, video initialization control param pointer
00078–0007B	=	Interrupt 1E, disk parameter table pointer
0007C–0007F	=	Interrupt 1F, graphics character table pointer
00080–00083	=	Interrupt 20, invokes DOS program termination
00084–00087	=	Interrupt 21, invokes all DOS function calls
00088–0008B	=	Interrupt 22, user-created, DOS-controlled interrupt routine invoked on program end
0008C–0008F	=	Interrupt 23, user-created, DOS-controlled interrupt routine invoked on keyboard break
00090–00093	=	Interrupt 24, user-created, DOS-controlled interrupt routine invoked at critical error
00094–00097	=	Interrupt 25, invokes DOS absolute disk read service
00098–0009B	=	Interrupt 26, invokes DOS absolute disk write service

0009C–0009F	=	Interrupt 27, ends program and keeps program in memory under DOS
000A0–000FF	=	Interrupts 28 through 3F, reserved
00100–00103	=	Interrupt 40, disk I/O (XT)
00104–00107	=	Interrupt 41, fixed disk parameters (XT)
00108–00123	=	Interrupt 41 through 48, reserved
00124–00127	=	Interrupt 49, keyboard supplement translation table pointer
00128–0017F	=	Interrupt 49 through 5F, reserved
00180–0019F	=	Interrupt 60 through 67, user-defined interrupts
Most data acquisition boards can be programmed to use anyone of the interrupts in the range of 60 through 67.		
Interrupt 67 is used by the expanded memory manager		
001A0–001FF	=	Interrupt 68 through 7F, not used
00200–00217	=	Interrupt 80 through 85, reserved for BASIC
00218–003C3	=	Interrupt 86 through F0, BASIC interpreter
003C4–003FF	=	Interrupt F1 through FF, not used

Table B.9
Interrupt vectors

00400–00401	=	Address of RS-232 adapter 1
00402–00403	=	Address of RS-232 adapter 2
00404–00405	=	Address of RS-232 adapter 3
00406–00407	=	Address of RS-232 adapter 4
00408–00409	=	Address of printer adapter 1
0040A–0040B	=	Address of printer adapter 2
0040C–0040D	=	Address of printer adapter 3
0040E–0040F	=	Address of printer adapter 4
0041D–00411	=	Equipment flag
00412	=	Manufacturing test indicator
00413–00414	=	Useable memory size in K
00415–00416	=	Memory in I/O channel for 64 K – planar PC
00417–00418	=	Keyboard status bits
00419	=	Alternate keyboard numeric input (future use)
0041A–0041B	=	Keyboard buffer head pointer
0041C–0041D	=	Keyboard buffer tail pointer
0041E–0043D	=	Keyboard buffer
0043E	=	Floppy disk seek status
0043F	=	Floppy disk motor status
00440	=	Floppy disk motor timeout
00441	=	Floppy disk status
00442–00448	=	Floppy disk controller status bytes
0449	=	CRT mode code
0044A–0044B	=	CRT column screen width
0044C–0044D	=	CRT regeneration buffer length

0044E–0044F	=	Starting address in regeneration buffer
00450–00451	=	Cursor position for CRT page 1
00452–00453	=	Cursor position for CRT page 2
00454–00455	=	Cursor position for CRT page 3
00456–00457	=	Cursor position for CRT page 4
00458–00459	=	Cursor position for CRT page 5
0045A–0045B	=	Cursor position for CRT page 6
0045C–0045D	=	Cursor position for CRT page 7
0045E–0045F	=	Cursor position for CRT page 8
00460–00461	=	Cursor mode
00462	=	Active page number
00463–00464	=	Address of current display adapter
0465	=	CRT mode
00466	=	Palette setting
00467–00468	=	Time count
00469–0046A	=	CRC register
0046B	=	Last input value
0046C–0046D	=	Low word of timer count
0046E–0046F	=	High word of timer count
00470	=	Timer revolver
00490–004CF	=	Used by MODE COM
00471	=	Break indicator
00472–00473	=	Reboot (Alt-Ctrl-Del) indicator
00474–00477	=	Fixed disk data area (XT)
00478	=	Printer 1 timeout (XT)
00478	=	Printer 2 timeout (XT)
0047A	=	Printer 3 timeout (XT)
0047B	=	Printer 4 timeout (XT)
0047C	=	RS-232 card 1 timeout (XT)
0047D	=	RS-232 card 2 timeout (XT)
0047E	=	RS-232 card 3 timeout (XT)
0047	=	RS-232 card 4 timeout (XT)
00480–00483	=	Additional keyboard buffer pointers (XT)
0484–004A8	=	EGA BIOS buffer
00484	=	Number of character rows
00485	=	Bytes per character
00487	=	Status byte
00488	=	Feature bits, DIP switches
004A8	=	Pointer save
004D0–004EF	=	Reserved
004F0–004FF	=	Intra-application communication area

Table B.10
BIOS data area

00500	=	Print-screen status
00504	=	Single-drive status (drive A or B)
00510–00511	=	BASIC's default data segment pointer
00512–00513	=	IP for BASIC's timer interrupt vector
00514–00515	=	CS for BASIC's timer interrupt vector
00516–00517	=	IP for BASIC's ctrl-break interrupt
00518–00519	=	CS for BASIC's ctrl-break interrupt
0051A–0051B	=	IP for BASIC's fatal-error interrupt
0051C–0051D	=	CS for BASIC's fatal-error interrupt
00600–XXXXX	=	DOS and 'other things'

Table B.11
DOS and BASIC data area

7FFFF	=	Top of 512 K
80000–9FFFF	=	AT, 128 K RAM expansion area
9FFFF	=	Top of 640 K, end of memory expansion area

Table B.12
RAM expansion area

A0000–AFFFF	=	Enhanced graphics adapter (EGA) screen buffers*
B0000–B7FFF	=	Monochrome adapter of EGA
B0000–B0FFF	=	Monochrome screen buffer
B1000–B7FFF	=	Reserved for screen buffers
B8000–BBFFF	=	Color/graphics adapter (CGA) or EGA
B8000–BBFFF	=	CGA Buffer
BC000–BFFFF	=	CGA/EGA screen buffers
C0000–C3FFF	=	EGA BIOS

Table B.13
CRT screen buffers

C4000–C7FFF	=	ROM expansion area
C8000–CCFFF	=	Fixed disk control (XT)
CD000–CFFFF	=	User PROM, memory-mapped I/O
D0000–DFFFF	=	User PROM. recommended 'LIM' location
E0000–EFFFF	=	ROM expansion area, optional I/O for PC/XT

Table B.14
User area

F0000–FDFF	=	ROM BASIC
FE000–FFFD9	=	BIOS
FFFF0–FFFF4	=	First code executed after power-on
FFFF5–FFFFC	=	BIOS release date
FFFFE–FFFFF	=	Machine ID

Table B.15
ROM

100000–FFFFFF	=	I/O channel memory (PC/AT extended memory, 15 Mb maximum)

Table B.16
AT extended memory

000–00F	=	DMA controller (8237A)
020–021	=	Interrupt controller (8259A)
040–043	=	Timer (8253)
060–063	=	PPI (8255A)
080–083	=	DMA page register (74LS612)
0A0	=	NMI mask register
200–20F	=	Joystick (game controller)
210–217	=	Expansion unit
2F8–2FF	=	Serial port (secondary)
300–31F	=	Prototype card
320–32F	=	Fixed disk
378–37F	=	Parallel printer (primary)
380–38F	=	SDLC
3B0–3BF	=	Monochrome adapter/printer
3D0–3D7	=	Color/graphics adapter
3F0–3F7	=	Diskette controller
3F8–3FF	=	Serial port (primary)

Table B.17
IBM PC/XT I/O map

000–01F	=	DMA controller (8237A-5)
020–30F	=	Interrupt controller (8259A)
040–05F	=	Timer (8254)
060–06F	=	Keyboard (8042)
070–07F	=	NMI mask register, real-time clock
080–09F	=	DMA page register (74LS612)
0A0–0BF	=	Interrupt controller 2 (8259A)
0C0–0DF	=	DMA controller 2 (8237A)
0F0–0FF	=	Math co-processor
1F0–1F8	=	Fixed disk
200–207	=	Joystick (game controller)
258–25F	=	Intel 'above board'
278–27F	=	Parallel printer (secondary)
300–31F	=	Prototype card
060–36F	=	Reserved
378–37F	=	Parallel printer (primary)
080–38F	=	SDLC or bisynchronous communications (secondary)
3A0–3AF	=	Bisynchronous communications (primary)
3B0–3BF	=	Monochrome adapter/printer
3C0–3CF	=	EGA, reserved
3D0–3DF	=	Color/graphic adapter
3F0–3F7	=	Diskette controller
3F8–3FF	=	Serial port (primary)

Table B.18
IBM I/O map

B.6   Bus signal information

PC 62-pin connector (bracket end of board)

Ground	←	B1	A1	←	–I/O CHCK	I/O channel check. When low a device on the bus has detected a parity error
+Reset	←	B2	A2	↔	SD7	Data bit No 7
+5 volts	←	B3	A3	↔	SD6	Data bit No 6
+IRQ2/9	→	B4	A4	←	SD5	Data bit No 5
–5 volts	←	B5	A5	↔	SD4	Data bit No 4
+DRQ2	→	B6	A6	←	SD3	Data bit No 3
–12 volts	←	B7	A7	↔	SD2	Data bit No 2
–OWS	↔	B8	A8	←	SD1	[1] zero wait state. Fast bus memory devices pull this line low to prevent the CPU from inserting extra wait cycles zero for 16-bit devices and two for 8-bit devices.

+12 volts	←	B9	A9	↔	D0	Data bit No 1
Ground	←	B10	A10	←	I/O CHRDY	I/O channel ready. Pulled low by a device on the bus that needs more time. Never hold low for more than 10 clock cycles. Cycles are extended in integral multiples of CLK cycles.
–S<E<W	←	B11	A11	↔	AEN	0address enable. When high, the DMA controller has control of the address lines, data lines, mem R/W and I/O R/W.
–SMEMR	←	B12	A12	↔	SA19	0memory write. This command instructs memory devices on the bus to store the data present on the data bus. Active low.
–IOW	↔	B13	A13	↔	SA18 C	0I/O write command. This line instructs an I/O device on the bus to store the data present on the data lines.
–IOR	↔	B14	A14	↔	SA17 O	0I/O read command. This line tells an I/O device on the bus to drive its data onto the data us. Active low.
–DACK3	↔	B15	A15	↔	SA16 M	
S+DRQ3	↔	B16	A16	↔	SA15 P	
0–DACK1	↔	B17	A17	↔	SA14 O	
L+DRQ1	↔	B18	A18	↔	SA13 N	
D–DACK0	↔	B19	A19	↔	SA12 E	
ECLK	↔	B20	A20	↔	SA11 N	0systems clock. This is either a 50% duty cycle signal at 6 MHz (AT) or 33% duty cycle at 4.77 MHz (XT).
R+IR17	↔	B21	A21	↔	SA10 T	
+IRQ6	↔	B22	A22	↔	SA9	
S+IRQ5	↔	B23	A23	↔	SA8 S	
I+IRQ4	↔	B24	A24	↔	SA7 I	
D+IRQ3	↔	B25	A25		SA6 D	
E–DACK2	↔	B26	A26	↔	SA5 E	
+T/C	←	B27	A27	↔	SA4	0terminal count. This signal goes high when the terminal count for a DMA cycle is reached.
+BALE	←	B28	A28	↔	SA3	0address latch enable. This line indicates when the CPU has a valid address on the bus, when used with AEN. Active high.
+5 volts	←	B29	A29	↔	SA2	
OSC	←	B30	A30	↔	SA1	0oscillator. A 50% duty cycle clock signal with a frequency of 14.31818 MHz.
Ground	←	B31	A31	↔	SA0	

Pin B43 is IRQ2 for an XT. Pin B4 is IRQ9 for an AT, which is redirected as IRQ2.

Table B.19
Pin assignments PC 62-pin connector

−memcs16	→	D1	C1	→	SBHE	Imemory 16-bit chip select signals a 16-bit, one wait-state memory cycle.
−I/OCS16	→	D2	C2	↔	LA23	I/O 16-bit chip select signals a 16-bit one-wait state I/O cycle.
+IRQ10	→	D3	C3	↔	LA22 C	
+IRQ11	→	D4	C4	↔	LA21 O	
S+IRQ12	→	D5	C5	↔	LA20 M	
O+IRQ15	→	D6	C6	↔	LA19 P	
L+IRQ14	←	D7	C7	↔	LA18 O	
D−DACK0	←	D8	C8	↔	A17 N	
E+DRQ0	→	D9	C9	→	MEMRE	
R−DACK5	←	D10	C10	→	MEMWN	
+DRO5	→	D11	C11	↔	SD08 T	
S−DACK6	←	D12	C12	↔	SD09	
I+DRQ6	→	D13	C13	↔	SD10 S	
D−DACK7	←	D14	C14	↔	SD11 I	ODMA acknowledge lines used to acknowledge DMA requests (DRQ). DACK0 is dedicated to the DRAM REFRSH function. Active low.
E+DRO7	→	D15	C15	↔	SD12 D	
+5 Volts		D16	C16	↔	SD13 E	
−MASTER	→	D17	C17	↔	SD14	A processor or the I/O channel may use this signal with DRQ to gain control of the address, data and control lines of the bus. Control lines processor may need to assume responsibility to refresh system memory every 15 usec.
Ground		D18	C18	↔	SD15	

Symbols pointing *toward* the connector designate signals *into* the system board (from devices on the bus), and vice versa. ↔ indicates bi-directional signals.

NOTE: **In the PC and PC/XT, the bus consists only of the 62-pin portion. Some of the signal names are different, but the functionality remains the same.**

Table B.20
Pin assignments PC 36-pin connector

B.7 Card dimensions

The physical card dimensions for both the PC/XT and PC/AT standards are given below.

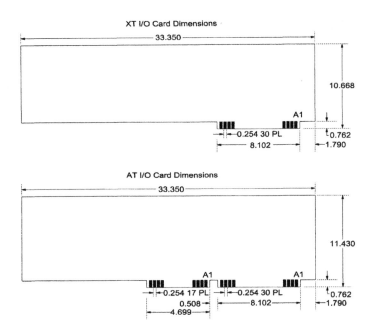

Figure B.1
Card dimensions for PC/XT and PC/AT

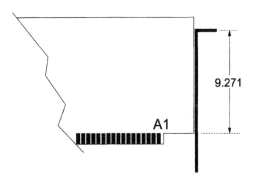

Figure B.2
Card bracket position

B.8 Centronics interface standard

The parallel printer or Centronics interface standard, which includes a 36-pin connector, does not normally present too many difficulties. The full signal definitions are given in Table B.23. This interface is used primarily to interface printers to computers or other intelligent devices. The interface has a limited distance because of its low-level +5 volt signals.

Signal name	Signal pin	Return pin	Signal definition
*DSTB	1	19	Low level pulse of 0.5 microseconds or more used to strobe the DATA signals into the printer. The printer reads the data at the low level of this signal. Ensure an –acknowledge has been returned before issuing the next –data strobe. –data strobe is ignored if the BUSY is high.
* DATA 1–8	2–9	20–27	8 data lines from the host. High level represents binary 1, low level represents binary 0. DATA 8 is the most significant bit. Signal must be high at least 0.5 microseconds before the falling edge of the – data strobe signal and held at least 0.5 microseconds after the rising edge.
Acknowledge	10	28	Low level pulse of 2 to 6 microseconds indicates input of a character into the print data buffer or the end of an operation.
BUSY	11	29	High level indicates the printer cannot receive data. Typical conditions that cause a high BUSY level are buffer full or ERROR condition.
PE (Paper Empty)	12		High level indicates that the printer is out of paper.
SLCT (Select)	13		High level indicates that the printer is ON LINE.
* AUTO FEED XT	14		Low level indicates LF (line feed) occurs after each CR (carriage return) code.
No connection	15		Reserved signal line.
Signal ground		16	Logic/signal ground level (0 volts).
Frame ground		17	Printer cabinet/frame ground line.
No connection			Reserved signal line.
Signal ground		19 - 30	Twisted-pair cable return lines.
* INIT initialize	31		Low level pulse of 50 microseconds or more resets the buffer and initializes the printer.
ERROR	32		Low level indicates the printer is OFF LINE, has a PAPER OUT or has sensed an ERROR condition.
Signal ground	33		Logic/signal ground level (0 volts).
No connection	34		Not used.
+5 volts regulate	35		Connected to the +5 volts source through a 3.3 kOhm resistor.
*SLCT IN (–select in)	36		Low level indicates the printer is placed ON LINE (selected) when the power is turned ON.

* means the signal is generated by the host system (e.g. the PC)

Table B.23
Centronics pin assignments

Appendix C

Review of the Intel 8255 PPI chip

This section contains brief qualitative information on the Intel 8255 programmable peripheral interface (PPI). Because of the chip's immense popularity in data acquisition boards, an entire appendix is devoted to it. For more detailed information on the 8255's operation, and the associated 8254 timer/counter (detailed in the following appendix), contact Intel for a copy of their data sheets for these chips.

The 8255 is used to interface real-world peripherals to the host computer bus. These may be switch sensors, relays, instruments with digital readouts and controls, industrial I/O mounting racks, other computer buses, and so on.

The chip has 24 programmable digital I/O lines. The 24 lines are divided into three 8-bit ports (named port A, port B and port C). The ports may be programmed either as two groups of 12 lines (group A and group B) or as 3 individual 8-bit ports. The ports/groups may be operated in three modes: simple I/O (mode 0), strobed I/O (mode 1) and bi-directional I/O (mode 2). The 8255 also allows single bits to be set or reset in port C.

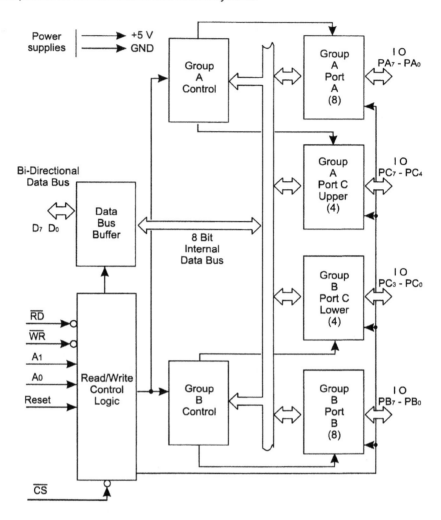

Figure C.1
8255 Block diagram

An 8255 occupies four consecutive addresses in the host computer's I/O address space. They are the data registers of ports A, B and C and the 8255 control register, as shown below

Offset	Write	Read
0	Port A (CIOA)	Port A (DIOA)
1	B (DIOB)	Port B (CIOB)
2	Port C (DIOC)	Port C (DIOC)
3	Control register (DIOCTRL)	

Table C.1
8255 Registers

Before any I/O can be performed, the mode and direction of the ports/groups of the 8255 must be set. This is done simply by writing one byte of configuration information to the control register of the 8255. The 8255 then operates in the specified mode until it is reset or new configuration information is written to the control register. The format of this register is shown below, followed by the format of the data registers and description on how to use the chip in the different modes.

C.1   DIO0CTRL – control register of the 8255

This register has two functions: to set the operating modes of the three ports in the chip, and to set and reset individual bits in port C. The function and bit names of the register therefore depend on the setting of bit 7. The layout of the register is shown below.

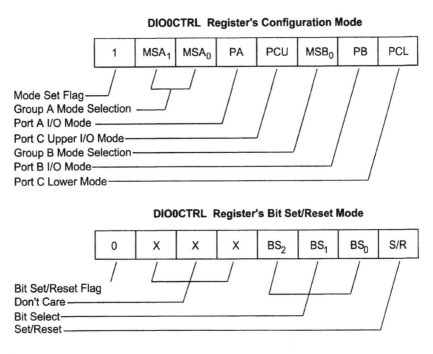

Figure C.2
DIO0CTRL 8255 control register layout

Bit 7, function select: If this bit is set when the register is written to, then the register is in configuration mode. If the bit is 0, then the register is in bit set/reset mode (see below). The functions of the remaining bits are described below; the functions depend on the setting of bit 7.

Configuration mode, bit 7 set

This allows the mode to be set for each of the two groups of the 8255 and the direction set for the individual ports. Note that the direction of port C is independently programmable in two 4-line nibbles.

Bits 6–5, group A mode select: These two bits set the mode of the group A ports. These are port A and the upper four lines of port C. The bit combinations are as follows:

MSA$_1$	MSA$_0$	Group A I/O mode
0	0	Mode 0, simple I/O
0	1	Mode 1, strobed I/O
1	X	Mode 2, bidirectional bus

Table C.2
Group A model select

Bit 4, port A I/O direction: If this bit is set, then port A functions as an input. If it is 0, then port A is configured as an output.

Bit 3, port C upper I/O direction: If this bit is set, then the upper four lines of port C function as inputs. If the bit is 0, then the lines become outputs.

Bit 2, group B mode select: This bit sets the mode of the group B ports. These are port B and the lower four lines of port C. The bit combinations are as follows:

MSA$_B$	Group B I/O mode
0	Mode 0, simple I/O
1	Mode 1, strobed I/O

Note that group B can only be used for simple or strobed I/O. Configuration mode, bit 7 set.

Table C.3
Group B model select

Bit 1, port B I/O direction: If this bit is set, then port B functions as an input. If it is 0, then port B is configured as an output.

Bit 0, port C lower I/O direction: If this bit is set, then the lower four lines of port C function as inputs. If the bit is 0, then the lines become outputs.

Bit set/reset mode, bit 7 clear

Bits 6–5: These bits have no effect in this function.

Bits 3–1, bit select: These bits select the bit in port C that is to be modified. A code of 000 selects port C line 0 to set or reset, 001 selects line 1 and soon up to 111 which selects line 7.

Bit 0, set/reset: This bit specifics the state into which the selected port C line will be placed. Writing a 1 will make the line go high and a 0 makes it go low. This operation has no effect on the other lines of port C.

C.2   DIOA – port A of the 8255 (offset 0, read/write)

This register is the data register of port A of the 8255. The port can be operated in simple I/O mode, strobed I/O mode or bi-directional bus mode, modes 0, 1 or 2.

A7	A6	A5	A4	A3	A2	A1	A0

Figure C.3
Port A of the 8255

The bits A7 (MSB) down to A0 (LSB) reflect the status of the port's I/O lines. Depending on the programmed I/O mode of the port, the lines may be inputs, outputs or bi-directional.

C.3   DIOB – port B of the 8255 (offset 1, read/write)

This register is the data register of port B of the 8255. The port can be operated in simple I/O mode or strobed I/O mode, modes 0, or 1.

B7	B6	B5	B4	B3	B2	B1	B0

Figure C.4
Port B of the 8255

The bits B7 (MSB) down to B0 (LSB) reflect the status of the port's I/O lines. Depending on the programmed I/O mode of the port, the lines may be inputs or outputs.

C.4   DIOC – port C of the 8255 (offset 2, read/write)

This register is port C of the 8255. It may operate in simple I/O mode, or some or all of its lines may be used as handshaking control lines for ports A and B when these ports operate in mode 1 or 2 and are therefore not available as I/O lines. It is more meaningful to refer to them by their functional names, summarized in the following table and described below that.

C7	C6	C5	C4	C3	C2	C1	C0

Figure C.5
Port C of the 8255

The bits C7 (MSB) down to C0 (LSB) reflect the status of the port's available I/O lines. Depending on the programmed 1/0 mode of the port, the lines may be inputs or outputs. The other lines may be handshakes or interrupt request lines.

Port C line	Simple I/O: mode 0	Strobed input: mode 1	Strobed output: mode 1	Bi-directional bus: mode 2
C7	I/O	I/O	/OBF$_A$	/OBF$_A$
C6	I/O	I/O	/ACK$_A$	/ACK$_A$
C5	I/O	IBF$_A$	I/O	IBF$_A$
C4	I/O	/STB$_A$	I/O	/STB$_A$
C3	I/O	INTR$_A$	INTR$_A$	INTR$_A$
C2	I/O	/STB$_B$	/ACK$_B$	I/O
C1	I/O	IBF$_B$	OBF$_B$	I/O
C0	I/O	INTR$_B$	INTR$_B$	I/O

The symbol '/' preceding a signal indicates that it is active low.

Table C.4
Port C line usage

When an 8255 is used in one of the handshaking modes, the /STB and IBF lines are used to synchronize input data transfers. The /OBF and /ACK lines are used to synchronize output transfers. The signals in the table above have the following functions:

Name	Type	Description
/STB	External input	Strobe: the external device, driving this line low loads data from the peripheral bus into the 8255 ports input latch.
IBF	External output	Input buffer full: the 8255 sets this line high to indicate to the external device that its data has been loaded into the 8255 port's input latch.
/ACK	External input	Acknowledge: the external device asserts this line low to indicate that the output data in the 8255 port has been read.
/OBF	External output	Output buffer full: the 8255 asserts this line low to indicate to the external device that there is data to be read from the port. This line can be used to strobe the data into the external device.
INTR	Internal output	Interrupt request: this signal becomes active (high) when the 8255 is requesting service from the host computer. For input operations, it indicates that there is data in the corresponding port to be read by the host. For output operations it indicates that the external device has read the data and thus the host may write another byte to the 8255 port. The appropriate interrupt enable bits must be set in the 8255 to allow this signal to reach the host computer.
/RD	Internal input	Read signal: this signal is generated by the control lines of the host computer. It should be activated when the program executes an input instruction from any 8255 register.
/WR	Internal input	Write signal: this signal is generated by the control lines of the host computer. It is activated when the program executes an output instruction to any 8255 register.

Table C.5
Port C line usage functions

The term external is used to refer to the external peripheral and internal to refer to the host computer bus. Input is an input signal or data to the 8255 while an output is a signal or line driven by the 8255.

The next section describes the three operating modes of the 8255 ports/groups and the bit set/reset operation from the software programmer's point of view. Reading from or writing to the ports typically takes the form of one of the following instructions:

Language	Port read	Port write
C	data = inp (addr)	outp (addr, data)
Pascal	data := port[addr]	port [addr] := data
BASIC	data = INP (addr)	OUT addr, data

Assembly language	mov al, data
	mov dx, addr
	in al, dx
		mov al, data
		mov dx, addr
		out dx, al

Where: addr **is the address of the 8255 register, in the host computer's I/O address map.**

data is the data byte read or written.

Table C.6
Instructions for reading or writing to the ports

C.5   Mode 0: simple I/O

This mode is used for simple input and output operations for each of the ports. No handshaking is required and no interrupts are generated. Data is simply read from or written to a selected port.

The following characterize mode 0:

- Two 8-bit ports (ports A and B) and two 4-bit ports (upper and lower nibble of port C)
- Any port can be configured for input or output
- Outputs are latched, inputs are not latched
- Data transfer by polled I/O

C.6   Mode 0 programming

To use the 8255 in mode 0:

- Write a single byte to the control register to set the 8255 into mode 0 with the three ports configured for the desired data direction
- Then read to or write from the I/O port corresponding to an 8255 port (port A, B or C) as many times as necessary to obtain or transfer the required amount of data

C.7   Mode 1: strobed I/O

In this mode data transfers are controlled by handshaking signals and hardware interrupts. Some of the port C lines are used for these control signals. Hence they take on different functions and names. The following characterize Mode 1:

- Two groups, group A and B. Each group consists of an 8-bit data port and three control lines
- Certain port C lines take on special functions
- The data ports can be either input or output ports

- Both inputs and outputs are latched
- One 2-bit simple I/O port
- Data transfer by interrupts or polled I/O

With both groups configured in mode 1, a single 8255 can read or write data 16-bits wide.

C.8   Mode 1 programming

To use the 8255 in mode 1 input with interrupts:

- Write a byte to the control register to configure the 8255 for mode 1 and the appropriate group for data input
- With the bit set/reset operation (see below), write a 1 to the interrupt enable flip-flop (INTE) of the desired port in the 8255
- The external device pulses the strobe (/STB) input line low. The trailing edge of this loads data into the input port
- The input buffer full (IBF) output line goes high to indicate that the data has been loaded into the input latch
- When the external device pulls the /STB line high, the interrupt request line (INTR) goes high. This indicates to the host system that there is data to be read from the 8255
- The computer reads the data using an interrupt service routine (ISR) and by doing so, automatically resets the INTR and the IBF signals
- The external device can now pulse the /STB low again to load another byte of data into the 8255

Whenever a group of the 8255 is in mode 1 input, the status of the handshaking lines and interrupt signals can be obtained by reading port C. The byte read contains the following information:

C7	C6	IBF$_A$	INTE$_A$	INTR$_A$	INTE$_B$	IBF$_B$	INTR$_B$

Figure C.6
Port C mode 1 input status information

The 8255 may alternatively be used in mode 1 and the data read by polled (program) transfer. This is done as follows:

- Write a byte to the control register to configure the 8255 for mode 1 and the appropriate group for data input
- The program continually monitors the appropriate IBF line by reading port C
- The external device pulls the /STB input line on the digital I/O connector low and this loads data into the input port
- The IBF output line goes high on the digital I/O connector to indicate that the data has been loaded into the input latch
- This also causes the corresponding IBF bit in port C to be set and this tells the program that it can now read the data
- Reading the data causes IBF to go low, thus the external device can now pull the /STB low again to load another byte of data into the 8255

The program could also enable the INTR line with the INTE flip-flop and then monitor INTR instead of the IBF line. In this case, the interrupts from the 8255 are not enabled in the host computer.

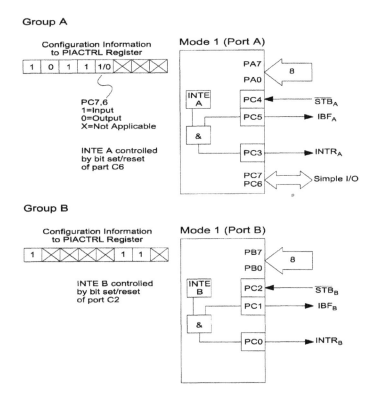

Figure C.7
8255 Group A and B as mode 1 inputs

To use the 8255 in mode 1 output with interrupts:

- Write a byte to the control register to configure the 8255 for mode 1 and the appropriate group for data output
- With the bit set/reset operation, write a 1 to the interrupt enable flip-flop (INTE) of the desired port of the appropriate 8255
- The 8255 interrupt request output (INTR) then goes high
- The host computer detects the INTR line is active. From an interrupt service routine (ISR) it writes a byte to the output port. This automatically resets the INTR line
- The output buffer full (/OBF) line goes low to indicate that there is data to be read by the external device from the 8255
- The external device pulses the acknowledge (/ACK) input low and then high again to indicate that it has read the data
- This makes INTR and /OBF go high again and the cycle may be repeated until all the required data has been written

Whenever a group of the 8255 is in mode 1 output, the status of the handshaking lines and interrupt signals can be obtained by reading port C. The byte read contains the following information:

/OBF$_A$	INTE$_A$	C5	C4	INTR$_A$	INTE$_B$	/OBF$_B$	INTR$_B$

Figure C.8
Port C mode 1 output status information

The 8255 may alternatively be used in mode 1 and the data written by polled (program) transfer. This is done as follows:

- Write a byte to the control register to configure the 8255 for mode 1 and the appropriate group for data output
- The program continually monitors the /OBF line by reading port C, waiting for it to go high. A high indicates that the last data written to the port has been read by the external device
- Then the program can write new data to the port
- The /OBF line goes low to indicate that there is data to be read by the external device from the 8255
- The external device pulses the /ACK input low and high to read the data.
- This makes the /OBF line go high again and the cycle may be repeated until all the required data has been written

The program could also enable the INTR line with the INTE flip-flop and then monitor INTR instead of the /OBF line. In this case, interrupts to the host computer are not enabled.

C.9 Mode 2: strobed bi-directional bus I/O

This mode provides a means for communicating with an external device using an 8-bit bus for both transmitting and receiving data. Both input and output handshaking signals similar to mode 1 are provided to maintain proper bus flow discipline. Hardware interrupts signal the host computer that the port needs attention.

The following characterizes mode 2:

- Only group A operates in mode 2
- One 8-bit bi-directional port, functions as both input and output
- Five of the port C lines take on special functions
- Both inputs and outputs are latched
- One 3-bit simple I/O port
- Data transfer by interrupts or polled I/O

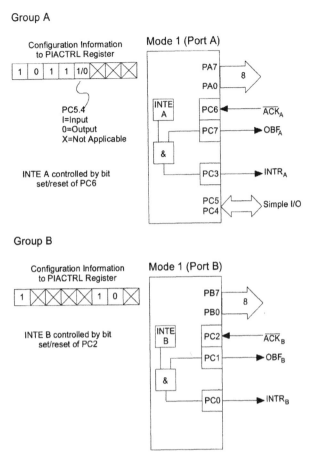

Figure C.9
8235 Group A and B as mode 1 outputs

C.10 Mode 2 programming

To use the 8255 in mode 2 with hardware interrupt transfer:

- Write a byte to the control register to configure the 8255 for mode 2 operation
- With the bit set/reset operation, write a 1 to the interrupt enable number 1 flip-flop (INTE$_1$) to enable output transfer interrupts. Write a 1 to the interrupt enable number 2 flip-flop (INTE$_2$) to enable input transfer interrupts. Both input and output interrupts may be enabled at the same time
- With both interrupt flip-flops enabled, the interrupt request line to the host computer is activated if the external device has strobed data into the 8255 input latch or if the external device has read the output data from the output latch
- The host computer detects the INTR line is active. The interrupt service routine (ISR) that services this interrupt determines whether it was an input or output interrupt by checking bit 5 (IBF$_A$) of the mode 2 status information from port C. (See below.) If the IBF line is high (bit 5 is set), then it is an input interrupt; otherwise it is an output interrupt
- The ISR simply reads the data from or writes data to the 8255

- This generates the appropriate handshake signals from the 8255
- The cycle continues when the next interrupt is generated

Whenever the 8255 is in mode 2, the status of the handshaking lines and interrupt signals can be obtained by reading port C. The byte read contains the following information:

/OBF$_A$	INTE$_1$	IBF$_A$	INTE$_2$	INTR$_A$	C2	C1	C0

Figure C.10
Port C mode 2 status information

The 8255 may alternatively be used in mode 2 with the data read and written by polled (program) transfer. This is done as follows:

- Write a byte to the control register to configure the 8255 for mode 2
- The program continually monitors both the IBF$_A$ and /OBF$_A$ lines by reading port C
- If the port C bit corresponding to IBF$_A$ is set, this indicates that the external device has written data into the input latch. The program must therefore read the data from the port A
- If the port C bit corresponding to /OBF$_A$ is set, this indicates that the external device has read the data written by the program to port A. It must therefore write more data to port A
- This process can be repeated until the required amount of data has been read and written

Interrupts to the host computer must not be enabled when data is transferred by polled I/O.

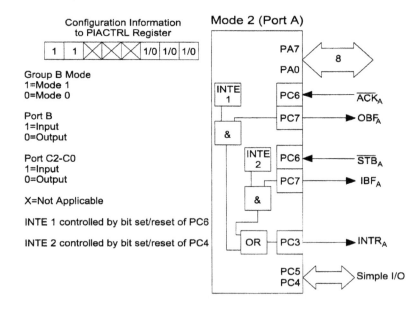

Figure C.11
8255 Group A in mode 2

C.11   Single-bit set/reset

Any of the eight bits of port C can be set or reset using a single output instruction to the DIOCTRL register. When port C is being used as status/control for port A or B any of these bits can be set or reset just as if they were data output ports. The format of the byte to write to the DIOCTRL register to set or reset a port C bit is shown below.

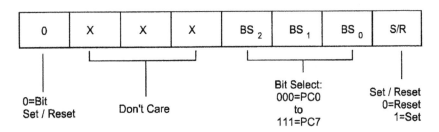

Figure C.12
DIOCTRL register – bit set/reset mode

C.12   Mixed mode programming

An 8255 is not constrained to operate in one mode only. For example, port A may operate in mode 2 and port B may then operate in either mode 1 or mode 0. For any combination, some or all of the port C lines are used for control or status. The remaining port C lines may be used in mode 0 either as inputs or outputs.

A read operation of port C returns all the port C lines except the /ACK and /STB lines. In their place will appear the status of the interrupt enable flip-flops ($INTE_X$). This is illustrated in Figures C.10 and C.11 above, and in the status information bytes that follow the figures.

A write operation to port C will only affect lines programmed as mode 0 outputs. To write to any port C output programmed as a mode 1 output or to change an interrupt enable flip-flop, the bit set/reset operation must be used.

Using the bit set/reset command, any port C line programmed as an output (including INTR, IBF and /OBF) can be written or an interrupt enable flag set or reset. Lines programmed as inputs (including /ACK and /STB) are not affected by this command.

Writing to these lines will affect the interrupt enable flags.

C.13   8255-2 mode 1 and 2 timing diagrams

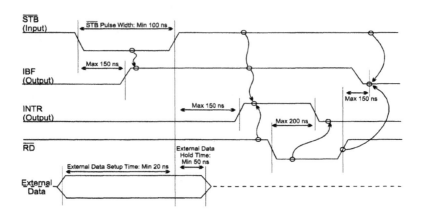

Figure C.13
Strobed input (mode 1)

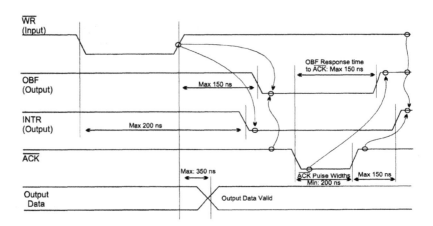

Figure C.14
Strobed output (mode 1)

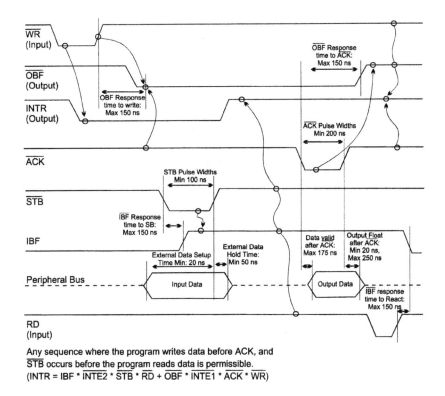

Any sequence where the program writes data before ACK, and $\overline{STB}$ occurs before the program reads data is permissible.
$(INTR = IBF * \overline{INTE2} * \overline{STB} * \overline{RD} + \overline{OBF} * \overline{INTE1} * \overline{ACK} * \overline{WR})$

Figure C.15
Bi-directional bus (mode 2)

Appendix D

Review of the Intel 8254 timer-counter chip

This section contains brief qualitative information on the Intel 8254 programmable timer-counter. Because of the chip's immense popularity in data acquisition boards, an entire appendix is devoted to it. For more detailed information on the operation on the 8254 and the associated 8255 PPI (detailed in the preceding appendix), contact Intel for a copy of their data sheets for these chips.

This appendix describes the architecture of the 8254, and then details the 8254 registers as seen from the host computer. Next, it describes programming the chip and lastly, it explains how the six different counting modes operate.

D.1   8254 architecture

The 8254 are a general-purpose 3-channel timer/counter device. Each timer counter is totally independent, and each may be programmed in different modes and data formats. Since all three timers are identical, the information provided here applies equally to each. The operation of a timer-counter (hereafter referred to simply as a timer) is as follows: The host computer writes a 16-bit word, called the *initial count,* to the timer. Every time the timer receives a clock pulse it decrements the count value.

Now, to the external system, a timer has three connections. They are:

- A clock pulse input
- A gate input
- A timer output

The behavior of the timer output depends on the timer's counting mode.

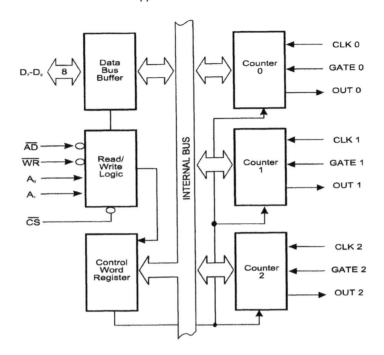

Figure D.1
8254 Block diagram

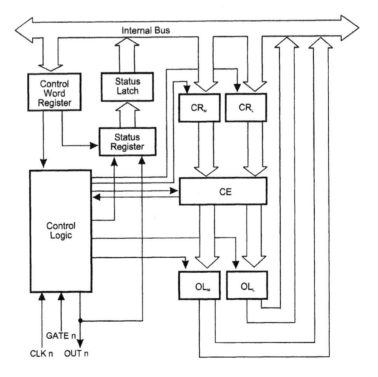

Figure D.2
Block diagram of a timer/counter

The timer/counter chip itself consists of a control word register, some logic circuitry, and the three counters.

Each counter consists of a 2-byte wide count register, a 16-bit counting element and a 2-byte wide output latch.

Count register (CR)

The count register stores the initial 16-bit count written to a counter. It consists of 2-byte wide registers, which are written to separately. When a counter is programmed with a control word, the count register is cleared. Both count register bytes are transferred (loaded) to the counting element simultaneously.

Counting element (CE)

The counting element is simply a 16-bit pre-settable synchronous down counter. It cannot be read from or written to directly. It is automatically loaded on specified conditions from data in the count register. The count value is always read from the output latch.

Output latch (OL)

The output latch normally follows the counting element. It consists of 2-byte wide registers, which are read from, separately. If a suitable counter latch command (see below) is sent to the counter, the current count value is latched in the output latch until it is read from the counter's data register (TC2, TCI or TC0). Thereafter, the output latch continues to follow the counting element.

D.2   8254 Registers

An 8254 occupies four consecutive addresses in the host computer's I/O address space. They are the data registers of timers 0, 1 and 2 and the 8254 control register, shown below.

Offset	Write	Read
0	Timer-counter 0 (TCO)	Timer-counter 0 (Ten)
1	Timer-counter I (TCI)	Timer-counter I (TCI)
2	Timer-counter 2 (TC2)	Timer-counter 2 (TC2)
3	Control register (TCCTRL)	-

Table D.1
8254 Registers

TCCTRL timer/counter control register (offset 3, write only)

The timer/counter control register is used to program, for each counter, the counting mode, the number of bytes to read/write and whether the counter counts in BCD or binary format. In addition, this register can be used to perform read-back commands and counter latch commands. Note that the function and bit names of this register differ according to whether configuration mode, counter latch command, or read-back command is selected with bits 7–6 (SC_1 and SC_0) and bits 5–4 (RW_1 and RW_0).

Read-back and counter latch commands, as well as the functions of the different counter modes are described below.

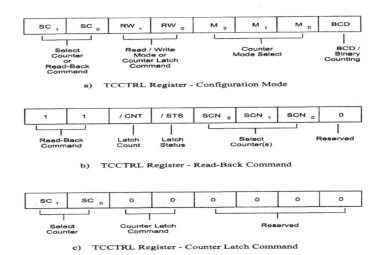

Figure D.3
TCCTRL register

The functions of the remaining bits are described below, depending on the setting of bits 7 and 6 and bits 5 and 4.

Configuration mode

Bits 7–6, select counter (SC): These two bits select the timer/counter to which the rest of the TCCTRL register bits will apply. The SC bits are defined as follows:

SC$_1$	SC$_0$	Operation performed
0	0	Select counter 0
0	1	Select counter 1
1	0	Select counter 2
1	1	Read-back command

Table D.2
Select counter

Bits 5–4, read/write mode (RW): These two bits select the read/write mode of the selected timer/counter. The RW bits are defined as follows:

RW$_1$	RW$_2$	Operation performed
0	0	Counter latch command
0	1	Read/write least significant byte only
1	0	Read/write least significant byte first
1	1	Read/write least significant byte first, then most significant byte

Table D.3
Read/write mode

Bits 3–1, counter mode select (M): These three bits select the operating mode of the timer/counter selected with bits 7-6. The Mode bits are defined as follows:

M_2	M_1	M_0	Operating mode
0	0	0	Mode 0
0	0	1	Mode 1
0	1	0	Mode 2
0	1	1	Mode 3
1	0	0	Mode 4
1	0	1	Mode 5

Table D.4
Counter mode select

Bit 0, counting mode select (BCD): This bit determines whether the selected counter is to count in BCD or binary format. A 0 specifies 16-bit binary counting and a 1 specifies 4 decade binary coded decimal counting.

Read-back command

When the read-back command is specified, the bit definitions of the TCCTRL register are:

Bits 7-6   These bits must both be set to 1 to invoke the read-back command.

Bit 5   Count (/CNT): Setting this bit to 0 causes the timer/counter chip to latch the count(s) of the counters selected with bits 3–1 of this register (see below).

Bit 4   Status (/STS): Setting this bit to 0 causes the timer/counter chip to latch certain status information from the counters selected with bits 3–1 of this register (see below). The format of the status byte is shown in Figure D.10.

Bits 3–1   Select counter (SCN): Setting one or more of these bits causes the corresponding counter to latch its count and/or status information, when the read-back command is issued. Setting SCN_2 latches counter 2 information, SCN_1 counter 1 and SCN_0 counter 0.

Bit 0   This bit performs no function and should be set to 0.

Counter latch command

The bit definitions of the TCCTRL register, when the counter latch command is specified, are as follows:

Bits 7–6   Select counter (SC): These two bits select the timer/counter whose count is to be latched.

Bits 5–4   These two bits must both be set to 0 to specify the counter latch command.

Bits 3–0   These four bits perform no function and must all be set to 0.

TCO – timer/counter 0 (offset 0, read/write)

This is the data register of the first timer/counter.

Before reading or writing from this register, a control word for this counter must be written to the timer/counter control register. Then 'reads' and 'writes' to this register must follow the format specified in the control word.

$TC0_7$	$TC0_6$	$TC0_5$	$TC0_4$	$TC0_3$	$TC0_2$	$TC0_1$	$TC0_0$

Figure D.4
TC0 register

The bits $TC0_7$ (MSB) down to $TC0_0$ (LSB) reflect the high byte or the low byte of the data 'read from' or 'written to' this counter.

TC1 – timer/counter 1 (offset 1, read/write)

This register is the data register of the second timer/counter.
Before reading or writing from this register, a control word for this counter must be written to the timer/counter control register. Then 'reads from' and 'writes to' this register must follow the format specified in the control word.

$TC1_7$	$TC1_6$	$TC1_5$	$TC1_4$	$TC1_3$	$TC1_2$	$TC1_1$	$TC1_0$

Figure D.5
TCI register

The bits $TC1_7$ (MSB) down to $TC1_0$ (LSB) reflect the high byte or the low byte of the data read or written to this counter.

TC2 – timer/counter 2 (offset 2, read/write)

This register is the data register of the third timer/counter.

Before reading or writing from this register, a control word for this counter must be 'written to' the timer/counter control register. Then 'reads from' and 'writes to' this register must follow the format specified in the control word.

$TC2_7$	$TC2_6$	$TC2_5$	$TC2_4$	$TC2_3$	$TC2_2$	$TC2_1$	$TC2_0$

Figure D.6
TC2 register

The bits $TC2_7$ (MSB) down to $TC2_0$ (LSB) reflect the high byte or the low byte of the data read or written to this counter.

D.3   Programming a counter

On power-up or reset, the state of the 8254 is undefined. Before any timer/counter operations can be performed, each timer to be used must be programmed with a control word, which is written to the control register. This sets the individual counters:

- Operating mode (Mode 0 to 5)

- Counting format (BCD or binary)
- Read/write format (LSB only, MSB only or LSB then MSB)

The programmed counter then operates in the specified format until it is reset or new configuration information is written to the control register. The format of this register when used for configuring a counter is repeated below for reference.

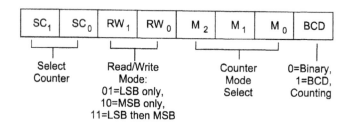

Figure D.7
TCCTRL register – configuration mode

Data transfer format

Using the control word, each counter may be programmed to transfer data from the host computer in one of three ways:

- Read/write least significant byte only
- Read/write most significant byte only or
- Read/write least significant byte first, then most significant byte

A new initial count may be written to a counter without affecting the counter's programming in any way. Counting will be affected as described in the mode definitions below.

Writing a 1-byte initial count simply consists of outputting the byte to the counter's data register. Writing a 2-byte count consists of writing the first byte (the least significant byte) to the counter's data register and then, at any time later, writing the second or most significant byte to the same data register.

Clock pulse input

The clock pulse input is the physical connection, where clock pulses are applied to a counter. A clock pulse is defined as a rising edge, then a falling edge, in that order, at a counter's clock input. New counts are loaded, and the counting element is decremented, on the falling edge of a clock pulse.

Gate input

Depending on the counter's mode, the gate input provides for enable/disable counting, count initiating (trigger), or setting/resetting the timer output.

D.4   Read operations

It is often necessary to read the value or status of a counter without disturbing the count in progress. There are three methods:

- A simple read operation
- A counter latch command or
- A read-back command.

The results of the read operation are read from the counter's data register, in the programmed format (LSB, MSB or LSB then MSB) of that counter.

This is called *reading a counter,* and may take 'one physical CPU read' instruction (LSB or MSB) or 'two physical CPU read instructions' (LSB then MSB).

Simple read operation

The simple read operation consists of reading the contents of the desired counter's data register. The clock input of the counter must be disabled for this to be successful, otherwise the count may be in the process of changing when it is read, returning a completely erroneous count value. The clock input may be disabled with the gate input or with external logic.

Counter latch command

A counter latch command is invoked by writing a special type of control word to the timer/counter's control register, TCCTRL.

The format of the control word for a counter latch command is diagrammed below.

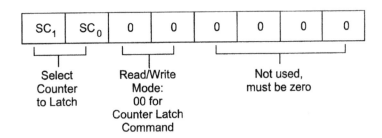

Figure D.8
TCCTRL register – counter latch command

Bits SC_1 and SC_0 *select* the counter whose count is to be latched. 00 selects counter 0, 01 selects counter 1 and 10 selects counter 2.

The selected counter's output latch, latches the count at the time the counter latch command is received. The count is held in the latch until it is read (or the counter reprogrammed). The count is then automatically unlatched and the output latch returns to following the counting element. Multiple counter latch commands may be used to latch more than one counter. Each latched counter holds its count until read. Counter latch commands do not affect the programming of the counter in any way.

If a counter is latched, any subsequent counter latch commands to the same counter, before the count has been read, will be ignored. When read, the count value returned will be the count at the time the first counter latch command was issued.

Read-back command

A read-back command is issued by writing a special type of control word to the timer/counter's control register, TCCTRL. Read-back commands may be used to latch one or more counter's *current count value* and/or *status* information.

The format of the control word for a read-back command is diagrammed below.

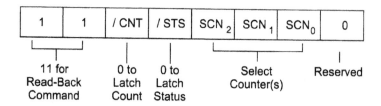

Figure D.9
TCCTRL register – read-back command

Setting any or all of bits 3 to 1 (SCN_2 to SCN_0) selects the counter(s) to which this command will apply. Setting the count (/CNT) bit to 0 causes the current count of the selected counter(s) to be latched, setting the status (/STS) bit to 0 causes the status byte of the selected counter(s) to be latched and setting both bits to 0 causes both the status and current count to be latched (see below).

Multiple counter latch

The read-back command is used to latch the current count of multiple counters in their respective output latches. This is done by setting the count bit to 0 in the control word for this command. This single command is functionally equivalent to multiple counter latch commands. Each counter's latched count is held until read (or the counter reprogrammed). That counter is automatically unlatched when read, but the other counters remain latched until they are read. If multiple read-back commands are issued to the same counter without reading the count, all but the first are ignored. The count returned is the current count at the time the first read-back command was issued.

Counter status information

The read-back command may also be used to latch certain status information of the selected counter(s), in their respective output latches. This is done by setting the status bit to 0 in the control word for this command.

The counter's status byte, when read, provides the information shown in the diagram below.

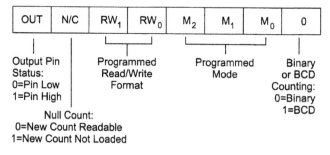

Figure D.10
TCX register – status byte

Bits M_2 to M_0 return a binary number corresponding to the counter's programmed mode.

The null count bit indicates if the last count written to the count register has been loaded into the counting element. If the count has not yet been loaded, then it cannot be read and the null count bit will be set to 1. Reading the count when the null count bit is set, will return the current count from the previous initial count, written to the count register. The exact time that the new count is loaded depends on the mode of the counter (see below), but the 'null count bit clear' indicates that the new count has been loaded.

Bit 7 (OUT) reflects the state of the counter's output pin. This provides software with the power to monitor this pin.

Latching both status and current count

Setting both the status (/STS) bit and the count (/CNT) bit to 0 causes both the status byte and the current count of the selected counter(s) to be latched simultaneously. This is functionally the same as issuing two separate read-back commands at once. If both the status and current count are latched, the first read operation of the counter's data register will return that counter's status byte, regardless of which was latched first. The next one or two reads, (depending on whether the counter has been programmed for one or 2-byte counts), return the latched count. Subsequent reads, return unlatched counts.

D.5   Counter mode definitions

The following six sections describe in detail the different counting modes of the 8254 timer/counter. In the descriptions that follow, *output* refers to the state of the output pin of the device, and *gate* refers to the state of the counter's gate input pin. Both signals from all three counters are available on the auxiliary connector.

The word *Trigger* is used to mean a rising edge at a counter's gate input, *Counter Loading* is the transfer of the initial count from the counter's count register to its counting element and *Clock Pulse* is a rising edge and then a falling edge at a counter's clock input.

Mode 0: interrupt on terminal count

After the mode byte is written to the control register, the output is low. Once an initial count has been written, the output remains low until the counter has counted down to zero. The output then goes high and remains high until a new count is written or the counter reprogrammed. The gate input inhibits counting when low, and enables counting when high.

After the control word and initial count have been written, the counter is loaded on the next clock pulse. This clock pulse does not decrement the count, so for an initial count of *N*, the output goes high $N+1$ clock pulses after the initial count was written.

If a new count is written to the counter, it will be loaded on the next clock pulse and counting will continue from the new count. If a two-byte count is written, the first byte disables counting and sets the output low. After the second byte is written, the full count is loaded on the next clock pulse. This allows the counting sequence to be synchronized by software. Again, the output goes high after $N+1$ clock pulses.

If the initial count is written when the gate is low, it will still be loaded on the next clock pulse. When the gate goes high, the output will go high, *N* clock pulses later.

Using the internal oscillator or bus clock, this mode can be used to generate a positive edge on the external output after a programmable time, or if the board is 'jumpered' for interrupts, to generate an interrupt after a programmable time.

This mode can also be used to count events or frequency.

Mode 1: hardware re-triggerable one-shot

After the counter is programmed the output will be high. Writing an initial count arms the counter and a subsequent trigger loads the counter. The output goes low on the next clock pulse and remains low until the counter reaches zero. The output then goes high and remains high until the next clock pulse after the next trigger.

An initial count of N results in a one-shot pulse N clock cycles long. The one-shot is re-triggerable; hence the output will remain low for N clock pulses after any trigger. The one-shot pulse can be repeated without rewriting the initial count to the counter. The gate input has no effect on the output.

If a new count is written to the counter during a one-shot pulse, the current one-shot pulse is not affected unless the counter is re-triggered. In that case, the counter is loaded with the new count and the current one-shot pulse continues until the new count expires.

Mode 2: rate generator

After the counter is programmed the output will be high. An initial count of N is loaded on the next clock pulse and when it has decremented down to 1, the output goes low for *one* clock pulse. The output then goes high, the counter automatically reloads the initial count and the process is repeated indefinitely. The sequence is repeated every N clock pulses.

The gate input enables counting when high, and inhibits counting when low. If the gate goes low during an output pulse, the output is set high immediately. A trigger reloads the initial count on the next clock pulse and the output goes low for one clock pulse after N clock pulses. Thus, the gate input can be used to synchronize the counter.

Writing a new count does not affect the current counting sequence. If a trigger is subsequently received before the end of the current period, the counter will be reloaded on the next clock pulse and counting will continue from the new count. Otherwise, the new count will be loaded at the end of the current cycle. In this mode, an initial count of 1 is invalid.

Mode 2 functions like a divide by N counter. It can also be used to generate an output frequency or a periodic interrupt.

Mode 3: square wave generator

Mode 3 is similar to mode 2 except for the duty cycle of the output. After the counter is programmed, the output will be high. An initial count of N is loaded on the next clock pulse. When half of the initial count has expired, the output goes low for the remainder of the count. The output then goes high, the counter automatically reloads the initial count and the process is repeated indefinitely. This results in a square wave with a period of N clock cycles.

The gate input enables counting when high, and inhibits counting when low. If the gate goes low when the output is low, the output is set high immediately. A trigger reloads the counter with the initial count on the next clock pulse. Thus, the gate input can be used to synchronize the counter.

Writing a new count does not affect the current counting sequence. If a trigger is subsequently received before the end of the current half-cycle of the square wave, the counter will be reloaded on the next clock pulse and counting will continue from the new count. Otherwise, the new count will be loaded at the end of the current half-cycle.

Mode 3 functions slightly differently for even and odd initial count values.

For even counts: the output is initially high. On the next clock pulse, the initial count is loaded. On subsequent clock pulses, it is decremented by *two*. When the count expires, the output toggles, and the counter is reloaded with the initial count. This process is repeated indefinitely.

For odd counts: the output is initially high. On the next clock pulse, the initial count *minus one* (an even number) is loaded. On subsequent clock pulses, it is decremented by *two*. One clock pulse *after* the count expires, the output goes low, and the counter is reloaded with the initial count minus one. Subsequent clock pulses continue to decrement the count by two. When the count expires, the output goes high again and the counter is reloaded with the initial count minus one. This process is repeated indefinitely. So for odd counts, the output is high for $(N+1)/2$ counts and low for $(N-1)/2$ counts, or high for one count longer than it is low.

Mode 3 is typically used to generate an output frequency.

Mode 4: software-triggered strobe

After the mode byte is written to the control register, the output is high. Once an initial count has been written, the output remains high until the counter has counted down to zero. The output then goes low for *one* clock pulse and then goes high again. The counting sequence is triggered by writing an initial count.

The gate input inhibits counting when low, and enables counting when high. It has no effect on the output.

After the control word and initial count have been written, the counter is loaded on the next clock pulse. This clock pulse does not decrement the count, so for an initial count of *N*, the output strobes low $N+1$ clock pulses after the initial count was written.

If a new count is written while counting, it will be loaded on the next clock pulse and counting will continue from the new count. If a two-byte count is written, the first byte written has no effect on counting. After the second byte is written, the full count is loaded on the next clock pulse. This allows the counting sequence to be re-triggered by software. Again the output strobes low after $N+1$ clock pulses.

Using the internal oscillator or bus clock, this mode can be used to generate a negative pulse on the external output after a programmable time. With an external clock source, it generates a pulse after a programmable number of events.

Mode 5: hardware-triggered strobe

This mode is similar to mode 4 except that the counting is triggered by a rising edge on the counter's gate input.

After the control word and initial count have been written, the output is high. The counter is loaded on the next clock pulse after a trigger is received. This clock pulse does not decrement the count. The output remains high until the counter has counted down to zero. The output goes low for *one* clock pulse and then goes high again. Therefore, for an initial count of *N*, the output strobes low $N+1$ clock pulses after the initial count was written.

The counting sequence is re-triggerable: a trigger causes the counter to be loaded with the initial count on the next clock pulse. The output will not strobe low until $N+1$ clock pulses after any trigger. The gate input has no effect on the output.

If a new count is written while counting, it will have no effect on the current count sequence. If a trigger is received after the new count is written but before the current count expires, the counter will be reloaded on the next clock pulse and counting will continue from there.

Using the internal oscillator or bus clock, this mode can be used to generate a negative pulse on the external output after a programmable time from an external trigger.

Operating mode	Low or going low	Rising	High
0	Disables counting	–	Enables counting
1	–	Initiates counting and resets output after next clock	–
2	Disables counting and sets output immediately high	Initiates counting	Enables counting
3	Disables counting and sets output immediately high	Initiates counting	Enables counting
4	Disables counting	–	Enables counting
5	–	Initiates counting	–

Table D.5
Summary of the functions of the gate inputs

D.6   Interrupt handling

Each timer/counter maybe configured to generate interrupt requests (IRQs) to the host computer. It is the same, in principle, as using any other device to generate PC interrupts:

- Program the counter and perform any other setup that has to be done
- Save the state (enabled or disabled) of the selected interrupt level on the PC system board and then disable it
- Save the old interrupt vector and install a new one that will point to the interrupt service routine that will service the timer interrupts
- Enable the interrupt level in the PC's interrupt controller

Now the software can continue with other tasks. The interrupt service routine must:

- Do whatever processing it needs to do
- Either chain another interrupt-routine if the interrupt line is being shared, or write an end-of-interrupt command to the PC system board interrupt controller

To clean up the interrupts after using them:

- Disable the interrupt level
- Restore the interrupt vector to what it was before
- Restore the interrupt level to its previous state

Note that some timer/counter modes are more suitable for generating interrupts (for example, mode 0) than others.

Appendix E

Thermocouple tables

The IPTS-68 standard defines thermocouple voltages as a function of temperature according to the following polynomial equation:

$$V=C_0+C_1T+C_2T^2+C_3T^3+\ldots\ldots+C_nT^n$$

Where:

V = thermocouple voltage in units of μV (10^{-6} V, or microvolts)
T = thermocouple temperature in °Celsius
$C_1, C_2, C_3, \ldots\ldots C_n$ = polynomial coefficients

Type B thermocouple

Number of ranges = 1
Range #1 0 to 1820°C
Order of polynomial = 8

Power of T	Coefficient
0	0.000000000000000/e+0000
1	−2.467 460 16200000E−0001
2	5.910211116900000E−0003
3	−1.4307 1234 300000E−0006
4	2.15091497500000E−0009
5	−3.17578007200000E−0012
6	2.40103674590000E−0015
7	−9.09281481590000E−0019
8	1 3299505l370000E−0022

Type BP Thermocouple

Number of ranges = 1
Range #1 0 to 1820°C
Order of polynomial = 8

Power of T	Coefficient
0	0.000000000000000E+0000
1	4.81936208460000E+0000
2	1.57022351980000E–0002
3	–2.28024180120000E–0005
4	3.12472605770000E–0008
5	–2.75501226440000E–0011
6	1.50248318750000E–0014
7	–4.448020196440000E–0018
8	6.12181360300000E–0022

Type BN thermocouple

Number of ranges = 1
Range #1 0 to 1820°C
Order of polynomial = 8

Power of T	Coefficient
0	0.000000000000000+0000
1	5.06610810080000E–0000
2	9.79202408090000E–0003
3	–2.13717056690000E–0005
4	2.90963456020000E–0008
5	–2.43743525730000E–0011
7	–3.73873871480000E–0018
8	4.79186308940000E–0022

Type E thermocouple

Number of ranges = 2
Range #1 –270 to 0°C
Order of polynomial = 13

Power of T	Coefficient
0	0.000000000000000E+0000
1	5.86958577990000E+0001
2	5.16675177050000E–0002
3	–4.46526833470000E–0004
4	–1.73462709050000E–0005
5	–4.87193684270000E–0007
6	–9.88965504470000E–0009
7	–1.09307673750000E–0010
8	–9.17845350390000E–0013
9	–5.25751585210000E–0015
10	–2.01696019960000E–0017
11	4.95021387820000E–0020
12	–7.01779806330000E–0023
13	–4.36718084880000E–0026

Range #2 0 to 1000°C
Order of polynomial = 9

Power of T	Coefficient
0	0.000000000000000E+0000
1	5.86958577990000E+0001
2	4.31109454620000E–0002
3	5.72203582020000E–0005
4	–5.40206680850000E–0007
5	0.54259221110000E–0009
6	–2.48500891360000E–0012
7	2.33897214590000E–0015
8	–1.19462968150000E–0018
9	2.55611274970000E–0022

Type J Thermocouples

Number of ranges = 2
Range #1 –210 to 760°C
Order of polynomial = 7

Power of T	Coefficient
0	0.000000000000000+0000
1	5.03727530270000E+0001
2	3.04254912840000E–0002
3	–8.5669750460000E–0005
4	1.33488257350000E–0007
5	–1.70224059660000E–0010
6	1.94160910010000E–0013
7	–9.63918448590000E–0017

Range #2 760 to 1200°C
Order of polynomial = 5

Power of T	Coefficient
0	2.97217517780000+0005
1	–1.50596328730000E+0003
2	3.20510642150000E+0000
3	–3.22101742300000E–0003
4	1.59499687880000E–0006
5	–3.12398017520000E–0010

Type JP thermocouples

Number of ranges = 1
Range #1 –270 to 760°C
Order of polynomial = 7

Power of T	Coefficient
0	0.000000000000000E+0000
1	1.79103202040000E+0001
2	4.66477610970000E–0003
3	–7.11724606090000E–0005
4	1.33722172380000E–0007
5	–1.50457626900000E–0010
6	1.53390150110000E–0013
7	7.52579474320000E–0017

Type JN Thermocouples

Number of ranges = 1
Range #1 –210 to 760°C
Order of polynomial = 7

Power of T	Coefficient
0	0.000000000000000E+0000
1	3.24624328230000E+0001
2	2.57607151740000E–0002
3	–1.44972898550000E–0005
4	–2.33915030000000E–0010
5	–1.97664327600000E–0011
6	4.07707598990000E–0014
7	–2.11338974270000E–0017

Type K thermocouples

Number of ranges = 2
Range #1 –270 to 0°C
Order of polynomial = 10

Power of T	Coefficient
0	0.000000000000000E+0000
1	3.94754331390000E+0001
2	2.74652511380000E–0002
3	–1.65654067160000E–0004
4	1.51909123920000E–0006
5	–2.45816709240000E–0008
6	–2.47579178160000E–0010
7	–1.55852761730000E–0012
8	–5.97299212550000E–0015
9	–1.26888-012160000E–0017
10	–1.13827973740000E–0020

Range #2 0 to 1372°C
Order of polynomial = 8

Power of T	Coefficient
0	−1.85330632730000E+0001
1	3.89183446120000E+0001
2	1.66451543560000E−0002
3	−7.87023744480000E−0005
4	2.28357855570000E−0007
5	−3.57002312580000E−0010
6	2.99329091360000E−0013
7	−1.28498487980000E−0016
8	2.22399743360000E−0020

Type KP Thermocouple

Number of ranges = 2
Range #1 −270 to 0°C
Order of polynomial = 12

Power of T	Coefficient
0	0.000000000000000E+0000
1	2.58357101330000E+0001
2	2.72021464150000E−0002
3	−3.83456376440000E−0004
4	−1.68410656320000E−0005
5	−4.46541645150000E−0007
6	−7.01614640110000E−0009
7	−7.01141755030000E−011
8	−4.57112620930000E−0013
9	−1.93669015050000E−0015
10	−5.13480975620000E−0018
11	−7.72685151860000E−0021
12	−5.02907385360000E−0024

Range #2 0 to 1372°C
Order of polynomial = 6

Power of T	Coefficient
0	0.000000000000000E+0000
1	2.58357101330000E+0001
2	2.61221522880000E–0002
3	–3.35533237550000E–0005
4	1.59014010170000E–0008
5	–6.03749339390000E–0013
6	–1.208750150000000E–0015

Type KN thermocouple

Number of ranges = 2
Range #1 –270 to 0°C
Order of polynomial = 12

Power of T	Coefficient
0	0.000000000000000+0000
1	1.36397230060000E+0001
2	2.63104723000000E–0004
3	2.17802309280000E–0004
4	1.53219743930000E–0005
5	4.21959974230000E–0007
6	6.76856722290000E–0009
7	6.85556478860000E–0011
8	4.51139628800000E–0013
9	1.92400134930000E–0015
10	5.12342695880000E–0018
11	7.72685151860000E–0021
12	5.02907385360000E–0024

Type R thermocouple

Number of ranges = 4
Range #1 −50 to 630.74°C
Order of polynomial = 7

Power of T	Coefficient
0	0.000000000000000+0000
1	5.28913950590000E+0000
2	1.3911099470000E+0002
3	−2.40052384300000E−0005
4	3.62014105950000E−0008
5	−4.46450193600000E−0011
6	3.84976918650000E−0014
7	−1.53726415590000E−0017

Range #2 630.74 to 1064.43°C
Order of polynomial = 3

Power of T	Coefficient
0	−2.64180070250000E+0002
1	8.04686807470000E+0000
2	2.98922937230000E−0003
3	−2.68760586170000E−0007

Range #3 1064.43 to 1665°C
Order of polynomial = 3

Power of T	Coefficient
0	1.49017027020000E+00003
1	2.86398675520000E+0000
2	8.08236311890000E−0003
3	−1.93384776380000E−0006

Range #4 1665 to 1767.6°C
Order of polynomial = 3

Power of T	Coefficient
0	9.54455599100000E+0004
1	−1.66425003590000E+0002
2	1.09757432390000E−0001
3	−2.22892169800000E−0005

Type S thermocouple

Number of Ranges = 4
Range #1 –50 to 630.74°C
Order of polynomial = 6

Power of T	Coefficient
0	0.000000000000000+0000
1	5.39957823460000E–0002
2	1.25197000000000E–0002
3	–2.2544821799700E–0005
4	2.84521649490000E–0008
5	–2.24405845440000E–0011
6	8.50541669360000E–0015

Range #2 –630.74 to 1064.43°C
Order of polynomial = 2

Power of T	Coefficient
0	02.98244816150000E+0002
1	8.23755282210000E+0000
2	1.64539099420000E–0003

Range #3 1064.43 to 1665°C
Order of polynomial = 3

Power of T	Coefficient
0	1.27662921750000E+0003
1	3.49709080410000E+0000
2	6.38246486660000–0003
3	–1.57224245990000E–0006

Range #4 1665 to 1767.6°C
Order of polynomial = 3

Power of T	Coefficient
0	9.78466553610000E+0004
1	–1.70502956320000E+0002
2	1.10886997680000E–0003
3	–2.24940708490000E–0005

Type T thermocouple

Number of ranges = 2
Range #1 −270 to 0°C
Order of polynomial = 14

Power of T	Coefficient
0	0.00000000000000E+0000
1	3.87407738400000E+0001
2	4.41239324820000E−0002
3	1.14052384980000E−0004
4	1.99744065680000E−0005
5	9.04454011870000E−0007
6	2.27660185040000E−0008
7	3.62474093800000E−0010
8	3.86489242010000E−0012
9	2.82986785190000E−0014
10	1.42813833490000E−0016
11	4.88332543640000E−0019
12	1.08034746830000E−0021
13	1.39492910260000E−0024
14	7.97958931560000E−0028

Range #2 0 to 400°C
Order of polynomial = 8

Power of T	Coefficient
0	0.00000000000000E+0000
1	3.87407738400000E+0001
2	3.31901980920000E−0002
3	2.07141836450000E−0004
4	−2.19458348230000E−0006
5	1.10319005500000E−0008
6	−3.09275818980000E−0011
7	4.56533371650000E−0014
8	−2.76168780400000E−0017

Type TP thermocouple

Number of ranges = 2
Range #1 −270 to 0°C
Order of polynomial = 14

Power of T	Coefficient
0	0.00000000000000E+0000
1	5.88026174000000E+0000
2	1.96585611920000E−0002
3	1.77122842010000E−0004
4	2.04796118410000E−0005
5	9.45106050990000E−0007
6	2.46395271480000E−0008
7	4.01667592050000E−0010
8	4.32562514960000E−0012
9	3.16195042210000E−0014
10	1.57848625730000E−0016
11	5.30107830900000E−0019
12	1.145496375100000E−0021
13	1.43860091110000E−0024
14	7.97958931560000E−0028

Range #0 to 400°C
Order of polynomial = 9

Power of T	Coefficient
0	0.000000000000000+0000
1	5.88062617400000E+0000
2	1.62014049810000E−0002
3	1.16368154490000E−0004
4	−1.63847540040000E−0006
5	9.48870459000000E−0009
6	−2.84437817350000E−0011
7	4.33143650190000E−0014
8	−2.64222483580000E−0017
9	−2.55611274970000E−0022

Type TN thermocouple

Number of ranges = 2
Range #1 –270 to 0°C
Order of polynomial = 13

Power of T	Coefficient
0	0.000000000000000+0000
1	3.28601476660000E+0001
2	2.44653712900000E–0002
3	–6.30704570300000E–0005
4	–5.05205273000000E–0007
5	–4.06520391200000E–0008
6	–1.87350864360000E–0009
7	–3.91934982500000E–0011
8	–4.60732739460000E–0013
9	–3.32082570160000E–0015
10	–1.50347922400000E–0017
11	–4.17752872630000E–0020
12	–6.51489067700000E–0023
13	–4.36718084880000E–0026

Range #2 0 to 1000°C
Order of polynomial = 9

Power of T	Coefficient
0	0.00000000000000E+0000
1	3.28601476660000E+0001
2	1.69887931740000E–0002
3	9.07736819560000E–0005
4	–5.56108081870000E–0007
5	1.543195960400000E–0009
6	–2.48380016340000E–0023
7	2.33897214590000E–00015
8	–1.19462968150000E–0018
9	2.55611274970000E–0022

Appendix F

Number systems

F.1 Introduction

All activities performed by the microprocessor of the PLC are in the binary form with two states '0' or '1'. The microprocessor memory generally has thousands of memory locations that are called words. Each word stores binary data in the form of binary digits orbits. 'Bits' is the shortened form of 'binary digits'. Each memory work generally consists of 16 bits, but could be 64 bits, 32 bits, 16 bits, or even 8 bits in length. (8 bits is generally referred to as a Byte or more correctly, as an Octet.)

PLCs use several numbering systems to convert user numbering information into binary digits for memory storage and control of outputs. Similarly, memory storage data is translated into the user selected numbering system for ease of interpretation.

The five most commonly used number systems in PLCs are:

- Binary
- Hexadecimal
- Octal
- Binary coded decimal (BCD)
- Binary coded octal (BCO)

After preliminary discussions on a generalized number system, these five systems will be discussed below. In addition, conversion between different systems will also be briefly examined.

F.2 A generalized number system

A number system is formed by allocating symbols to specific numerical values. Any group of symbols can be used with the total number of symbols for a number system called the base of the system. The three most common bases are:

- Binary with two symbols (0 and 1) and hence a base of 2
- Hexadecimal with sixteen symbols (0, 1, 2...9,A, B....F) and hence a base of 16

- Decimal with ten symbols (0, 1, 2...9) and hence a base of 10

When numbers with different bases are being used in the same descriptive text they sometimes have the subscript referring to the base being used, as in 3421.19_{10} for a decimal or base 10 number.

Numerical symbols have to be combined in a certain way to represent other combinations of numbers. The decimal numbering system has the structure laid out in Table F.1 for weighting each digit in the number 3421.19_{10} in a combination of numbers written together.

Exponential notation is used here, where for example: 10^2 means 100 and to 10^{-3} means 0.001.

Weight	10^4	10^3	10^2	10^1	10^0	.	10^{-1}	10^{-2}	10^{-3}	10^{-4}	10^{-5}
	0	3	4	2	1	.	1	9	0	0	0

Table F.1
Decimal weighting system

The most significant digit (or MSD) in this number is 3. This refers to the left most non-zero digit that has the greatest weight (10^3 or 1000) assigned to it.

The least significant digit (or LSD) in this number is 9. This refers to the right most non-zero digit that has the least weight (10^{-2} or 0.01) assigned to it.

This represents the number calculated below:
$$0 \times 10^4 + 3 \times 10^3 + 4 \times 10^2 + 2 \times 10^1 + 1 \times 10^0 + 1 \times 10^{-1} + 9 \times 10^{-2} + 0 \times 10^{-3} + ...$$

F.3 Binary numbers

Binary numbers are commonly used with computers and data communications because they represent two states – either ON or OFF. For example, the RS-232-C standard has two voltages assigned for indicating ON (–5 volts, say) or OFF (+5 Volts, say). Any other voltages outside a narrow band around these voltages are undefined.

The word 'bit', referred to often in the literature, is a contraction of the words 'binary digit'.

The same principles for representing a binary number apply as in the previous section. For example, the number 1011.1_2 means the following using Table F.2:

Weight	2^4	2^3	2^2	2^1	2^0	.	2^{-1}	2^{-2}	2^{-3}	2^{-4}	2^{-5}
	0	1	0	1	1	.	1	0	0	0	0

Table F.2
Binary weighting system

This translates into the following number:
$$........0 \times 2^4 + 1 \times 2^3 + 0 \times 2^2 + 1 \times 2^1 + 1 \times 2^0 + 1 \times 2^{-1} + 0 \times 2^{-2} + ...$$
The most significant bit (MSB) in the above number is the left most bit and is 1 with weighting of 2^3. The right most bit is the least significant bit (LSB) and is valued at 1 with a weighting of 2^{-1}.

F.3.1 Conversion between decimal and binary numbers

Table F.3 gives the conversion between decimal and binary numbers. Note that the binary equivalent of decimal 15 is written in binary form as 1111 (using 4 bits). This 4 bit binary grouping will have significance in hexadecimal arithmetic later. As expected binary 0 is equivalent to decimal 0.

Decimal number	Binary equivalent
0	0
1	1
2	10
3	11
4	100
5	101
6	110
7	111
8	1000
9	1001
10	1010
11	1011
12	1100
13	1101
14	1110
15	1111

Table F.3
Equivalent binary and decimal numbers

The procedure to convert from a binary number to a decimal number is straightforward. For example, to convert 1101.01_2 to decimal, use the weighting factors for each bit to make the conversion.

$1101.01_2 = 1 \times (2^3) + 1 \times (2^2) + 1 \times (2^1) + 1 \times (2^0) + 1 \times (2^{-1}) + 1 \times (2^{-2})$

This is equivalent to:

$1101.01_2 = 1 \times (8) + 1 \times (4) + 1 \times (2) + 1 \times (1) + 1 \times (\frac{1}{2}) + 1 \times (\frac{1}{4})$

This then works out to:

$1101.01_2 = 8 + 4 + 0 + 1 + 0.25$

$1101.01_2 = 13.25$

The conversion process from a decimal number to a binary number is slightly more complex. The procedure here is to repeatedly divide the decimal number by two, until the quotient (the result of the division) is equal to zero. Each of the remainders forms the individual bits of the binary number.

For example, to convert decimal number 43_{10} to binary form:

2	43 remainder 1 (LSB)
2	21 remainder 1
2	10 remainder 0
2	5 remainder 1
2	2 remainder 0
2	1 remainder 1 (MSB)
	0

Table F.4
Illustration of decimal to binary conversion

This translates a number 43_{10} to 101011_2

F.4   Hexadecimal numbers

Most of the work done with computers and data communications systems is based on the Hexadecimal number system, as it is easy to translate a binary number into a hexadecimal equivalent. This has a base of 16 and uses the sequence of symbols:
0, 1, 2, 3, 4, 5, 6, 7, 8, 9, A, B, C, D, E, F
Hence the number of $FA9.02_{16}$ would be represented as below in Table F.5:

Weight	16^4	16^3	16^2	16^1	16^0	.	16^{-1}	16^{-2}	16^{-3}	16^{-4}	16^{-5}
	0	0	F	A	9	.	0	2	0	0	0

Table F.5
Hexadecimal weighting structure

This translates into the following number:
$0 \times 16^4 + 0 \times 16^3 + F \times 16^2 + A \times 16^1 + 0 \times 16^{-1} + 0 \times 16^{-2} + \ldots\ldots$
$= 15 \times 16^2 + 10 \times 16^1 + 9 \times 1 + 2 \times 1/16^2$
$= 15 \times 256 + 10 \times 16 + 9 \times 1 + 2\ /256$
$= 4009.0078125$
The most significant digit (MSD) in the above number is the left most symbol and is *F* with weighting of 16^2. The right most symbol is the least significant digit (LSD) and is valued at 2 with a weighting of 16^{-2}.

F.4.1   Conversion between binary and hexadecimal

The conversion between binary and hexadecimal is effected by modifying Table F.3 to the table following:

Decimal	Hexadecimal	Binary
0	0	0000
1	1	0001
2	2	0010
3	3	0011
4	4	0100
5	5	0101
6	6	0110
7	7	0111
8	8	1000
9	9	1001
10	A	1010
11	B	1011
12	C	1100
13	D	1101
14	E	1110
15	F	1111

Table F.6
Relationship between decimal, binary, and hexadecimal numbers

As can be seen from the table, the binary numbers are grouped in fours for the largest single digit hexadecimal character or symbol. A similar approach, of grouping bits in fours, is followed in expressing a binary number as a hexadecimal number.

In converting the binary number 1000010011110111_2 to its hexadecimal equivalent, the following procedure should be adopted. First, break up the binary number into groups of four commencing from the least significant bit. Then equate the equivalent hex symbol to it (derived from Table F.6 above).

1000010011110111 becomes:

1000	...	0100	...	1111	...	0111_2
8		4		F		7_{16}

Or $8\ 4\ F\ 7_{16}$

In order to convert a hexadecimal number back to binary the procedure used above must be reversed.

For example, in converting from C9A4 to binary this becomes:

C	...	9	...	A	...	4_{16}
1100	...	1001	...	1010	...	0100_2

Or 1100100110100100_2

F.5   Octal

Some of the older computer systems used octal for their coding. This uses the sequence of symbols:

0, 1, 2, 3, 4, 5, 6, 7
Hence, the number 23471.6_8 would be represented as below:

Weight	8^4	8^3	8^2	8^1	8^0	.	8^{-1}	8^{-2}	8^{-3}	8^{-4}
	2	3	4	7	1	.	6	0	0	0

Table F.7
Octal weighting structure

This translates into the following decimal number
$2 \times 8^4 + 3 \times 8^3 + 4 \times 8^2 + 7 \times 8^1 + 1 \times 8^0 + 6 \times 8^{-1} = 10041.75_{10}$

F.6 Binary coded decimal

The Binary coded decimal approach can be used to convert decimal numbers into binary form and assigns four (4) binary digits to each decimal digit.

For example, 4 decimal numbers could be represented as:

Weight	Most significant digit (MSD)				Least significant digit (LSM)							
	2^3 2^2 2^1 2^0				2^3 2^2 2^1 2^0				2^3 2^2 2^1 2^0			2^3 2^2 2^1 2^0
	0 1 0 1				0 1 1 1				0 1 0 0			0 0 0 1

This is represented as four decimal numbers indicated as follows:
$0 \times 2^3 + 1 \times 2^2 + 0 \times 2^1 + 1 \times 2^0 + 0 \times 2^3 + 1 \times 2^2 + 1 \times 2^1 + 1 \times 2^0 \quad 0 \times 2^3 + 1 \times 2^2 + 0 \times 2^1 + 0 \times 2^0 + 0 \times 2^3 + 0 \times 2^2 + 0 \times 2^1 + 1 \times 2^0$
$= 5\ 7\ 4\ 1_{10}$

There is a certain amount of wastage in this coding system as the maximum 4 bit binary combination for BCD is 1001_2 or 9_{10}. The binary combinations 1010 to 1111 are unused (and illegal) in a BCD encoding system.

BCD notation is useful for applications where absolute precision is required (unlike floating point notation which gives a high precision but no guarantee of absolute precision). Unfortunately, a specialized method of arithmetic operations has to be built into the system, as normal binary arithmetic is inadequate.

F.7 Binary coded octal systems

The binary coded octal system uses 8 bits to represent 3 digit octal numbers from 000_8 to 377_8. This approach is not often used today.

The largest binary coded octal number could be represented as:

Weight	Most significant digit (MSD)			Least significant digit (LSM)				
	2^1	2^0		2^2	2^1	2^0		2^2 2^1 2^0
	1	1		1	1	1		1 1 1

This is represented as three octal numbers
$1 \times 2^1 + 1 \times 2^0 \quad 1 \times 2^1 + 1 \times 2^0 \quad 1 \times 2^2 + 1 \times 2^1 + 1 \times 2^0$
$= 3\ 7\ 7_8$

F.8 Internal representation of information

As has been discussed previously, there are two kinds of information that must be represented within the PLC:

- Program information
- Data information

There are two types of data:

- Numeric
- Alphanumeric or characters

F.8.1 Numeric data

Numeric data can be further subdivided into

- Integers
- Floating point numbers

Each of these types must have the sign of the number encoded as well. There are a few methods of encoding this numeric data:

Sign magnitude

In this example, the most significant bit represents the sign bit. Hence, 2 would be represented in an 8-bit notation as:

0 0 0 0 0 0 1 0

–2 would be represented as:

1 0 0 0 0 0 1 0

Although easy to visualize the encoding of the number, it is not popular because of the complexity of performing arithmetic operations.

One's complement

The approach here is to take the mathematical complement of the number to derive its negative value. For example, the number 2 would have the representation of:

0 0 0 0 0 0 1 0

The representation for –2 would be (complementing the above byte):

1 1 1 1 1 1 0 1

(The MSB is still the sign bit and in this example, it is 1 indicating a negative number.)

Unfortunately with this approach, the rules of normal binary arithmetic do not work if the signs are different.

Two's complement

This is probably the most effective approach and its calculation is as follows:

- Take the number and complement each bit.
- Add '1' to the least significant bit.

For example:

'+2' is represented as:	0000	0010
'–2' is represented as (1's complement)	1111	1101
(2's complement)	1111	1110

A check to see whether this is correct is to add +2 to –2 (to achieve)

Add '+2' to '–2'   0000 0010
                    1111 1110
                    0000 0000

The carry 1 at the end of the operation is thrown away

Fractional numbers

The principle of floating point notation is to allocate parts of the word to the mantissa, the mantissa's sign, the exponent and the exponent's sign.

The mantissa must be normalized to a number between 2^{-1} and 1

In the representation here, eight (8) bits are allocated to the exponent, including sign and 32 bits to the mantissa including sign:

0	1	7	8	9	31
Sign	Exponent		Sign	Mantissa	

F.8.2 Alphanumeric data representation

Two standards have evolved for encoding alphanumeric symbols. These are the ASCII and EBCDIC notations. ASCII is probably the most popular, with 128 possible unique characters. Although 7 bits are used to represent all the ASCII characters, the eighth bit is used to encode the parity information. This ensures full utilization of the 8-bit byte. There is an extended ASCII version available which uses full 8-bits. This is not very common with PLCs.

F.9 Binary arithmetic

Addition

Knowledge of binary addition is useful although it can be cumbersome. It is based on the following four combinations of adding binary numbers:

0 0   1   1
0 1   0   1
0 1   1   0   and carry 1

The carry 1 (or bit) is the only difficult part of the process. This addition of the individual bits of the number should be done sequentially from the LSB to the MSB (as in normal decimal arithmetic).

An example of addition is given below.

1010001001_2
0011101010_2
1101110011_2

Subtraction

The most commonly used method of binary subtraction is to use 2's complement. This means that instead of subtracting two binary numbers (with the attendant problems such as having 'carry out' bits), the addition process is applied.

For example, take two numbers and subtract the one from the other as follows:

12	which is equivalent to:	1100
–4	Subtrahend	– 0100
8	Result	1000

The two's complement (as discussed earlier) is found by first complementing all the bits in the subtrahend and then adding 1 to the least significant bit.

Complementing the number results in 0100 becoming: 1011

Adding 1 to the least significant bit gives a two's complement number of: 1100.

Add 1100_2 to 1100_2 as follows:

```
    1100
    1100
    1000    carry 1
```

(Drop the carry 1 to achieve the same result as above.)

Exclusive–OR (XOR)

Exclusive–OR is a procedure very commonly used with binary numbers in the error detection sequences of data communications. The result of an XOR operation on any two binary digits is the same as the *addition* of two digits *without the carry bit.*

Consequently, this operation is sometimes also called the Modulo–2 adder. The truth table for XOR is shown below.

Bit 1	Bit 2	XOR
0	0	0
0	1	1
1	0	1
1	1	0

Table F.8
Exclusive OR truth table

Appendix G

GPIB (IEEE-488) mnemonics and their definitions

AC	Addressed Command
ACDS	Accept Data State
ACG	Addressed Command Group
ACRS	Acceptor Ready State
AD	Addressed
AH	Acceptor Handshake
AH1	Complete Capability
AH10	No Capability
AIDS	Acceptor Idle State
ANRS	Acceptor Not Ready State
APRS	Affirmative Poll Response State
ATN	Attention
AWNS	Acceptor Wait for New Cycle State
C	Controller
CACS	Controller Active State
CADS	Controller Addressed State
CAWS	Controller Active Wait State
CIDS	Controller Idle State
CPPS	Controller Parallel Poll State
CPWS	Controller Parallel Poll Wait State
CSBS	Controller Standby State
CSNS	Controller Service Not Requested State
CSRS	Controller Service Requested State
CSWS	Controller Synchronous Wait State
CTRS	Controller Transfer State
DAB	Data Byte
DAC	Data Accepted
DAV	Data Valid

DC	Device Clear
DCAS	Device Clear Active State
DCIS	Device Clear Idle State
DCL	Device Clear Local
DD	Device Dependent
DIO	Data Input/Output
DT	Device Trigger
DTAS	Device Trigger Active State
DTIS	Device Trigger Idle State
END	End
EOI	End or Identity
EOS	End of String
F	Active False
(F)	Passive False
GET	Group Execute Trigger
GTL	Go to Local
GTS	Go to Standby
IDY	Identify
IFC	Interface Clear
IST	Individual Status
L	Listener
LE	Extended Listener
LACS	Listener Active State
LADS	Listener Addressed State
LAG	Listener Address Group
LIDS	Listener Idle State
LLO	Local Lockout
LOCS	Local State
LON	Listen Only
LPAS	Listener Primary Addressed State
LPE	Local Poll Enable
LPIS	Listener Primary Idle State
LTN	Listen
LUN	Local Unlisten
LWLS	Local with Lockout State
M	Multiline
MLA	My Listen Address
MSA	My Secondary Address
MTA	My Talk Address
NBA	New Byte Available
NDAC	Not Data Accepted
NPRS	Negative Poll Response State
NRFD	Not Ready for Data
NUL	Null Byte
OSA	Other Secondary Address
OTA	Other Talk Address

PACS	Parallel Poll Addressed to Configure State
PCG	Primary Command Group
POFS	Power Off State
PON	Power On
PP	Parallel Poll
PPAS	Parallel Poll Active State
PPC	Parallel Poll Configure
PPD	Parallel Poll Disable
PPE	Parallel Poll Enable
PPIS	Parallel Poll Idle State
PPR	Parallel Poll Response
PPSS	Parallel Poll Standby State
PPU	Parallel Poll Unconfigure
PUCS	Parallel Poll Unaddressed to Configure State
RDY	Ready
REMS	Remote State
REN	Remote Enable
RFD	Ready for Data
RL	Remote/Local
RPP	Request Parallel Poll
RQS	Request Service
RSC	Request System Control
RSV	Request Service
RTL	Return to Local
RWLS	Remote with Lockout State
SACS	System Control Active State
SCG	Secondary Command Group
SDC	Selected Device Clear
SDYS	Source Delay State
SE	Secondary
SGNS	Source Generate State
SH	Source Handshake
SIAS	System Control Interface Clear Active State
SIC	Send Interface Clear
SIDS	Source Idle State
SIIS	System Control Interface Clear Idle State
SINS	System Control Interface Clear Not Active State
SIWS	Source Idle Wait State
SNAS	System Control Not Active State
SPAS	Serial Poll Active State
SPD	Serial Poll Disable
SPE	Serial Poll Enable
SPIS	Serial Poll Idle State
SPMS	Serial Poll Mode State
SR	Service Request
SRAS	System Control Remote Enable Active State
SRE	Send Remote Enable
SRIS	System Control Remote Enable Idle State
SRNS	System Control Remote Enable Not Active State
SRQ	Service Request
SRQS	Service Request State
ST	Status

STB	Status Byte
STRS	Source Transfer State
SWNS	Source Wait for New Cycle State
T	Active True
(T)	Passive True
TE	Extended Talker
TACS	Talker Active State
TADS	Talker Addressed State
TAG	Talk Address Group
TCA	Take Control Asynchronously
TCS	Take Control Synchronously
TCT	Take Control
TIDS	Talker Idle State
TON	Talk Only
TPAS	Talker Primary Addressed State
TPIS	Talker Primary Idle State
U	Uniline Message
UC	Universal Command
UCG	Universal Command Group
UNL	Unlisten
UNT	Untalk

Index

Printed and bound by CPI Group (UK) Ltd, Croydon, CR0 4YY

03/10/2024

01040338-0012